Lateral Pressure Reduction
on
Earth-Retaining Structures
Using Geofoam

Front-cover graphic: A whimsical rendition of what is arguably the iconic rigid earth-retaining structure...a classical free-standing gravity retaining wall...combined with a simplistic active earth pressure wedge within the retained soil. This example of 'geo-humor' from the writer's files was used on a flyer promoting a now-forgotten (by the writer) conference that was most likely held in Europe in the circa-1970s timeframe.

Rear-cover photos

Upper left: One of the earliest (circa 1990) uses of EPS-block geofoam to both reduce settlements adjacent to and lateral pressures on below-grade walls at a site underlain by compressible soils was at the Carousel Center (now called Destiny USA) shopping mall in Syracuse, NY. Note that this project occurred before stair-stepping of EPS blocks into the retained soil was considered a necessary design detail for such applications. A years-later expansion project uncovered some of these blocks which were found to be in excellent physical and mechanical condition and reused in the new construction. [credit: BASF]

Upper right: This circa-2000s project at the Tamarack ski resort in Idaho showcases the all-weather constructability of EPS-block geofoam, in this case to reduce lateral pressures on a permanent soldier-pile-and-lagging wall constructed as part of an abutment for one of the resort structures. [credit: unknown, via J-U-B Engineers and Terracon, both of Boise, ID]

Center left: This shows the replacement Puente Cayumapu bridge near Valdivia, Chile under construction on 7 August 1997. This bridge is of integral-abutment design and the EPS-block geofoam was used for part of the approach embankment to both reduce lateral pressures on the abutments as well as minimize settlement of the finished roadway as the site is underlain by compressible soils. Of note is that this site is in an area with high seismic potential as it is near the epicenter of the 1960 earthquake that is often credited as being the most powerful earthquake ever recorded, with M_w = 9.4 - 9.6. [credit: John S. Horvath]

Center right: A relatively rare example of using EPS-block geofoam to reduce lateral pressures on a flexible earth-retaining structure, in this case an anchored bulkhead at a port in Japan. [credit: Expanded Polystyrol Construction Method Development Organization (EDO) - Japan]

Lower left: One of the earliest (circa 1990s) project uses of the *GeoTech GeoInclusion*™...a panel-shaped geocomposite consisting of resilient (elasticized) EPS-block, glued polystyrene porous block, and nonwoven geotextile that collectively act as a compressible inclusion for lateral pressure reduction while providing positive drainage of groundwater...was around the basement walls of a high-rise building in Kansas City, MO. [credit: GeoTech Systems Corporation]

Lower right: The first known (circa 1998) use of the *GeoTech GeoInclusion*™ to reduce seasonal increases in lateral pressures on a semi-integral abutment (integral-backwall) bridge was on the U.S. Route 60 crossing of the Jackson River in Alleghany County, VA. This project was instrumented and monitored over a period of years by the Virginia Transportation Research Council on behalf of the Virginia DOT and Federal Highway Administration. The VaDOT found the installation to be successful in achieving design goals and subsequently used the same design on other, similar projects including the VA Route 18 crossing of Blue Spring Run in the same county. [credit: GeoTech Systems Corporation]

Lateral Pressure Reduction
on
Earth-Retaining Structures
Using Geofoam

A Scholarly Monograph
by

John S. Horvath, Ph.D., P.E., Life Member.ASCE

Consulting Professional Engineer
d/b/a
John S. Horvath Consulting Engineer
Scarsdale, New York, U.S.A.

and

Professor of Civil Engineering (retired)
Manhattan College
School of Engineering
Civil and Environmental Engineering Department
Bronx, New York, U.S.A.

Published
by
John S. Horvath Consulting Engineer
Scarsdale, New York, U.S.A

Lateral Pressure Reduction on Earth-Retaining Structures Using Geofoam
by John S. Horvath

This document is a scholarly monograph written solely for the purpose of dissemination of technical knowledge and information for consideration by licensed design professionals and academicians educated in the field of civil engineering. As such, it is not a textbook, design manual, or standard, and should not be used or referenced as such. Furthermore, no liability is expressed or implied as to the suitability or accuracy of any information presented herein for any specific application in engineered construction or academic research. Use of any of the methods presented herein remains solely the liability of the licensed design professional using these methods.

Published March 2018
by
John S. Horvath Consulting Engineer
Scarsdale, New York, U.S.A.
linkedin.com/in/jshce

Library of Congress Control Number: 2018902878

ISBN-13: 978-1-7320953-0-4

Contents

Preface

This is my second monograph related to geofoam. It marks my return to the roots of my initial interest in geofoam that began more than three decades ago and subsequently led to my broad research into cellular geosynthetics in general, and geofoam in particular, that along the way resulted in my first monograph, *Geofoam Geosynthetic*, in 1995.

In this era of questioning the source of funding of scientific research...with the implication that researchers, who are, after all, human, may be predisposed, perhaps subliminally if not overtly, to slanting the outcomes of their research to those who are 'paying the bills' as it were...it is important that I state at the outset that this latest monograph, as with my 1995 monograph, is entirely self-funded and self-published both in terms of my time as well as all production costs. Furthermore, the two U.S. patents that I hold related to technologies presented in this monograph have expired so the methodologies they cover are entirely in the public domain at the present time and I derive no financial benefit from their use. Consequently, there is no hidden agenda in this monograph to promote any geotechnology or commercial product that benefits me financially, either directly or indirectly. My desire and goal are simply sharing of knowledge.

To provide some background as to how this new monograph came to be, when I began my full-time academic career at Manhattan College in August 1987 it was, in part, to be able to do the kind of ongoing, long-term research and concomitant writing and publication that proved to be impossible in my then full-time employment in the geotechnical engineering consulting industry. As with anyone entering full-time academia, I immediately began to look for an area of research that was of intellectual interest to me. However, I had to deal with the realities of being at an institution such as Manhattan College that was and had always been primarily an undergraduate educational school with relatively heavy teaching loads for faculty (eight courses per academic year when I started) and grading (all done by the faculty, even for laboratory courses); no release time for unfunded research; minimal research laboratories and research support of any kind; no promise of Master-level graduate (or even qualified undergraduate) students to assist with research; and not even a guarantee of being granted a paid sabbatical leave. Thus, I knew from Day One that I had to find a research topic that was not only uncrowded with other researchers but one where solo research of an analytical nature was possible. This was because I knew when I started at Manhattan that there would never be any way that I could compete in a field already occupied by faculty at research-oriented universities with all the financial, human, and laboratory resources that implies.

By coincidence, around the time (mid-1987) that I was transitioning from full-time engineering practice to full-time academia, I read the first of several technical papers and reports that I would encounter over the next few years on the topic of using what the authors called the *boundary-yielding concept* that eventually came to be known as the *compressible-inclusion* (a term coined by me some years later) function of *geofoam* (another term independently coined by me years later) to reduce lateral earth pressures on rigid earth-retaining structures. Specifically, I read a case history paper dealing with the basement wall of a high-rise commercial building in Philadelphia, Pennsylvania. At the time, I was very familiar with the traditional design of such walls from my then-fifteen years of industry experience. The compressible-inclusion concept fascinated me and caused me to explore it further. Being practice-oriented, I was especially interested in finding a commercially available material and product that could be used for this purposes in routine practice in the U.S. It was through new contacts with the U.S. expanded polystyrene (EPS) industry that I eventually found and cultivated that I learned about the broader uses of EPS, a subject that wound up occupying much of my time and led to the aforementioned publication of my 1995 monograph.

The use of geofoam products, especially those composed of block-molded EPS (EPS-block), has increased dramatically worldwide in recent decades. Most of this usage has involved the obvious and intuitive *lightweight-fill* function and has been the result, in no small part, of the publication of not only my 1995 monograph but also two comprehensive U.S. National Cooperative Highway Research Program (NCHRP) reports, in 2004 and 2013, for which I was a co-Principal Investigator and co-author.

Despite these significant successes in technology transfer, other potentially useful geofoam functions...compressible inclusion in particular...remain significantly under-recognized and underutilized in routine practice. Although there has been some increase in academic research into the compressible-inclusion function, particularly under dynamic loading (as what others have termed a *seismic buffer* or, much less commonly, *seismic isolation*), I have been distressed to see that much of this research has been, in my opinion, poorly crafted and interpreted as the principal investigators did not appear to have a sound and correct understanding of the relevant geofoam material behaviors.

Because of both the continued underuse and, worse in my opinion, incorrect use of the compressible-inclusion function of geofoam, I felt that a comprehensive monograph focusing on this function was long overdue. Although this function was addressed in my 1995 monograph as well as in numerous technical papers that I authored both prior to and subsequent to publication of my monograph, I felt that there was still a need to collect and present all relevant information in one resource, hence this current document.

It is important to note that this new monograph, despite its substantial length, is not the final word on the subject. Far from it, actually. Although the basic concepts of how geofoam can be used to reduce lateral pressures from both retained soil and surface surcharges on a variety of earth-retaining structures are well understood, there is still much research and observation in practice to be done to both verify, and modify if necessary, existing analytical methods as well as develop new analytical methodologies for application variations that can arise in practice. However, to borrow a line from the film "Wag The Dog":

"A good plan today is better than a perfect plan tomorrow."

Consequently, I feel that it is better to write and publish a monograph documenting what we know about compressible inclusions at present rather than wait until some vague time in the future when the knowledge-base of the profession is greater. Hopefully this monograph will stimulate not only greater usage of the compressible-inclusion function but more and better research that will ultimately extend the knowledge-base for the benefit of all concerned.

Finally, it is worth noting that even though the focus of this monograph is on using geofoam to reduce earth and other forces in a nominally horizontal (lateral) direction, for the sake of completeness I have included a brief introductory and overview chapter that addresses the use of geofoam to achieve force reductions in a vertical direction.

John S. Horvath, Ph.D., P.E., Life Member.ASCE
Scarsdale, New York, U.S.A.
March 2018

Chapter 1

Introduction

1.1 BACKGROUND

The construction of *earth-retaining structures* (ERSs), especially free-standing *gravity retaining walls* of various types that are in the ERS sub-category of *rigid earth-retaining structures* (RERSs), is one of the oldest human endeavors as the building of a wall to retain soil for some purpose is arguably intuitive if not instinctive. This human endeavor was, and to some extent still is, facilitated by the fact that nature provides an abundance of rock fragments of the size now classified as *cobbles* (75 to 300 millimetres/3 to 12 inches in nominal diameter) and *boulders* (> 300 mm/12 in) on or close to the ground surface in many parts of the world. This provides the basic building blocks (literally) of such walls, both within plain view and readily accessible to human hands.

Except for the fact that early on humans advanced wall-building technology by learning how to shape pieces of sound rock into prismatic blocks to help satisfy their material needs rather than just rely on the size and shape of what nature provided, relatively little changed with regard to RERS construction over the millennia of human existence until the latter centuries of the second millennium CE[1] when Coulomb, Rankine, and others developed analytical methodologies that have been applied to ERSs of all types ever since. This allowed the design of ERSs to be turned from purely an art with an unpredictable outcome (in terms of satisfactory long-term performance of the ERS) to a science with a reasonably consistent, reliable outcome. That these analytical methods, which are simplistic approximations of actual behavior and are now more than 200 years old in some cases, are basically sound is attested by the fact that they still form the basis of ERS analysis and design in routine practice in the present. Nevertheless, their shortcomings and deviation from reality are discussed in this monograph.

These analytical advances of the 18th and 19th centuries were followed and complemented by the ascendancy of Portland-cement concrete (PCC) into consistent, global commercial use by the early 20th century. PCC impacted and influenced the design of RERSs in a way that persists to the present as it removed the need to build RERSs solely out of natural or quarried stone.

However, the real revolution in ERS technology occurred in the last few decades of the 20th century with the more or less contemporaneous development of several improved and completely new ERS technologies that can be either competing or complementary depending on the specific application. These geotechnologies include:

- Pre-fabricated structural bins that are assembled (in some cases), arranged, and filled with soil on-site to create a modernized version of the time-honored gravity retaining wall. These have proven to be more cost-effective to use in many cases than traditional gravity retaining walls constructed of either natural stone or PCC.

- *Mechanically stabilized earth walls* (MSEWs) that utilize layers of *geosynthetic tensile reinforcement* embedded within the retained soil. This was a revolutionary, completely

---

[1] In this context, CE means *Common Era* and is used as an alternative to the traditional term AD.

new concept at the time of its development in that it effectively turned the retained-soil mass into its own gravity retaining wall and dispensed with the need for a relatively massive RERS.

- *Segmental retaining walls* (SRWs) that utilize precast-PCC blocks and synergistically combine the behaviors of the traditional gravity retaining wall and newer MSEW concept in many cases.

The net result of this technological revolution of the last several decades is that there are now more alternatives than ever when a design professional is confronted with the need to construct an ERS. As will be seen, the concepts presented in this monograph will add to this list of choices.

1.2 PROLOGUE

The new ERS technologies such as MSEWs and SRWs have become widely known and used in practice over the last several decades and, as a result, have had a significant effect on ERS design for new construction worldwide. Relevant to this monograph is the fact that another new geotechnology called *geofoam*, a type of *cellular geosynthetic* (which includes *geocombs* as well), independently emerged and evolved in the final decades of the 20th century as well. Thus, it is only logical to wonder how these different geotechnologies might be combined in a complementary fashion to yield synergistic results.

The writer has been intimately and extensively involved with researching, writing, educating, and consulting about cellular geosynthetics in general, and polymeric[2] geofoams in particular, since 1987. In the last several decades, many design professionals and others in the construction and construction-materials industries have become aware of some geofoam materials, products, and applications, especially the use of *expanded polystyrene* (EPS)[3] in its generic block-molded form (referred to as *EPS-block geofoam*[4]) as lightweight fill for earthworks such as embankments and other types of fills on soft ground as well as for slope stabilization. However, most stakeholders involved in engineered construction remain unfamiliar with not only other lightweight-fill applications for which EPS-block can be used but other geofoam materials, products, functions, and applications as well.

In the writer's opinion, at the top of the list of under-recognized and underutilized geofoam functions and applications is their use with ERSs or even as ERSs. The overall,

---

[2] This term is used throughout this monograph in lieu of the colloquial term 'plastic' as plastic has a well-established, material-behavior meaning in engineering mechanics that predates the invention of polymeric materials.

[3] EPS is one of two types of *rigid cellular polystyrene* (RCPS), the other being *extruded polystyrene* (XPS), that collectively have long been the predominant types of polymeric foam used in construction applications, both engineered and otherwise. It is important to note that in some cases the RCPS terminology is used to the exclusion of EPS and XPS mentioned individually. One significant example of this is in several material standards promulgated by ASTM International (formerly the American Society for Testing and Materials). The reasons for this are strictly business-political, not technical, in nature. Unfortunately, this can be confusing to those who might be new to using these standards.

[4] There are many different types of polymeric and non-polymeric geofoam materials as well as a variety of geofoam products made from a given material. Thus, it is important when a specific geofoam material and product is being discussed to use specific terminology and not just refer to it simply as 'geofoam'. This is no different than when dealing with other types of geosynthetics such as geotextiles where the manufacturing process (woven versus non-woven) and polymer chemistry of the thread used both need to be stated explicitly.

primary benefit of using geofoam with traditional RERSs such as gravity retaining walls is to reduce lateral[5] earth and surface-load ('surcharge') pressures[6] relative to what would be acting on the ERSs in the absence of geofoam.

There are two distinct purposes and concomitant benefits for doing this depending on whether a new (i.e. proposed) or existing ERS is involved:

- For newly built ERSs, there can be a net reduction in the overall capital construction and possibly life-cycle operation and maintenance (O&M) costs relative to a baseline condition of no geofoam usage. In some cases, it is even possible to eliminate the traditional ERS entirely and have the geofoam act as its own ERS (called a *geofoam wall*) provided that the ground behind the geofoam is either self-stable or stabilized in some fashion.

- For existing ERSs that may either be exhibiting signs of physical distress or be in acceptable condition but have an inadequate margin of safety under current design standards and associated loads (e.g. an ERS that was not designed originally for seismic loads that needs to be upgraded to resist such loads), the reduction of costs associated with renovating, rehabilitating, or upgrading such structures occurs because it is often possible to achieve the desired level of improvement simply by using geofoam and without any alteration of the original ERS. The cost of replacing some of the retained soil with geofoam is generally less than other, more-conventional alternatives that involve alteration of the original ERS.

This latter application with existing ERSs is particularly noteworthy because the other modern ERS technologies of MSEWs and SRWs noted previously are, for the most part, limited to new construction with minimal potential, economical use with existing ERSs, especially RERSs.

As will be seen, the use of geofoam with both new and existing ERSs in general, and RERSs in particular, can involve two distinctly different geofoam functions:

- the well-known, aforementioned lightweight fill and

- the less-well-known *compressible inclusion*[7].

---

[5] The term 'lateral' is used synonymously for 'horizontal' throughout this monograph, primarily when it applies to the direction of a normal stress ('pressure'). This is in consonance with colloquial usage in U.S. engineering practice when dealing with all types of ERSs.

[6] The term 'pressure' is used synonymously for 'stress' (more specifically, normal stress) throughout this monograph in consonance with colloquial usage in U.S. engineering practice when referring to horizontal normal stresses from the ground ('earth') acting on some ERS. Note, however, that the use of the term pressure in this context is, strictly speaking, terminologically incorrect as pressure in a civil engineering context is, by definition, a synonym for normal stresses only when dealing with a fluid (gas or liquid). However, the dominant (in U.S. practice) use of the phrase 'lateral earth pressure' as opposed to 'horizontal earth stress' likely derives from an observation made in the earliest days of modern geotechnical engineering that in many applications such stresses can be reasonably approximated and visualized as varying linearly with depth which is the same as the depth-wise variation of fluid pressure. In fact, many textbooks, handbooks, etc., especially older ones, refer to lateral earth pressures behind an ERS as being 'equivalent fluid pressures'.

[7] In the multiple late-1980s to early-1990s published works by A. McGown, K. Z. Andrawes, and their colleagues in the U.K., this geofoam function was referred to as the *boundary-yielding concept*.

In general, these two functions are competitive as opposed to complementary in most ERS applications. However, each function has its benefits and advantages when used with ERSs, and part of the design process is selecting which geofoam function provides the more cost-effective alternative in a given project application.

1.3 OBJECTIVE AND SCOPE

In the writer's opinion, geofoam has the potential to significantly impact, if not revolutionize, ERS, especially RERS, design not only for new construction but for the renovation, rehabilitation, and performance-upgrading of existing ERSs as well. This latter capability is particularly noteworthy given the enormous worldwide inventory of existing ERSs. Furthermore, as noted previously, there are situations where the use of geofoam may remove the need for a traditional ERS entirely and form its own ERS as a geofoam wall. Note that this is in the same, broad, conceptual manner in which MSEWs have replaced traditional RERSs by having the retained soil effectively retain itself through the creative use of internal tensile reinforcement. However, despite these significant, revolutionary, game-changing potentials, the use of geofoam with or as ERSs has not yet received anywhere near the attention, exposure, and use in practice that could make this happen (Horvath 1999b, 1999c, 1999d, 2001d).

It is the writer's opinion that the primary reason for this mismatch between potential and actual use to date is primarily educational in nature and not because of some underlying technical deficiencies in geofoam technology. Thus, the genesis and primary purpose of this monograph is to address this knowledge gap and summarize in one document the current state of knowledge of using geofoam either alone or synergistically with other geosynthetic materials for functional applications related to ERSs.

The writer wants to emphasize that the focus of this monograph is on the current state of knowledge which, in many ways, is far from perfect. Although both the lightweight-fill and compressible-inclusion functions for lateral pressure reduction on RERSs have been explored since at least the 1980s, there is still much to be done for both functions to verify, improve, and, as necessary, revise the current understanding of both behavior and analytical methodologies. Therefore, this monograph is intended to hopefully serve as a baseline and starting point for much-needed additional research of both functions.

As will be seen, this monograph is limited to *polystyrene* (PS)-based geofoam such as the aforementioned EPS-block and materials related to or derived from EPS-block for the simple reason that PS-based materials have proven to be the geofoam materials of choice for all ERS applications encountered or envisioned to date. However, as will be seen, the analysis and design methodologies presented in this monograph are completely general in nature so that their use is not limited to PS-based geofoam. Rather, the methodologies presented herein can be used, in principle, with any solid or nominally solid material whose basic mechanical (stress-strain-time-temperature) properties are known.

To accomplish these educational goals, this monograph is intended first and foremost to be a resource for sharing state of knowledge, practice-oriented technical information for use by licensed design professionals educated in civil engineering. It focuses on practical, usable design alternatives that to date have not yet entered the mainstream of geotechnical engineering textbooks and similar reference publications which means that relatively few design professionals in practice are even aware of the entire suite of design alternatives available to them no less use them.

However, an important secondary objective of this monograph is its use as a resource document for academic research and education as proved to be the case for the writer's first monograph (Horvath 1995b)[8] that dealt with the broad aspects of geofoam and was well-received internationally. Although in recent years there has been some academic research by others into the use of geofoam with RERSs, primarily under seismic loading, much of this work has, unfortunately, contained what the writer believes are some significant errors concerning interpretation of the load-bearing behavior of PS-based geofoam materials (Horvath 2010b). In the writer's opinion, these errors have, as a minimum, diminished the utility of the conclusions drawn by the researchers who performed the work. Consequently, to provide maximum benefit to researchers who might be interested in exploring and/or extending concepts presented in this monograph, included herein is considerable detail concerning the theoretical bases for the design alternatives presented along with an extensive list of relevant publications that includes both cited references and a supplemental bibliography.

Because this monograph is intended to be as informational and informative as possible, names of specific business entities and tradenames of their products are mentioned throughout this monograph. This has been deemed by the writer to be both desirable and necessary for the stated educational purposes of this monograph. However, in no way should this be interpreted as an endorsement by the writer of any named business or product, nor any guarantee or warrantee, expressed or implied, that a specific material or product will perform as desired in a specific project application. Liability for any application discussed in this monograph rests solely with the licensed design professional(s) involved in the design, construction, and/or operation of that application.

As it turns out, many of the geofoam materials, products, and concepts used for lateral earth and surface-surcharge pressure reduction on ERSs are also useful for reducing vertical earth pressures in two broad categories of applications:

- over *underground conduits* such as pipes, culverts, and small 'box' tunnels. This is simply a modern, geosynthetics-based version of a concept referred to as the *Induced Trench Method* or *Imperfect Ditch Method* that was noted and used more than a century ago, before the dawn of modern soil mechanics (Spangler and Handy 1982); and

- beneath *structural slabs*[9] built over expansive soil or rock.

Although these vertical-pressure applications are beyond the primary scope of this monograph, given that they are also lesser-known geofoam applications and have much in common with ERS applications, a chapter discussing the basic concepts of each of these two applications is included in this monograph. This material on vertical-pressure reduction is to provide the reader with basic knowledge that can be useful to those involved with the design and construction of underground conduits, especially under high fills, as well as those who deal with regionally significant problems involving expansive earth materials.

---

[8] All cited references are listed in a separate section at the end of this monograph. Note that there is an additional separate section that contains a selected bibliography of relevant publications that were either not accessed by the writer or not cited explicitly in this monograph.

[9] Structural slabs are defined herein as reinforced-PCC slabs constructed on the ground but not relying on the underlying ground (also called the *subgrade*) for long-term, permanent support of either the slab itself and/or loads applied to the top of the slab. This distinguishes structural slabs from *slabs-on-grade* that rely on the underlying subgrade for all long-term support. Structural slabs are typically framed into and supported on *grade beams* which are, in turn, framed into and supported by foundations (usually deep foundations).

1.4 OVERVIEW AND ORGANIZATION

The content of this monograph was developed and organized assuming no prior knowledge of using PS-based geofoam with ERSs. This was done intentionally so that this document might be useful to the broadest possible audience.

The remaining contents of this monograph are as follows:

- Chapter 2 discusses the fundamental aspects of the overall problem of ERS types and behavior including basic terminology used throughout this monograph; the lateral earth pressure states that can develop with ERSs; and the types of geosynthetics that can be used to reduce lateral earth and surface-surcharge pressures on ERSs.

- Chapter 3 provides an introduction and overview of the two distinctly different functional ways (lightweight fill and compressible inclusion) in which geofoam products, used either alone or synergistically together with other geosynthetics, are used to reduce lateral pressures on ERSs.

- Chapter 4 contains a detailed treatment of design of traditional non-yielding and yielding ERSs using the lightweight-fill function.

- Chapter 5 contains a detailed treatment of design of traditional non-yielding and yielding ERSs using the compressible-inclusion function.

- Chapter 6 deals with self-yielding ERSs. These are addressed separately from other ERSs as not only are they a unique category of ERSs but they are the only ERS application that can involve complementary use of both the compressible-inclusion and lightweight-fill functions simultaneously.

- Chapter 7 discusses additional geosynthetic functions that geofoam products can provide that are relevant to ERS applications. These additional functions are not always necessary or even desirable, but in some cases they are inherent in the geofoam product used so must be considered during design.

- Chapter 8 provides guidance on selecting geofoam and other geosynthetic products for the applications discussed in this monograph.

- Chapter 9 contains suggestions for future research and development.

- Chapter 10 contains an introduction to and overview of using the compressible-inclusion function to reduce vertical earth pressures over underground conduits and beneath structural slabs.

- Appendix A contains a summary of various lateral earth pressure solutions and other relevant analytical theories used as the basis for the analytical methodologies presented in this monograph. Most of this information has been published in various books and scholarly journals over the years but is presented here to serve as a convenient single-source reference and using uniform notation consistent with that used throughout this monograph.

- <u>Appendix B</u> contains a summary of the different parameters that have been proposed for use as metrics for quantifying the reduction in lateral pressures on ERSs under both gravity and seismic loading.

- <u>Appendix C</u> contains analytical methodologies and concomitant solutions for lightweight-fill partial-depth placement applications for ERSs under yielding conditions. This is original research by the writer, never published before, for a problem that has never been solved before.

- <u>Appendix D</u> contains a discussion of the use of 'true' vs. 'engineering' stress and strain with compressible-inclusion applications.

- <u>Appendix E</u> contains background information on the finite-element (FE) software used for various analyses in this monograph. The results for some baseline analysis cases are also presented in this appendix.

- <u>References</u> cited in this monograph, including Internet links for accessing a copy of a document wherever possible.

- <u>Supplemental Bibliography</u> of uncited references that are relevant to the contents of this monograph.

This page intentionally left blank.

Chapter 2

Problem Definitions and Fundamentals

2.1 TERMINOLOGY

Precise, correct use of technical terminology by a design professional is a necessity, not a luxury. It is an important, essential...though often underappreciated...component of engineered construction, especially in professional work-products such as consulting reports and construction specifications where written words have significant legal implications in terms of professional liability and construction contracts respectively. Toward that end, some basic terminology relevant to the subject matter presented in this monograph is defined in this chapter. This is especially necessary for the geofoam-related aspects of this monograph as the writer's experience to date indicates that there is currently widespread imprecise and sometimes grossly incorrect terminology used in practice, at least in the U.S., that has led to unnecessary construction claims and even construction-related litigation.

Note that every effort is made throughout this monograph to define and use correct terminology based on the writer's assessment of current, colloquial practice in the U.S., with past, regional, and international variations noted where deemed relevant for historical or informational purposes.

2.2 EARTH-RETAINING STRUCTURES

2.2.1 Types

This monograph focuses primarily on RERSs such as traditional retaining walls, bridge abutments, and below-grade (basement) walls of buildings as the greater potential for using geofoam to reduce lateral pressures is with such structures as opposed to with *flexible earth-retaining structures* (FERSs) as typified by sheet-pile walls and bulkheads. Furthermore, the geotechnology of using geofoam to reduce lateral pressures on ERSs is much more advanced and proven in practice for RERSs compared to FERSs.

Note that the terms 'rigid' and 'flexible' are used here in a relative context. A RERS is defined as an ERS constructed of such materials and geometry and in such a manner that it can be assumed to be *non-deforming*, at least under service loads[10] and up until the point of a collapse-type failure (the *Ultimate Limit State*, ULS) of the overall ERS-ground system. In this context, 'non-deforming' means that the ERS does not change or distort its physical shape to any meaningful or significant extent, at least under service loads.

It is important to note that even though a RERS is assumed herein to be inherently non-deforming it may be capable of rigid-body *displacement*, primarily in the lateral direction, if not restrained either structurally or geometrically. In this context, 'displacement' means movement in some *mode* or combination of modes relative to the

---

[10] Consistent with current usage, the term 'service loads' is synonymous with the older term 'working loads' and means the best estimate of actual loads under some specified load case.

original position of the RERS. Displacement modes applicable to RERSs are discussed later in this chapter.

Although the use of geofoam with FERSs appears to have less technical benefit and is substantially less developed analytically compared to RERSs, this monograph does address the subject for the sake of completeness. By definition, FERSs are inherently deformable in the lateral direction under service loads, at least compared to RERSs (Horvath 1997a). Note that this deformation is separate from and in addition to any lateral displacement of the overall structure or portions thereof.

Finally, as noted in Chapter 1, there are situations where with certain geofoam products arranged in certain geometric configurations no ERS of any kind is required to restrain the geofoam. Such applications, where the geofoam is essentially its own ERS, are referred to colloquially as geofoam walls.

2.2.2 System Components

Figure 2.1 illustrates some of the generic terminology and notation used throughout this monograph to describe physical components of and locations on an overall system consisting of an ERS plus surrounding ground. This terminology is not limited to free-standing gravity retaining walls (a common type of RERS) as used in this figure for illustrative purposes only.

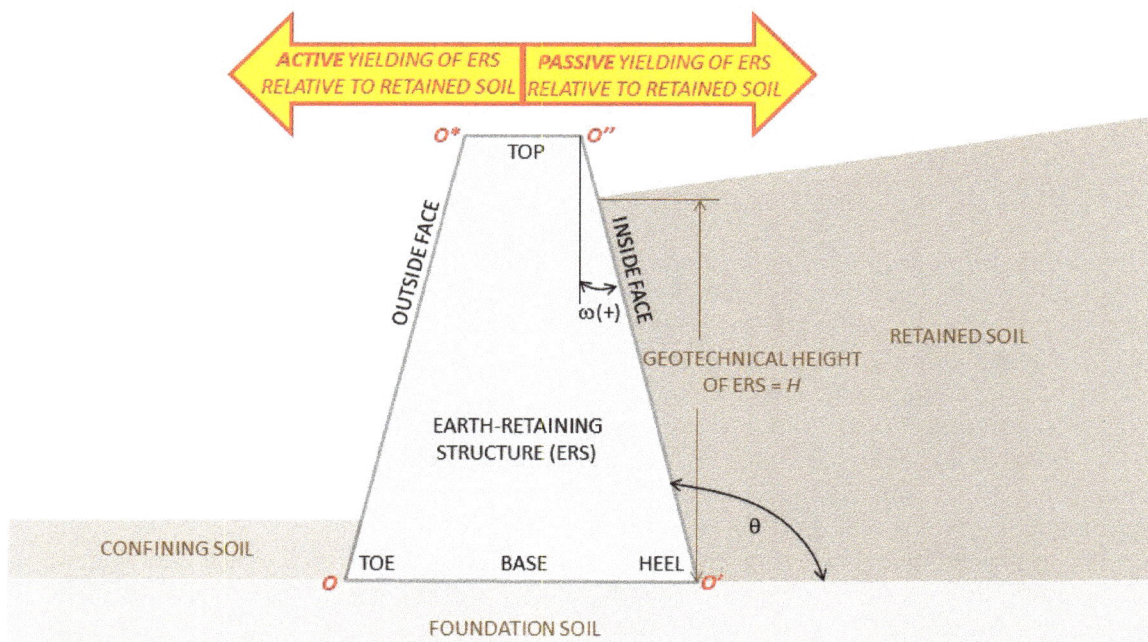

Figure 2.1. Basic ERS Geometry and Definitions.

Note that it is assumed in this monograph that whatever the particular type of ERS, its inside face is always planar. This is because the case of a non-planar inside face (usually stepped in geometry) turns out to be a surprisingly complex problem to deal with rigorously and correctly, with widely varying degrees of accuracy of approximate methods that have been used or proposed for use in routine practice (Horvath 1990a). However, although the inside face is planar, it is not necessarily vertical as $0° < \theta < 180°$[11] is considered allowable in principle for the traditional lateral earth pressure theories used throughout this monograph.

Note also that although θ is always positive in sign, the sign of the angle ω (= θ - 90°) as defined in Figure 2.1 can be positive (as shown) or negative, a detail that is important in some applications presented in this monograph.

Note too that the *geotechnical height* of the ERS, *H*, which is the dimension used to calculate lateral pressures from the retained soil and surface-surcharge on the surface of the retained soil, is distinctly different from the actual or physical height of the ERS which is used to calculate the weight (under static loading) or mass (under dynamic loading) of the ERS. Textbooks, handbooks, and similar documents seldom make this distinction (the top of the ERS is often shown flush with the surface of the retained soil, something that would rarely be done in practice for any number of technical and legal reasons) which is important in practice in almost all cases. Also, the geotechnical height of an ERS is always a vertical distance measured relative to the heel (Point O' in Figure 2.1) of the ERS. This is true even if the inside face of the ERS is not vertical, i.e. $\theta \neq 90°$, and/or the base of the ERS is inclined relative to the horizontal.

Note further that, in general, the confining, foundation, and retained soils may each be different in composition and each may be classified as *backfill* or *fill*. The terms 'backfill' and 'fill' as used in this monograph are not synonymous although these terms are often used synonymously (and incorrectly) in U.S. practice, even by design professionals and construction contractors who know, or should know, better.

More to this point, 'backfill' is material used to replace existing material (that may or may not be soil) at a site that was excavated as part of the ERS construction process. The backfill material may be the excavated material if it is satisfactory for the purpose or it may be a different material that is usually imported to the site or from elsewhere on the site. Furthermore, backfill may or may not be placed up to the original ground-surface elevation of the removed material.

On the other hand, 'fill' is material placed above the original surface elevation to an elevation greater than that which existed prior to construction. Note that fill might be placed on backfill or directly on original ground. Fill material on a given project may or may not be the same as backfill material.

These terminological distinctions between backfill and fill are not trivial, semantic exercises. Even though it is not uncommon in practice for backfill and fill materials to be similar, if not identical, in nature on a given project, they are often bid contractually at different unit prices (i.e. per cubic metre or cubic yard) because of significantly different conditions involved in their respective placement. For example, backfilling a relatively

---

[11] The SI (*Système International d'Unites*) version of metric units is primary throughout this monograph although equivalents in *Imperial* (a.k.a. *U.S. customary* or *English*) units are given as well in most cases. The one exception to this is that angles are stated only in *degrees*, which is technically the Imperial unit for angular measurement, as opposed to the SI angular-measurement unit of *radians*. This is to be consistent with geotechnical engineering practice worldwide. Furthermore, this monograph follows the recommended SI-unit practice of spelling the basic unit of length measurement as *metre* to avoid confusion with the word 'meter' meaning a measurement device or instrument of some type and purpose. Note that this preferred spelling of metre in an engineering/scientific context has nothing to do with the general interchangeability of '-er' and '-re' word endings between American and British/Canadian English in broader, non-technical, colloquial usage.

narrow utility trench can be relatively expensive on a unit-volume basis because of the attendant safety and productivity issues of working in a confined space below the adjacent ground surface. On the other hand, placing fill material over a large, open area across a site can usually be done very efficiently by comparison so the unit-volume cost basis would be expected to be less even if it is the same material used for backfilling.

Separate from construction considerations, there can be important distinctions related to design that makes the distinction between backfill and fill potentially significant. Because placing fill by definition creates a new ground surface that is higher than the original ground surface ('raises the grade' is a common colloquial expression), the resulting vertical stresses on the underlying foundation soils will generally be greater than those that existed originally. Depending on the nature of the foundation soils, this can result in significant settlement of the ERS-ground system that needs to be considered.

2.2.3 Analysis Versus Design

It is important to distinguish between *analysis* and *design*, terms that are also often used (incorrectly) synonymously. 'Analysis' is arguably the more fundamental of the two as, in its most basic form, it is the fundamental deterministic exercise of many engineering disciplines and involves performing a single, explicit calculation or sequence of calculations involving one or more unknown parameters to get a specific, repeatable outcome ('the answer'), i.e. the value(s) of the unknown parameter(s).

A simple example in civil engineering is the case of an existing beam with the problem geometry (span length and supports), all material properties, and loading known. One analysis for that system might involve calculating the maximum deflection of that beam under the specified conditions using a specific theoretical solution or analytical methodology. There would be only one calculated outcome for such an analysis, no matter how many times or by whom it was performed. Assuming that the theoretical solution or analytical methodology used was acceptably accurate for the application, the calculated outcome would be unambiguously considered 'the' correct answer.

On the other hand, using the same basic problem scenario, a 'design' would be when there is no existing beam but one is proposed to span the known distance and carry certain loads. In principle, there are numerous combinations of beam materials and geometries that could be used to satisfy the technical need. There is even the potential to use an entirely different structural concept (e.g. a truss or cable-supported structure) to satisfy the design requirements. In short, there is no, one explicit 'correct answer' in this case. However, after considering other factors including, but not limited to, applicable design codes, costs, schedule, and compatibility with other project needs such as aesthetics and other client inputs, a licensed design professional would be expected to come up with the overall optimum result. Note that among the many design tasks a design professional would undertake during the design process would be to calculate the maximum deflection of different design alternatives. So, beam-deflection <u>analysis</u> in this case is but one component of the overall <u>design</u> process.

In summary then, analysis involves an explicit, straightforward (in principle at least) calculation for some known or assumed condition to produce an explicit answer that is presumed to be 'correct'. In practice, analysis is used both to assess the condition of an existing structure as well as a component of the design process for a proposed (newly built) structure.

As applied to the contents of this monograph, because the technologies presented herein are new to most potential users, the vast majority of applications in the near future will be to design new installations of these technologies as opposed to analyzing the condition of existing structures built with these technologies. Therefore, the emphasis of this monograph is on the overall design process using these technologies. However, because analysis is an inherent component of any design process, the necessary analytical methodologies are identified and explained where necessary as well.

2.2.4 New Versus Existing

As noted previously, one of the many desirable features offered by geofoam when used to reduce lateral pressures on ERSs is that technical and economic benefits can be realized with both new construction as well as existing structures. Thus, all of the basic concepts and methodologies presented and discussed in this monograph are equally applicable to both new and existing ERSs, at least in principle.

However, the practical realities of working with an existing ERS, specifically, being able to excavate and replace existing material retained by the ERS within the physical and operational constraints that may exist at a given site, sometimes means that the actual geofoam-based design that is ultimately implemented is based on what is reasonably and economically achievable, not what might be the maximum desirable from a purely technical perspective if one were building the same ERS new. More specifically, practical limitations sometimes cause geofoam configurations to be used that deviate from a basic or ideal configuration that would normally be used for new construction or reconstruction if there were no site constraints.

For obvious reasons, it is impossible to anticipate and address in this monograph every potential deviation from the ideal that might ever arise in practice. Therefore, this monograph focuses primarily on the basic ideal geofoam configurations that would be used with few exceptions for all new construction and would be desirable for use with existing ERSs in the absence of site and project constraints that dictate otherwise. However, there is some additional, general discussion of how these basic configurations are likely to be altered in practice when dealing with existing ERSs and how this affects the analyses performed.

2.2.5 Yielding (Lateral-Displacement Restraint) Conditions

2.2.5.1 Introduction and Overview

With reference again to Figure 2.1, a common term used primarily in the published literature generated by academicians relative to all types of ERSs (but RERSs in particular) is *yielding*. In this context, the term yielding is synonymous with 'lateral displacement', more specifically, the <u>potential</u> for lateral displacement of an ERS that may or may not be actualized under service loads ('potential' cannot be emphasized too strongly). Throughout this monograph, the symbol Δ is used for yielding (= lateral displacement).

Note that either 'displacement' or 'movement' is arguably a better term to use than yielding in this context as either word more clearly describes the physical process involved. Furthermore, the term yielding has older, well-established, more widely known and used definitions with significantly different meaning in both solid and particulate mechanics (typically with regard to material stress-strain behavior and strength) and should arguably be reserved for those traditional definitions. Nevertheless, the term yielding is used throughout the remainder of this monograph exclusively for the above-stated displacement-related definition to be consistent with its extensive historical usage in the published literature with ERSs in general and RERSs in particular.

Historically, yielding has been used to define two broad categories of RERSs:

- *non-yielding* which is when the RERS is restrained either structurally or geometrically against any lateral displacement and

- *yielding* which is when the RERS is unrestrained either structurally or geometrically against lateral displacement[12].

These assumed conditions are used for two broad purposes in both the analysis of existing RERSs and the design of new ones:

- to define the *earth pressure state* (*active* or *at-rest*) within the retained soil that is assumed to exist for both geotechnical and structural purposes[13], at least when using traditional *Allowable Stress Design* (ASD)[14]-based methodologies with safety factors and

- to determine whether or not geotechnical (as opposed to structural) failure mechanism(s) of the RERS are physically possible as the primary mechanism(s) of ULS-type failure of the overall RERS-ground system.

It is important to point out that whether a RERS is classified as being non-yielding or yielding is not so much determined by the inherent nature of the RERS itself (although this is true in some cases) but by the physical and/or geometric constraints placed on a RERS in a specific application. Stated another way, the same basic type of RERS might be non-yielding in one application but yielding in another solely by virtue of different physical and/or geometric constraints between the two applications.

This point is strongly emphasized here as the writer's experience is that both textbooks/design manuals and design professionals alike often do not recognize and appreciate the fact that the yielding vs. non-yielding decision is almost always governed by the specific application and not the type of RERS itself. Consequently, this subtlety will be explained further in the following sections.

Also defined and explored in following sections are the newly identified third category of RERSs called *self-yielding* as well as a discussion of the various *modes* of yielding.

Finally, discussed in a following section is the issue of yielding of FERSs vs. RERSs. Historically, this subject has not received any extensive, formal attention but for reasons that will become clear later in this monograph this issue needs to be addressed up front.

2.2.5.2 Modes of Yielding

Before discussing non-yielding, yielding, and self-yielding RERSs further, it is very important to discuss one of the most significant discoveries in recent decades related to yielding RERSs. This is the fact that the *mode* of yielding is an important parameter for

---

[12] A special case or exception to this general rule illustrated subsequently occurs when displacement may be partially restrained structurally so that yielding can occur but is confined to a single physical mechanism (mode).

[13] As will be seen, there can be sound, logical reasons why different earth pressure states might be assumed for geotechnical and structural calculations on the same RERS.

[14] This was called *Working Stress Design* (WSD) in the past.

understanding the behavior of RERSs in general and especially the lateral earth pressures (in terms of both magnitude and distribution) that develop behind them in particular. In this context, mode means the physical mechanism and concomitant displacement pattern by which RERS yielding occurs.

The potential modes of yielding for a RERSs are:

- *Rigid-body translation* (a.k.a. *translation* or, colloquially, *sliding*) which is single-value (i.e. Δ = constant) displacement of the entire RERS as an assumed rigid body along an assumed planar surface that is defined by the spatial orientation of the base of the RERS. For most RERSs, the base is horizontal as shown in Figure 2.1 so the direction of sliding will also be horizontal. However, with newer bin-type gravity retaining walls or SRWs, it is possible and sometimes desirable to place the bottom layer of bins or blocks that define the base of the wall at a slight angle relative to the horizontal so that the toe (Point O in Figure 2.1) of the RERS is slightly higher than the heel (Point O' in Figure 2.1). In such cases, sliding along an inclined surface is not only kinematically possible but likely so should be considered where appropriate.

- *Rigid-body rotation* (a.k.a. *rotation*) which is angular displacement of the entire RERS as an assumed rigid body about some pre-defined fixed point on the RERS. This mode is further sub-divided depending on whether the fixed point of rotation is at the base (specifically the toe, Point O in Figure 2.1) or top (either Point $O*$ or Point O'' in Figure 2.1) of the RERS. These sub-modes are referred to as *rotation-about-bottom* and *rotation-about-top* respectively. In practice, rotation-about-bottom is the more common by far and is sometimes referred to colloquially as *overturning* although this term is better applied only to the ULS, i.e. rotation of a RERS about its toe to the point that it overturns completely. Rotation-about-top can occur only when there is some partial or limited physical restraint applied to the RERS at its top but the RERS is otherwise free to yield (this is the special case or exception to the general definition of a yielding RERS noted previously). This restraint is generally structural in nature and nowadays is most commonly found to occur with *integral-abutment bridges* (Horvath 2000, 2004c, 2005).

Note that in reality, an actual RERS may yield in one mode or some combination of modes. However, in the latter case, experience indicates that one mode usually predominates.

Historically (and to the present in most codified or standardized design methodologies used in routine practice), yielding mode was not recognized as being a relevant or necessary variable to consider when analyzing or designing a RERS. This is understandable as the classical lateral earth pressure theories for RERSs such as Coulomb's traditional solution or a modern-day (circa mid-20th century) 'exact' solution such as Caquot-Kerisel's (which is essentially an evolutionary improvement to Coulomb) ignored yielding mode completely in their theoretical development. These solutions assume that as long as the RERS yields in some unspecified fashion and magnitude, the same (at least in terms of the magnitude of resultant earth force) limiting active earth pressure state will develop within the retained soil. The triangular (i.e. linearly increasing with depth in hydrostatic or equivalent-fluid-pressure fashion) distribution of lateral earth pressures that is typically assumed to represent these earth pressure states is also assumed to be independent of yielding mode.

By comparison, both theory and research, including the instrumented observation of full-scale RERSs, indicate clearly that the actual shape of the lateral earth pressure distribution is parabolic although the exact geometry is strongly dependent on yielding mode (Tschebotarioff 1973, Fang and Ishibashi 1985, Handy 1985, Harrop-Williams 1989). This behavior appears to be due to the fact that the retained soil is not a rigid block of material as assumed in the theoretical formulations of classical earth pressure solutions (Coulomb and its modern derivatives such as Caquot-Kerisel) but undergoes deformation as yielding occurs that induces the phenomenon of arching to develop within the retained soil (Handy 1985, Harrop-Williams 1989). The significant impact of this actual behavior is that the resultant earth force typically acts somewhat above the one-third-of-geotechnical-wall-height point that is implied by the traditional triangular lateral earth pressure distribution. Depending on the particular geotechnical or structural analysis being performed, this can produce a more-critical condition than that obtained with traditional solutions.

In the writer's experience (Horvath 1997c), the influence of yielding mode on the magnitude and distribution of lateral earth pressures continues to be largely unrecognized and ignored to the present, at least in U.S. practice, although it has been broadly recognized and proven by research for almost as long as modern soil mechanics has existed. The writer finds this disconnect between reality and practice surprising as even in the earliest days of modern soil mechanics the fact that the simplistic triangular distribution of lateral earth pressures did not always exist with ERSs in general was well known and established (Terzaghi 1943) so should have always been part of even the most basic geotechnical knowledge-base taught on the undergraduate level. Subsequent research and publications as cited above have only improved our understanding of the actual distribution of lateral earth pressures and produced analytical methodologies to allow its being considered in even routine practice. Thus, while it is defensible for a licensed design professional to say that yielding mode was considered in an analysis or design but the simpler traditional solutions used instead because they were judged to be more conservative in their outcomes and easier to use, it is indefensible in the writer's opinion in this day and age to ignore yielding mode altogether.

In any event, the reasons that the issue of yielding mode is not only noted but highlighted and emphasized here are twofold. First, the parabolic lateral earth pressure distributions that result from an understanding of yielding modes are observed with certain geofoam applications with RERSs. Thus, a basic knowledge and understanding of yielding modes is essential for some of the concepts presented in this monograph.

Second, as will be seen, simplified analytical methods developed for some geofoam applications presented and discussed in this monograph utilize empirical values of lateral displacement of the RERS, Δ. These empirical values are defined very specifically depending on yielding mode.

2.2.5.3 Non-Yielding

A non-yielding RERS is defined as being laterally restrained either physically or geometrically against all the modes of yielding. Physical restraint is the more common condition and is defined as some system of relatively fixed reaction points external to the RERS that prevents its lateral displacement in any mode. Typically, this is the result of some structure or components of a structure external, but mechanically connected, to the RERS.

Geometric restraint is due to the three-dimensional geometry of the RERS itself preventing displacement. For example, the simple gravity retaining wall shown in Figure 2.1 may look as though it is free to displace laterally and thus be considered to be in the yielding category. However, if this RERS were curved in plan view, this requisite lateral displacement could not occur for all cross-sections of the wall simultaneously so the wall would actually be non-yielding in this case. This simple example illustrates the need to always consider the three-dimensional setting of a proposed (new) or existing RERS before making a final determination of its potential to yield.

Regardless of the source or cause of restraint, the hallmark of non-yielding RERSs is that they are typically designed (or at least should be designed) assuming that the at-rest earth pressure state exists within the retained soil under service loads, with due consideration, where appropriate, of residual compaction-induced stresses as well (Duncan et al. 1991, 1993; Clough and Duncan 1991; Filz 2003).

Examples of non-yielding RERSs include:

- below-grade (basement) walls of buildings, where floors and slabs that frame into or otherwise abut the wall provide physical restraint;

- abutments for traditional jointed bridges, where the bridge deck or superstructure provides physical restraint by virtue of the bearings that rest on the abutments;

- water and wastewater treatment tanks[15], where their complex, as opposed to linear, geometry in plan view inherently prevents the wall(s) enclosing such tanks from displacing laterally simultaneously;

- navigational locks[16], where the relatively massive nature of their sidewalls makes them inherently non-yielding; and

- otherwise free-standing retaining walls that are either physically or geometrically restrained.

Note that a non-yielding RERS cannot, by definition, ever suffer a geotechnical failure as its primary ULS mechanism, at least with regard to the traditional geotechnical failure mechanisms of sliding and overturning. Nor is the geotechnical *Serviceability Limit State* (SLS) mechanism of non-positive bearing (contact) stresses along the base of the RERS likely to ever be a concern. Rather, any ULS or SLS will always be primarily structural in nature, at least initially, and usually involve flexural failure of the RERS itself. Note, however, that the structural ULS of the RERS will have geotechnical consequences, i.e. the soil retained by the RERS will almost always slump downward after the RERS has collapsed structurally.

2.2.5.4 Yielding

A yielding RERS must always satisfy two conditions:

- There are no inherent physical or geometric constraints on the overall RERS-ground system that would inherently preclude at least one of the three modes of yielding defined previously from occurring. As a result, the RERS has the potential (which, in a given application, may or may not be actualized for reasons discussed subsequently) for more-or-less unlimited translation and/or rotation without any inherent constraint other than overall force and moment equilibrium within the RERS-ground system.

---

[15] In some cases, such structures can actually behave in the category of self-yielding that is discussed later in this chapter.

[16] See the prior footnote for water and wastewater treatment tanks.

- The nature of the RERS is such that it yields solely as a response or reaction to the retained soil, i.e. it is the inherent force of the retained-soil mass alone that is governing or driving the yielding of the RERS.

The meaning and significance of the latter condition will become clear in the following section.

The hallmark of a yielding RERS is that it is typically designed, at least geotechnically[17], assuming that the active earth pressure state can develop within the retained soil under service loads, possibly with a consideration of residual compaction-induced stresses as well (Duncan et al. 1991, 1993) although this appears to be rarely done in routine practice based on the assumption that such compaction-induced stresses are inherently self-correcting as they will dissipate, or at least have the potential to dissipate if necessary, as a result of RERS yielding. The classical example of a yielding RERS is a free-standing gravity or cantilever retaining wall that is not physically or geometrically restrained and is thus capable of displacing laterally as a result of the loads placed on it by the retained soil plus any surface surcharge on the surface of the retained soil.

It is important to note that a yielding RERS has, in principle, only the potential for lateral displacement sufficient to mobilize the active earth pressure state within the retained soil. Whether or not the RERS actually yields sufficiently to actualize this potential and mobilize the active state over the full height of the RERS in a given application is dictated by the relative magnitudes of driving and resisting forces acting on the RERS under service load conditions. In fact, observation of instrumented retaining walls indicates that the active earth pressure is usually not fully developed and mobilized under service load conditions for a properly engineered RERS.

The fact that a well-performing yielding RERS may not actually yield sufficiently in-service to mobilize the active earth pressure state over it full geotechnical height derives from the fact that under service load conditions a 'safe' (i.e. stable) RERS will always have an excess of resisting forces compared to driving forces in order to provide the requisite margin of safety against the geotechnical ULS for the overall RERS-ground system. As a result, the actual lateral earth pressures may remain at their post-construction level (usually assumed to be the at-rest state plus some compaction effects) or, more likely, be somewhere between the initial at-rest and assumed final active states.

Field measurements of actual instrumented RERSs indicate that the actual distributions of lateral earth pressures that develop tend to be relatively complex. Specifically, the earth pressure states that develop are not depth-wise uniform but tend to vary between the at-rest and active states as a function of depth along the inside face of the RERS. This is because the lateral displacement of an RERS is not necessarily uniform with depth due to combinations of translation and rotation-about-bottom modes of yielding.

This behavioral nuance concerning the actual vs. assumed (in design) earth pressure state existing behind a yielding RERS is neither often discussed in the literature nor considered routinely in practice. Whether this is good or bad is actually a complex assessment.

On the one hand, it is irrelevant as far as the geotechnical behavior of the overall RERS-ground system is concerned because geotechnical safety is based on assumed

---

[17] In this context, 'geotechnically' means with regard to calculating what is traditionally considered to be the geotechnical safety of the overall RERS-ground system with regard to translation, rotation, bearing stresses and capacity, and global stability of the RERS as an assumed rigid body. This is distinctly different from the structural design of the RERS itself.

conditions that exist under the ULS, not service, conditions. So, if a RERS is stable geotechnically under loads greater than would exist at the ULS (due to excess geotechnical capacity or 'safety' being built into the design) then so be it. Because the overall behavior of a yielding RERS is driven by geotechnical conditions, the RERS can simply yield an additional amount if necessary (within limits of course) to mobilize additional shear strength within the retained soil and reduce loads from the retained soil if need be.

Bottom line, a yielding RERS designed geotechnically assuming the active earth pressure state within the retained soil will, all things considered, always perform adequately in service because sufficient reserve geotechnical capacity has been built into the system. This was true under the traditional ASD approach and remains true under newer *Load and Resistance Factor Design* (LRFD)[18] approach that intentionally 'factors up' the active earth pressures to build a margin of safety (actually performance reliability) into the system.

Note, however, that the situation is very different with regard to the <u>structural</u> behavior of a yielding RERS. The reason is that for the reasons explained above, a yielding RERS may actually have lateral earth pressures at or close to the at-rest, as opposed to active, state acting on it day in, day out indefinitely during the service life of the RERS. This means that the structural components are theoretically overstressed day in, day out on an open-ended basis. However, in this case the structural components cannot self-reduce this overstress (unless there is substantial displacement or deformation of the structural components) because the yielding of the RERS, which is the only way in which the earth loads can reduce from the at-rest to active earth pressure state, is dictated solely by geotechnical stability of the RERS as explained above.

Stated another way, the <u>geotechnical</u> failure mechanisms (sliding, overturning, etc.) have a natural, built-in 'safety valve' against overstress as the RERS can yield any time it needs to reduce the earth loads from the original, as-constructed at-rest state to the active state. However, the <u>structural</u> failure mechanisms (overstressing of the RERS materials) have no such built-in load-reduction capability and thus no corresponding safety valve. As a result, the actual day in, day out margin of safety for the <u>structural</u> components of the RERS would actually be less than what the designer planned on and anticipated based on using active earth pressures for structural design as is typically done in practice. If such a situation exists, then the RERS may be at risk for structural failure, at least the SLS (e.g. excessive cracking of PCC) if not the ULS (e.g. yielding of reinforcing steel), occurring before the geotechnical ULS even though the designer thought the opposite case exists (or at least did so 'on paper'). This is especially true given the fact that the geotechnical safety of yielding RERSs is generally greater than thought because it is common in practice to use conservative assumptions concerning soil properties used in design.

This raises the issue that it may actually be rational to design a yielding RERS for at-rest lateral earth pressures structurally but active lateral earth pressures geotechnically. Alternatively, given the global migration away from traditional ASD with safety factors to LRFD with load and resistance factors, it may be logical to increase the load factors on the structural components to allow for the fact that the structural components are likely to be 'overloaded' routinely and continuously in practice without any natural, built-in mechanism for load reduction as exists with the geotechnical components. Again, the basis-in-logic for this is the fact that a yielding RERS will typically have lateral earth pressures closer to the at-rest state acting on it during its design life due to geotechnical margins of safety built into the design. And while the overall RERS can yield as necessary during its life to reduce loads

---

[18] Formerly known as *Ultimate Strength Design* (USD).

as necessary, the structural components of the RERS have no such built-in pressure-relief mechanism other than to suffer the ULS.

2.2.5.5 Self-Yielding

It is now recognized that in addition to the traditional conditions of non-yielding and yielding RERSs there is a third condition, defined and studied only in recent years, that will be referred to in this monograph as *self-yielding*. RERSs that fall into this category are usually physically connected to an external structural element (referred to collectively and generically in this monograph as the *superstructure*) as often occurs with non-yielding RERSs[19]. However, there are some fundamental differences.

To begin with, unlike in the case of a non-yielding RERS whose lateral displacement is effectively prevented because the superstructure to which it is attached is not only effectively rigid but non-displacing as well, in the case of a self-yielding RERS the attached superstructure displaces more or less continuously as a function of time for the life of the structure under service-load conditions. Typically, this occurs with repeated reversals of direction on a periodic, cyclic basis that is driven by seasonal variations in atmospheric temperature. This means that the RERS also displaces laterally continuously with time due to its physical connection(s) with the displacing superstructure. Furthermore, this lateral displacement of the RERS reverses direction periodically, alternating between moving away from and into the retained soil as illustrated in Figure 2.1, a feature that is unique compared to yielding RERSs where the lateral displacement only occurs monotonically in one direction, i.e. away from the retained soil.

The term 'self-yielding', which was coined by the writer, derives from the fact that in this case the source of the yielding of the RERS is the RERS itself. Thus, the important, in fact defining, feature of a self-yielding RERS is that the lateral displacement of the RERS is not a reaction or response to loading from the retained soil (as in the traditional case of a yielding RERS) but is the direct result of displacement of the attached superstructure. In fact, the cause-effect roles between RERS and retained soil are reversed in the case of a self-yielding RERS so that it is now the retained soil adjacent to the RERS that is reacting (in terms of its lateral displacement and concomitant earth pressure state) to the RERS displacement rather than the other way around as in the case of a yielding RERS. And because the superstructure and its attached self-yielding RERS are constantly displacing with time this means that the earth pressure state within the retained soil adjacent to the self-yielding RERS is also constantly varying with time, both increasing and decreasing repeatedly as a result of the typically periodic, cyclic nature of the superstructure-plus-RERS displacement.

In summary, the truly unique characteristic of a self-yielding RERS is that it causes or initiates lateral displacement on its own (by virtue of the displacing superstructure to which it is connected) and does not displace as a reaction to the retained soil. In fact, it is the RERS that causes reactions in the retained soil and not the other way around as happens in traditional RERS applications, both non-yielding and yielding. Complicating the overall problem is that all these cyclic displacements and variations in lateral earth pressure are occurring continuously and with constantly reversing directions as a function of time for the entire service life of the overall structure. As will be seen, the cyclic changes in lateral earth pressures result in complex geomechanical behavior of the retained soil.

---

[19] Self-yielding RERSs can also exist on their own, without being connected to a superstructure, in situations where the RERS is its own superstructure. Examples are given at the end of this section.

The most common, classic example of a self-yielding RERS is the broad category of what are referred to as *jointless bridges* (Horvath 2000, 2004c, 2005). These are bridges, typically used for roads, that eschew the traditional use of expansion joints and, in most cases, the bearings between the superstructure and RERSs (abutments in this case) on which the superstructure is supported. The predominant type of jointless bridge is the *integral-abutment bridge* (IAB)[20] in which the bridge deck is the superstructure that is both physically and structurally connected at each end to an abutment that is the self-yielding RERS. Thus, in the IAB design, bridge bearings are eliminated in addition to expansion joints. Because the superstructure (bridge deck in this case) expands and contracts with natural, ambient air-temperature changes, it forces the abutments connected to it to move. Displacement of the abutments includes both translation and rotation-about-bottom modal components although research and experience indicate that the latter rotational modal component tends to predominate as all the induced displacement is concentrated at the top of the RERS where the bridge deck connects with the abutment.

The other type of jointless bridge is the *semi-integral-abutment bridge* (SIAB)[21] design in which the superstructure (bridge deck) is extended horizontally compared to the traditional jointed-bridge design to cover the top of each abutment and actually butt directly against the retained soil. The traditional bridge-bearing detail is retained in the SIAB design but is moved to the top of what it usually a relatively short 'stub' type of abutment so that the bridge deck can move horizontally relative to the top of the abutment as a result of seasonal temperature variation and concomitant length changes. With the SIAB design, the abutment remains more or less spatially fixed (so is a traditional non-yielding, not self-yielding, RERS) and only the ends of the bridge deck (which function as a RERS) displace translationally toward and away from the retained soil.

For the sake of completeness, it is worth noting that self-yielding RERSs can exist on their own without a connected superstructure to 'drive' them. Experience indicates that this typically involves a RERS that is essentially its own superstructure and contains an open body of liquid on one side. Such applications typically involve either a linear RERS (as in the case of navigational locks) or a RERS that is circular in plan view (as in the case of a water or wastewater treatment tank). Some of these applications are discussed in detail by England (1994), England and Dunstan (1994), and England et al. (1995).

These publications by England and colleagues also contain a very detailed presentation of the complex geomechanics behavior that occurs within the retained soil adjacent to a self-yielding RERS. In particular, a phenomenon called *ratcheting* is defined and explained. This is a mechanism that, because of the nonlinear behavior of soil-structure systems, results in a self-yielding RERS not load-cycling in a linear manner, i.e. always returning to the position it started from. As a result, lateral earth pressures within the retained soil tend to accumulate after each cycle (typically annual) of temperature-induced displacement and thus exhibit a net increase over the life of the RERS. This can result in SLS or even ULS structural failure of the RERS at some point in the future unless the RERS is designed for these long-term earth loads that can approach the classical passive earth pressure state.

In any event, even though self-yielding RERSs constitute a fascinating, increasingly common type of RERS (by virtue of the increasing use of IABs), the focus of this monograph is on RERSs under the traditional conditions of non-yielding and yielding as the use of geofoam for reducing lateral pressures with these types of RERSs is sufficiently developed

---

[20] IABs are also referred to in the literature as (alphabetically) *integral bridges, integral bridge abutments, rigid-frame bridges,* and *U-frame bridges.*

[21] SIABs are also referred to in the literature as *integral-backwall bridges.*

so that they could and should be used in routine practice. However, recent research (Horvath 2000, 2004c, 2005) has indicated that several geofoam materials and products have great potential benefit when used with self-yielding RERSs, especially IABs. Although this potential has been demonstrated by numerical modeling as well as limited application in practice, the state of knowledge has not yet advanced to the point where relatively simple but reasonably reliable design methods are available for such applications. Nevertheless, a basic, introductory treatment of applications involving self-yielding RERSs is presented in Chapter 6.

2.2.5.6 Non-Rigid Yielding (Flexible Earth-Retaining Structures)

As stated previously, the focus of this monograph is on RERSs as with very few exceptions all of the known research and application to date involving geofoam for lateral pressure reduction have involved such ERSs. The relatively few applications to date involving FERSs are discussed at the end of Chapter 4. However, it is the qualitative behavior of FERSs, which are always yielding in nature due to their inherent flexibility and ability to deform, that actually turns out to be of some potential interest with regard to geofoams and RERSs. Thus, the qualitative behavior of FERSs merits some discussion.

It is obvious that by virtue of their inherent rigid nature the yielding of RERSs can be defined by a single value of Δ, either that of the entire RERS in translation mode or the peak value at either the top or bottom of the RERS in the relevant rotation mode. As will be discussed at length later in this monograph, this fact has been used historically (e.g. Clough and Duncan 1991) to develop empirical correlations for the minimum magnitude of Δ necessary to mobilize either the active or passive earth pressure state with conventional RERSs.

More importantly and of significant relevance to this monograph is that such correlations have figured prominently in at least some applications of geofoam to reduce lateral pressures on RERSs that date back to the 1980s. Specifically, correlations with Δ have been used as the cornerstone of a simplified analytical methodology for the 'sizing' (determining the minimum required thickness) of geofoam when used as a compressible inclusion. As will be seen, in this case Δ is assumed to be the magnitude of the compression of the geofoam. This is because it is the geofoam that is yielding in cases where the RERS does not.

However, more-recent research conducted by the writer during preparation of this monograph indicated some flaws in this reasoning as it applies to geofoam and RERSs. This is because the depth-wise pattern of yielding (geofoam compression in this case) associated with these geofoam applications is highly non-uniform and thus qualitatively similar to that found not with RERSs but with FERSs such as sheet-pile walls/bulkheads and the facing of MSEWs with non-rigid facing systems such as wrapped geotextiles.

Unfortunately, extending the knowledge gained over the years with FERSs is of little to no benefit for the geofoam applications discussed in this monograph. To begin with, when dealing with FERSs, Δ is not typically correlated with lateral earth pressure state which tends to be very complex for FERSs and thus does not lend itself to simplistic classification as 'active' or 'passive' as in most cases involving RERSs. Rather, Δ of FERSs is typically correlated with another parameter or metric such as damage to adjacent structures (for braced excavations) or relative tensile-reinforcement stiffness (for MSEWs). Furthermore, even if it were possible to correlate earth pressure state with Δ for a FERS, when dealing with non-uniform patterns of Δ it becomes a subjective 'judgment call' as to whether any empirical correlations with mobilized earth pressure states should be based on a maximum value of Δ, a depth-wise average value, or something else.

In summary, as will be seen later in this monograph, for certain geofoam applications involving lateral earth pressure reduction it is desirable to correlate Δ with expected behavior in terms of the magnitude of earth pressure reduction. In view of the preceding discussion, it will be necessary to investigate how this is done given that Δ is depth-wise variable for all of these applications.

2.2.6 Vertical Displacements

2.2.6.1 Introduction

Vertical displacements (*settlement* and *heave* or *rebound*), especially *differential* (relative) *settlement* between an ERS and adjacent soil on each side of it, have historically been ignored in practice with respect to having any influence on lateral earth pressures that develop from the retained soil acting on an ERS. This is true for both RERSs and FERSs. However, a combination of theory, research, and experience has demonstrated conclusively that differential vertical displacements <u>always</u> influence lateral earth pressures and thus should always be considered, at least qualitatively even if not quantitatively, for every ERS application in both practice and research.

The physical mechanism by which vertical displacements influence lateral earth pressures is fundamentally the same for both yielding and non-yielding ERSs. Specifically, it is the Coulomb-type ('dry') sliding friction that occurs along the interface between the ERS and adjacent soil, especially the inside face of an ERS and retained soil. However, the implications of this *soil-wall friction*, as it is usually called, are different for yielding and non-yielding ERSs. Therefore, separate discussions are required for each category of ERSs.

2.2.6.2 Effect On Yielding Earth-Retaining Structures

Whenever active and passive earth pressure coefficients from some lateral earth pressure solution (Coulomb, Caquot-Kerisel, etc.) are selected for analysis or design involving yielding ERSs, there is always an implicit assumption made about vertical displacements. This fact appears to be insufficiently emphasized in textbooks as well as significantly unrecognized and underappreciated by design professionals in both engineering practice and academia alike. Thus, simply using earth pressure coefficients without appreciating the surprising significance vertical displacements within an ERS-ground system can have on the magnitude of these coefficients is a potentially significant and serious error. This was clearly demonstrated decades ago by Tschebotarioff (1973) for a case history application and will be illustrated here as well.

To begin with, using a classical lateral earth pressure solution (specifically, Coulomb's active state which is presented and discussed in detail in Appendix A), Tschebotarioff demonstrated that the relative vertical displacement between an ERS and its retained soil (more specifically, the theoretical soil wedge that forms at failure within the retained soil) influences lateral earth pressures, all other variables (i.e. problem geometry and soil shear-strength parameters) being equal. This is because Coulomb's solution for the

24

coefficient of active earth pressure, K_a (see Equation A.4 in Appendix A, with all parameters defined in Figure A.1[22] in Appendix A), depends on the sign of the soil-wall friction angle, δ.

For the active earth pressure state, δ is defined as <u>positive</u> when the theoretical failure wedge within the retained soil moves <u>downward</u> relative to the ERS. This is by far the more common case in actual applications because as an ERS displaces outward and away from the retained soil (toward the left in Figure 2.1), that soil tends to slump downward behind the ERS. This is the condition design professionals tend to 'assume' by default (i.e. by not giving the matter any explicit thought in most cases) and use routinely.

On the other hand, δ is defined as <u>negative</u> when the failure wedge within the retained soil moves <u>upward</u> relative to the ERS (actually what happens in reality is that the ERS moves downward relative to the retained soil due to differential settlement but the net effect is the same). While this condition is not common in practice, Tschebotarioff (1973) illustrated a case history for an early bin-type gravity retaining wall located on Long Island in the State of New York where just such a situation occurred due to compressible foundation soils and the fact that the PCC bin wall was overall denser than the retained soil so settled more. This condition had apparently not been considered during design of the wall and the consequence was that both the geotechnical and structural ULS occurred in the ERS-ground system.

That such a failure can happen merely by sign reversal of δ can be illustrated by a simple example calculation using realistic soil properties and problem parameters. Referring to Figure A.1, using Equation A.4, and assuming a yielding RERS with:

- a vertical inside face ($\theta = 90°$),

- horizontal ground surface on the retained soil ($i = 0°$), and

- strength properties of $\phi = 35°$ and $\delta = \pm20°$

the resulting Coulomb values of K_a are:

- 0.245 for the more common +δ case and

- 0.366 for the less common -δ case

which is an increase of 49%. This means that the resulting active earth force for the (uncommon) -δ case would also be 49% greater than that of the (much more common) +δ case which is a sufficiently large relative increase that could effectively eliminate most, if not all, of both structural and geotechnical excess capacity ('safety') incorporated into a design whether it be using the traditional ASD or newer LRFD methodologies.

In addition, note that the vertical component of the resultant active earth force for the (uncommon) -δ case would act <u>upward</u> on the inside face of the ERS as opposed to <u>downward</u> for the for the (much more common) +δ case due to simple vector geometry. This which would have the effect of further reducing safety for the geotechnical failure mechanisms, especially overturning of the ERS, as the vertical component of the resultant active earth force goes from being a stabilizing force to a destabilizing force.

---

[22] Note that the base of the ERS shown in this figure is inclined relative to the horizontal. This is simply the more general case compared to the generic ERS shown previously in Figure 2.1.

Although not illustrated here, it is relevant to note that the coefficient of passive earth pressure, K_p, is also significantly affected by the sign of δ. In this case, the more common scenario in actual applications is for a passive soil wedge to move <u>upward</u> relative to the ERS as the ERS displaces laterally into the retained soil (toward the right in Figure 2.1). This is defined as the -δ case. If, on the other hand, the passive soil wedge moved <u>downward</u> relative to the ERS this would be the +δ case.

All other things being equal, the magnitude of K_p for the less common +δ case is always smaller than for the more common -δ case. Note that in most ERS applications where passive earth pressures can reliably develop they produce a resisting force within the overall ERS-ground system. In the common FERS application of anchored bulkheads, for example, the passive resultant force is essential to the stability and safety of the overall ERS-ground system. Thus, if ERS-ground friction were to be the less common +δ case, the net result would generally be much-reduced stability and safety to the overall ERS-ground system.

Note that this issue of the relative sign of the ERS-ground friction angle as applied to the passive earth pressure case is independent of and separate from the issue of the relative accuracy of different passive earth pressure solutions. This latter issue is addressed in Appendix A.

The conclusion drawn from this discussion is that the relative vertical displacement between a yielding ERS (either rigid or flexible) and the soil it retains should always be considered explicitly in each ERS application in practice, even if only on a qualitative as opposed to quantitative basis. While the majority of applications would be expected to have the typical +δ for the active earth pressure state and -δ for the passive, the consequences of the atypical, but possible, cases of sign reversal of δ for both the active and passive earth pressure states are sufficiently troublesome in terms of their potential failure outcomes to warrant this explicit consideration as a routine matter of course.

There are also indications that more-careful attention to vertical displacements could explain, at least in part, various observed behaviors such as the complex lateral earth pressures distributions found with anchored bulkheads (Horvath 1990a, 2014). So, while this monograph focuses consideration on modes of yielding (i.e. lateral displacement), where appropriate and relevant vertical displacements of the ERS-ground system are considered as well.

2.2.6.3 Effect On Non-Yielding Rigid Earth-Retaining Structures

While vertical displacements can be important for both RERSs and FERSs under yielding conditions, from a practical perspective the effect of vertical displacements on non-yielding ERSs is limited to RERS. As noted previously, the issue here is again related to the Coulomb-type frictional mechanism between the inside face of the RERS and the retained soil. However, the cause and consequences of this soil-structure friction are completely different compared to yielding ERSs.

The usual argument made with non-yielding RERSs is that because, by definition, no lateral displacement of either the RERS or retained soil occurs, no failure wedge develops within the retained soil. Consequently, because there is no soil wedge to displace up or down relative to the RERS, there can be no friction between the RERS and retained soil as, again by definition, mobilization of soil-structure friction requires relative displacement between the two problem components.

However, this seemingly logical argument is based on simplistically 'wishing' both the RERS and retained soil into place. In reality, the retained soil is placed as backfill and/or fill after construction of the RERS[23]. Even if the retained soil is placed in compacted lifts, there will still be some slight vertical compression and concomitant settlement of the retained-soil mass soil relative to the RERS as additional material is placed above it. Note that this vertical compression within the retained soil is independent of and in addition to any compression of the foundation soil on which the retained soil is placed.

This progressive compression and concomitant settlement of the retained soil relative to the RERS (typically larger in magnitude toward the bottom of the RERS as opposed to the top) will usually be sufficient to mobilize at least some soil-structure friction. As is well known from decades of research and field observations related to deep foundations, only relatively modest differential displacements of the order of 3 mm (⅛") are required for full mobilization of shaft friction between a deep-foundation element and adjacent soil. Differential settlements of such magnitude between a non-yielding RERS and adjacent retained soil are well within the realm of possibility, even if compression of the foundation soil is removed from the discussion.

As a result, significant soil-wall friction between a non-yielding RERS and adjacent retained soil would be expected to develop as a matter of course in the vast majority of situations. This has been confirmed by research (Horvath 1990a, Filz and Duncan 1997, Filz 2003). Of course, additional compression due to settlement of the foundation soil, especially if the RERS is supported on deep foundations, would only further guarantee that these soil-structure frictional stresses will develop.

The question, then, is what to do or make of these frictional stresses between the RERS and retained soil. Because the retained soil does not develop a failure wedge as in the previously discussed case of a yielding ERS, there is, theoretically at least, no issue concerning the sign of this friction and how it influences lateral earth pressure coefficients. In addition, because a non-yielding RERS cannot, by definition, fail initially by any geotechnical failure mechanism[24], the resulting resultant vertical force acting along the inside face of the RERS due to soil-wall friction does not factor into any stability analyses. However, presence of these soil-wall shear stresses does affect the stress field along the RERS-retained soil interface, including the orientation of the principal stresses, so as a minimum this behavior is something to be considered during any advanced analytical assessment that includes numerical analyses of the RERS-retained soil system.

2.2.7 Retained-Soil Conditions

The majority of the research and applications to date where geofoam has been used either alone or with other geosynthetics to reduce lateral pressures on ERSs have involved retained soils that are referred to in this monograph as 'normal soils'. These are defined as being:

- predominantly granular in nature;

---

[23] Note that good practice typically calls for a non-yielding RERS such as the basement wall of a building to be fully braced in terms of lateral support (typically floor systems) before any backfilling/filling adjacent to the RERS commences. This is because a RERS that was designed to be braced in the long term cannot generally tolerate loading while temporarily unbraced without suffering either the geotechnical or structural ULS.

[24] It is possible in principle for a non-yielding RERS to suffer a classical bearing capacity failure due to being constructed on weak foundation soil. However, it is assumed that this geotechnical failure mechanism would be addressed during design and eliminated as a possibility by using deep foundations to support the RERS as necessary to prevent such an occurrence.

- overall coarse-grain in gradation;

- always drained in terms of volume-change and shearing behavior;

- no or negligibly small time-dependent pore-pressure effects, i.e. consolidation effects are negligible; and

- no or negligibly small volume-change issues related to suction, especially matric suction, so that unsaturated pore pressures can be reasonably modeled as being zero so that effective stresses and total stresses are equal.

Thus, the information presented in this monograph will focus on such conditions.

However, as will be seen, one of the more significant potential applications for some of the geotechnologies presented in this monograph is with soils that are predominantly fine-grain in nature, especially those with significant clay-mineral content that are considered 'expansive' or 'swelling' in nature, i.e. where matric-suction effects that exist under partial-saturation conditions are significant and cannot be ignored. Historically, such soils were often referred to as 'problem soils' in the literature and considered to be localized, regional anomalies or exceptions to 'normal soils' and normal-soil behavior.

However, due to decades of pioneering research and publication by Fredlund (Fredlund and Rahardjo 1993) and others, it is now widely appreciated that 'problem soils' are actually geographically widespread and thus significant. This has led to a wider appreciation and understanding of what has been termed *unsaturated soil mechanics* (alternatively *partially saturated soil mechanics*) in which the aforementioned 'normal soils' are a subset. Therefore, some consideration is given to these materials as well.

2.3 GEOSYNTHETICS

2.3.1 Introduction

Geosynthetics have been one of the revolutionary, game-changing technological developments in engineered construction beginning in the latter decades of the 20th century. As noted in Chapter 1, one the biggest impacts of geosynthetics has been on the way ERSs are conceptualized, designed, and constructed. In fact, current ERS alternatives such as MSEWs and SRWs simply would not exist without geosynthetics and that is certainly true of the subject of this monograph.

2.3.2 Geofoam

2.3.2.1 Basics

The primary category of geosynthetic considered in this monograph is *geofoam* which has been recognized and accepted (at least by most) as a geosynthetic only relatively recently (circa 1990s) compared to traditional planar geosynthetics such as geotextiles, geogrids, and geomembranes. However, the predominant materials used for what we now call geofoam have actually been used for geotechnical applications since at least circa 1960 which is well before most of these traditional planar geosynthetic products even existed or the term geosynthetics was even coined. Thus the 'new' geosynthetic called geofoam is arguably one of the oldest, if not <u>the</u> oldest, geosynthetic.

It is assumed that readers and users of this monograph have at least a fundamental working knowledge of geofoam as can be found in Horvath (1995b) so basic, general information about geofoam is not presented here. The information presented in this monograph is limited to and focused on those specific geofoam materials and associated products that have been used or have the potential to be used for applications involving ERSs.

Experience indicates that of all the different geofoam materials and various products derived from those materials that are or have been available commercially, the aforementioned expanded polystyrene (EPS), in particular in its generic, commodity block-molded[25] product form called EPS-block, is the overall, all-around geofoam material and product of choice for use to achieve lateral pressure reduction on ERSs. A detailed presentation and discussion of all aspects of the EPS block-molding process can be found in Horvath (2011). An equally detailed presentation and discussion of U.S-sourced standards for EPS-block used in geofoam applications can be found in Horvath (2012).

Although EPS-block is the overall geofoam material and product of choice for the applications discussed in this monograph, there are some functional applications where it is not the optimum alternative from either a technical or cost perspective. There are two additional polystyrene-based geofoam materials, both related to EPS-block and both capable of being produced by the same molder (manufacturer) that makes EPS-block (although not all EPS molders make these products), that are used either alone or in combination as a single geocomposite product in some ERS applications:

- *resilient* (a.k.a. *elasticized*[26]) EPS (hereinafter referred to as *resilient-EPS* or simply R-EPS) that was actually developed commercially, in the U.S. at least, specifically for some of the applications discussed in this monograph and

- *glued polystyrene porous block* (a.k.a. *brown board* due to the color that the binding agent used to glue the *prepuff* from the first stage of EPS manufacturing together imparts to the final material and hereinafter referred to as GPS-PB) that was developed originally as a unique geosynthetic material specifically for use with RERSs as a thermally insulated sheet-drain product[27]. By pure coincidence, this material, at least as

---

[25] EPS can be molded (manufactured) in two ways: by block molding or *shape molding* (sometimes referred to as *custom-shape molding*). The ubiquitous white plastic foam coffee cup is perhaps the most familiar consumer-product example of a shape-molded EPS (*EPS-shape*) product. The availability and concomitant use of EPS-shape geofoam products, which to date have always been proprietary in nature and thus tend to be available only regionally and from limited sources, is relatively uncommon and globally very limited at present. Consequently, for simplicity in the remainder of this monograph, any reference to EPS will be assumed to refer only to EPS-block.

[26] This material, which is essentially factory-modified very-low-density EPS-block, is discussed in further detail in Chapter 8. Although the term 'elasticized' was used historically for this material, especially in western Europe where it evolved originally as a construction material for non-geofoam applications, the writer coined and prefers the term 'resilient' as it more accurately describes the behavior of this material in geofoam applications relative to normal EPS-block. Therefore, the term 'resilient' is used for this material throughout the remainder of this monograph.

[27] The 'glued' distinction is not trivial in this case. There is, reportedly, a material called *molded polystyrene porous block* (MPS-PB) in which the prepuff is molded into a final material that has the same open texture as GPS-PB. Because no binding agent is used as with GPS-PB, the final MPS-PB material retains the inherent white color of the prepuff so is **[continued on following page]**

manufactured in the U.S., has similar mechanical (stress-strain-time) properties as R-EPS. Thus GPS-PB, whether used alone or with R-EPS as part of a geocomposite product called a *GeoInclusion*™, has an additional functional application as a compressible inclusion for RERSs. This material is discussed further in Chapters 7 and 8.

2.3.2.2 Functional Uses With Earth-Retaining Structures

Concomitant with the ascendency of the use of geosynthetics in engineered construction has been the universal realization and appreciation that the only rational way to design with geosynthetics in any and all applications is the concept of *design by function*[28]. Geosynthetics are, or at least should be, used only when they can clearly provide a function or functions the natural ground alone cannot provide or at least cannot provide technically and/or economically to the extent desired or required in that application. For example, geogrids used with MSEW provide the function of *reinforcement* in tension. Tensile strength is inherently lacking in a particulate material such as uncemented soil.

The *design by function* concept may seem both obvious and intuitive but there have been cases, much more prevalent in the early years of geosynthetics usage compared to the present, where a geosynthetic was used in a situation where it had no functional benefit. In the past, this caused some to question the value of geosynthetics because there did not seem to be any benefit to doing so. In reality, the fault was due to misapplication by the end user and was not inherent in the geosynthetic itself.

As an aside, it is relevant to the content of this monograph to note that geosynthetic misuses in the present tend to be not so much function related as it is users not correctly understanding material properties and behavior of a geosynthetic material or product when used for a given function. The writer has found this to be a particular problem with EPS-block geofoam in general (Horvath 2010a) and with applications involving ERSs in particular (Horvath 2010b).

In any event, as will be discussed throughout this monograph, many geosynthetic materials and products inherently provide more than one function at the same time (even if not all of them are used or even desired in a given application) in which case the geosynthetic is called *multifunctional*. Note that this inherent multifunctionality of a single geosynthetic material or product is distinct from the case of a geosynthetic product that is a factory-manufactured composite (*geocomposite*) of two or more different materials, at least one of which is a geosynthetic. Geocomposites are by design always inherently multifunctional. Note also that there can be applications where two or more different geosynthetics are used together in the same overall application to have some combined post-construction, synergistic multifunctionality but are placed separately in the field (and not necessarily physically connected) as opposed to being a true geocomposite.

One of the most significant aspects of geofoam in general, and EPS-block in particular, is that not only are they inherently multifunctional but, with few exceptions, they provide functions no other geosynthetic can provide. In fact, most functions now identified and associated with geofoam did not even exist in the world of geosynthetics until geofoam was identified as a legitimate geosynthetic-material category in the early 1990s (Horvath 1995b). Unfortunately, some geosynthetic 'purists' continue to not recognize geofoam as a

---

referred to colloquially as *white board*. MPS-PB is not known to be or ever have been manufactured in the U.S. MPS-PB is believed to have been manufactured in then-West Germany at one time although the current availability is unknown.

[28] 'Function' as used here and throughout this monograph is in the geosynthetics context of being the technical role provided by a geosynthetic material or product.

type of geosynthetic and therefore do not list geofoam functions along with the traditional (planar) geosynthetic functions of reinforcement, separation, filtration, etc.

As noted in Chapter 1, there are two distinctly different primary geosynthetic functions that geofoam in general can provide when used with ERSs for the purpose of lateral pressure reduction:

- as a *lightweight-fill*[29] material replacing part of the ground (soil, rock, or non-earth material) that would normally be retained by an ERS and

- as a *compressible inclusion* between an ERS and adjacent ground.

As discussed in greater detail throughout this monograph, generic EPS-block geofoam in the form of prismatic blocks is the geofoam material and product of choice for all lightweight-fill functional applications and ERSs are no exception. EPS-block has also been used for compressible-inclusion functional applications but is almost always technically inefficient (and thus needlessly costly) and can even be technically ineffective in that role. The related geofoam materials of R-EPS and GPS-PB in panel-shaped product form are by far the preferred materials and products for compressible-inclusion applications, at least with ERSs. These issues are discussed in more detail in Chapter 8.

In concept, the overall <u>qualitative</u> benefit, in that some degree of lateral pressure reduction on an ERS can be achieved, is the same with both of these functional categories. Thus, it is important to recognize that these two geofoam functions are, in general, competitive as opposed to complementary in nature when the primary goal is lateral pressure reduction on an ERS. Therefore, on most projects where ERS lateral pressure reduction using geofoam is explored as a design alternative there should always be an economic comparison between these two functional applications to see which is the overall more cost-effective on that particular project. Note that this assessment should always consider the other functions, desired or otherwise, that the geofoam product provides.

The key differences between the two functional applications that should be kept in mind when making such a project-specific assessment are:

- Product availability. EPS-block as used for the lightweight-fill function is a generic, commodity product that is literally available almost anywhere in the world and, in larger countries, usually from multiple suppliers. However, relatively large volumes of material are required to achieve meaningful benefits with ERSs which means that shipping and associated costs can be an issue in relatively remote areas. On the other hand, only relatively small volumes of the specialized geofoam materials R-EPS and GPS-PB are required for significant benefits when using the compressible-inclusion function. However, these are either proprietary or limited-manufacture materials which means that their geographic availability can be limited.

- Overall design issues and analytical methodologies are completely different.

---

[29] In earlier publications, the term *ultralightweight* was sometimes used in lieu of 'lightweight' to emphasize the fact that the density of EPS-block is not only two orders of magnitude lower than that of normal earth materials (soil and rock) but also an order of magnitude lower than the density of other non-geofoam lightweight-fill materials such as foamed-PCC that have been and are used in practice. Thus EPS-block geofoam is unique in the larger world of lightweight-fill materials, at least in terms of its density.

- The <u>quantitative</u> benefit in terms of the relative (from the baseline case of no geofoam) reduction in lateral pressures acting on the ERS that can be achieved under a range and combination of loading conditions applicable to a given application, e.g. gravity vs. seismic, without or with a surface surcharge, etc. In determining this quantitative benefit, the physical volume that must be occupied by the geofoam in order to achieve the desired level of benefit must always be considered. This is particularly important for application with an existing ERS as in many cases various physical site constraints will effectively determine the upper bound of the volume that can be occupied by geofoam.

- For applications involving new construction, the overall net capital construction cost associated with the quantitative benefit that must simultaneously consider both the cost <u>reduction</u> for the ERS itself but concomitant cost <u>increase</u> due to use of more-costly geosynthetics in lieu of normal earth backfill or fill behind the ERS.

- For applications involving the performance upgrade or improvement of existing ERSs, the overall net capital construction cost associated with the quantitative benefit that must simultaneously consider the cost value of the net improvement to ERS performance vs. the cost of the geosynthetics compared to the net cost of other design strategies to improve performance of the ERS, including complete replacement of the ERS. Again, the overall influence of site construction constraints that will impact all design strategies to varying degrees is emphasized as this consideration often looms large in practice and will obviously vary greatly from project to project making generalizations difficult.

- The project-specific need and benefit from other geosynthetic functions provided by a specific geofoam material and product which will vary significantly on a project-by-project basis. For example, the writer is familiar with projects (the Carousel Center/Destiny USA shopping mall in Syracuse, NY shown on the back cover is one) where soft-ground conditions adjacent to a building (which was supported on deep foundations) meant that using the lightweight-fill function for lateral pressure reduction was the clear preference because not only would lateral pressures on the below-grade building walls be greatly reduced but settlements of the paved parking areas adjacent to the building could be readily controlled because of the low vertical stresses imposed on the existing subgrade by the EPS-block geofoam placed adjacent to the walls.

Further to the final point, it will be seen that EPS-block and related geofoams can, depending on the specific materials and product(s) used, also and simultaneously provide the following additional functions either alone or in some combination:

- *thermal insulation,*

- *fluid (gas and liquid) drainage,* and

- *noise and small-amplitude vibration damping.*

These are discussed further in Chapter 7.

While the inherent multifunctionality of EPS-block and related geofoam materials is overall beneficial because it increases their cost effectiveness (i.e. one product provides multiple technical benefits in a given application), it should always be remembered that there is, in principle, always a downside to geosynthetic multifunctionality. This is because one of the issues concerning geosynthetic multifunctionality that is not always recognized and considered by design professionals is the fact that a geosynthetic product's functions cannot be selectively turned on and off and so are omnipresent throughout the installed life of that product. Because of this, a geosynthetic product can sometimes provide functions that are not only unnecessary in a given application but also undesirable because they can be detrimental in that application, even to the point of causing failure of some or all of the overall structure at some point in time. This is also discussed further in Chapter 7 with specific regard to the geofoam products considered in this monograph.

Note that this is not intended to imply that a multifunctional geosynthetic with a potentially negative function or functions should not be used in a given application. Quite the opposite. The point made here is that <u>all</u> construction materials have their pluses and minuses, and due diligence by design professionals in practice simply means that both positives and negatives of a material and product are always considered in a given application so that the negatives do not create problems or are at least reasonably contained and managed at a risk level that is acceptable to the owner of the structure. For example, given time and the proper environmental conditions, steel oxidizes (rusts); PCC shrinks, cracks, and can deteriorate; and structural timber shrinks and can rot in addition to being inherently flammable. None of these behaviors has prevented these materials from being widely used in construction worldwide for a very long time. Rather, the negatives of a construction material are or should be simply factored into the long-term operation and maintenance (O&M) of the structure.

With this in mind, the point made here is that when using EPS-block and related geofoams for lateral pressure reduction on ERSs, the design professional should always remember that in addition to providing the desired primary function of either lightweight fill or compressible inclusion, depending on the specific material and product used one or more of the additional functions noted above will also be provided. In general, these additional functions are at least benign if not desirable in most applications. However, experience has shown that in certain applications involving EPS-block used as lightweight fill, the excellent thermal-insulative properties of the EPS can result in the phenomenon of *differential icing* of an overlying pavement surface (Horvath 1995b, 2001b; Stark et al. 2004a). This particular outcome is typically at the top of the list of undesirable and potentially hazardous phenomena that are potential outcomes of using geofoam.

2.3.3 Other

Although PS-based geofoam products are the primary and predominant geosynthetic discussed in this monograph, as will be seen there are applications where other geosynthetics can be used to either complement and/or interact synergistically with the geofoam product to effect desirable end results that no one type of geosynthetic could effect if used alone. The additional geosynthetic functions provided and the products that provide them are:

- *tensile reinforcement* in the traditional manner of *mechanically stabilized earth* (MSE) using either metallic[30] (strip or mesh) or polymeric (primarily geogrid although geotextiles are possible) sheet-like reinforcements and

---

[30] Historically and to the present, metallic elements in the form of strips or wire mesh used as tensile reinforcement in MSE and MSEW applications are not considered **[continued on following page]**

- *filtration* by a geotextile (typically non-woven) that is incorporated into a manufactured drainage geocomposite utilizing a special, high-permeability geofoam material (GPS-PB) as its core. Such products are discussed further in Chapter 8.

---

geosynthetics by geosynthetics 'purists'. This appears to be due to the traditional view of geosynthetics as being limited to planar, polymeric materials such as geotextiles, geomembranes, and geogrids. The writer long ago adopted a broader (and arguably more logical) definition of geosynthetics as being any manufactured (i.e. synthetic) material used on or in the ground in some functional application normally associated with geosynthetics (tensile reinforcement in this case). Note, however, that this extended definition of geosynthetics can be overdone and misused in practice. For example, the writer has seen EPS blocks used within a building's superstructure for, say, stadium-type seating inside a movie theater or auditorium referred to by the EPS industry promoting the application as 'geofoam' when this clearly is incorrect.

This page intentionally left blank.

Chapter 3

Overview of the Design Process

3.1 FUNDAMENTAL OBJECTIVES

Before addressing specific analysis and design methodologies, it is useful to understand the broad conceptual issues that shape these methodologies in general and how they are presented in this monograph in particular.

To begin with, it is important to remember that the overall goal of all engineered construction is to *design to prevent failure*, whether this involves a proposed new structure or an existing structure that will be rehabilitated, upgraded, or renovated for some existing or different use. 'Failure'...nowadays often referred to synonymously as the 'limit state' as noted in Chapter 1...in a civil engineering context has a broad definition as being the *loss of function*.

In this context[31], function means the primary operational purpose(s) for which a structure is to be built or already exists. Thus, failure as it applies to engineered construction always includes both *serviceability* failure (the Serviceability Limit State, SLS) due to excessive displacement and/or deformation creating operational issues for the structure and/or its users as well as the obvious *ultimate* or *collapse* failure (Ultimate Limit State, ULS) due to a physical collapse of all or part of a structure.

Note that the SLS tends to be the more common type of failure nowadays in engineered construction compared to the ULS, especially for new construction where sophisticated analytical methodologies allow for materials to be accurately analyzed to their performance limits. Note also that what defines the SLS in a given project application can often be highly subjective and variable depending on the specific intended use of a structure. For example, a warehouse where the floors have settled differentially may be acceptable for some uses but not others depending on structure-specific variables such as what kind of products and how high they are stacked on the floors; what equipment might be used to traffic within the warehouse; etc. So, depending on the intended use of the warehouse, a certain magnitude of differential settlement might exceed the SLS in one case but not in another.

Note that assessing the SLS for a structure is typically achieved by analyzing behavior under service loads to ensure that displacements and deformations of the overall structure as well as individual structure components as appropriate are within pre-defined limits for the intended use of the structure.

For ULS mechanisms, design-to-prevent-failure involves satisfying the following generic, fundamental equation for either the overall system or individual system components for each potential ULS mechanism that can occur for a structure or its components:

---

[31] Note that 'function' as used here is completely unrelated to its specific use and definition relative to geosynthetics as discussed in Chapters 1 and 2 although it has the same broad meaning of role or purpose. It is unfortunate that this word has two completely different contextual uses in this monograph but this simply reflects the reality of the use of this word in practice. The vast majority of usage of the term function, in this monograph at least, is in the previously defined geosynthetics context discussed in Chapter 2.

$$capacity\ (resistance) > demand\ (loads) \qquad \textbf{(3.1)}$$

where the excess capacity or resistance implied in this equation represents a *margin of safety* against occurrence of the ULS.

The need to incorporate an acceptable margin of safety into new structures as well as ensure that existing structures maintain an adequate margin of safety throughout their service life is universally accepted. How to do this can vary from the traditional safety-factor concept based on service loads and reduced material strengths in the traditional ASD/WSD methodology to some combination of factored loads and possibly reduced material strength as well in the alternative LRFD/USD approach. Because different methods for incorporating a margin of safety exist in practice, the analytical methodologies presented in this monograph are all illustrated in terms of service loads for both consistency and simplicity throughout this document. Thus, these loads can either be used as-is when using ASD/WSD or factored as desired when using LRFD/USD.

Within this overarching primary objective of designing to prevent failure that drives all modern civil engineering practice, there is a secondary objective that is unique to the subjects covered by this monograph. This is the desire to develop relatively simple analytical models and concomitant analytical methodologies that are amenable for use in routine practice for design.

As is well known, geotechnical engineering practice has been revolutionized in recent decades by the ability to use sophisticated numerical-analysis tools, primarily the finite-element (FE) method (FEM), on a routine basis using commercially available software and computers available to anyone. However, FE analysis still requires an investment in software costs, user education and experience, and time so that there is still a need for simpler analytical methodologies that can be used for smaller projects as well as preliminary feasibility assessments on larger projects. Consequently, a significant, recurring theme throughout this monograph is the development, presentation, and assessment of various simplified analytical models and methodologies that have been proposed over the years and to the present.

3.2 OVERALL SYSTEM DESIGN

3.2.1 Basic Concepts

When the above-defined concept of failure is applied to ERSs, there are four potential failure scenarios that must be considered for each specific project application, at least in principle. This is because the SLS and ULS must each be considered separately for both the geotechnical and structural components of the ERS-ground system. It is important to recognize that, theoretically at least, any one of these four failure scenarios may be the most critical and thus govern overall design of an ERS-ground system.

That having been said, there is obviously extensive history and practical experience with human construction of ERSs in general. Based on this history and experience, certain rules-of-thumb have evolved in routine practice that provide guidance as to which of the four potential failure scenarios (geotechnical vs. structural, SLS vs. ULS) is likely to govern the performance of different types of ERSs. These rules-of-thumb are incorporated where relevant into the detailed analytical methodologies presented in this monograph. Nevertheless, while these practical guidelines are useful and form the basis of routine, efficient design practice, the fact remains that four potential failure scenarios exist with any

ERS. Thus, due diligence requires that this not be forgotten by design professionals, especially when performing analyses to assess the stability of an existing ERS.

Finally, it is important to note that the various loads and loading diagrams presented and discussed in this monograph are unfactored, i.e. they are based on service loads only, as would be used for a traditional ASD-based design. When used for designs based on LRFD, the magnitudes will need to be increased (factored) by the appropriate load factors which may vary for gravity and seismic loads. Given the many and ever-changing LRFD codes that exist globally, it is a practical impossibility to consider LRFD issues explicitly in this monograph.

3.2.2 Rigid Earth-Retaining Structures

3.2.2.1 Non-Yielding

For non-yielding RERSs, the geotechnical SLS and ULS are not a consideration as such ERSs are, by definition, constrained against the primary geotechnical ULS modes (sliding and overturning). It is further assumed that the secondary geotechnical ULS modes of bearing capacity and global stability are non-issues as well. Therefore, only structural failure needs to be considered[32]. This is normally done by a structural engineering specialist using a project-specific design earth pressure diagram developed by a geotechnical engineering specialist. Thus, the focus of this monograph is solely on the development of such pressure diagrams when geofoam is used to reduce lateral pressures. The actual structural design is beyond the scope of this monograph as it is very standard reinforced-concrete design that is adequately and better addressed in texts and other documents on the subject.

3.2.2.2 Yielding

A yielding RERS is a more complex problem to deal with as the potential exists for both geotechnical and structural SLS and ULS. Thus, all four failure scenarios need to be considered, at least in principle, during the design phase of every project.

However, experience indicates that only the strength-based geotechnical and structural ULS need to be considered in routine practice. This is because the typical geotechnical displacements and structural deformations of a RERS under service loads are negligibly small in magnitude as long as there is an adequate margin of safety against the ULS in each case. In fact, the methodologies used for either analyzing or designing both the geotechnical and structural components of a yielding RERS in routine practice under both gravity and seismic loads do not explicitly consider or calculate displacements and are based on the geometry of the undisplaced, undeformed structure (what is referred to as a *linear analysis* in structural engineering).

---

[32] The fact that the structural ULS of a non-yielding RERS will typically trigger the geotechnical ULS within the retained soil is irrelevant to the design process. This is because the geotechnical ULS is a consequence of the structural ULS, not the initiator of the primary failure. Only those mechanism that initiate failure need be addressed explicitly in design although the geotechnical consequences of any failure should always be considered, at least qualitatively.

38

3.2.2.3 Self-Yielding

Self-yielding RERSs are the most behaviorally complex type of RERS identified to date and the subject of ongoing research, especially IABs which are by far the most common types of self-yielding RERSs. Although some guidelines for IAB design (the abutments of an IAB are the RERSs in this case) in particular have been developed by various researchers and engineering organizations such as state departments of transportation (Horvath 2000, 2004c, 2005; IAJB 2005), there is no one methodology or even a small group of methodologies that have been widely accepted and disseminated to the point that they appear in the standard geotechnical engineering textbooks, at least in the U.S. Based on the writer's more than three decades of experience as an educator in academia, as a result of this dearth of treatment in undergraduate geotechnical engineering textbooks, relatively few geotechnical engineers graduate with any basic awareness of, no less analytical knowledge about, self-yielding RERSs.

To date, the published analytical methodologies used for the design of IAB abutments tend to be conceptually and qualitatively identical to those for non-yielding RERSs such as the abutments of conventional bridges, with the only difference being quantitative in nature. Specifically, it is assumed that the only potential failure mode for an abutment of an IAB is the structural ULS so that only a structural design based on an assumed lateral earth pressure diagram is necessary. No geotechnical ULS is assumed possible due to physical constraints placed on the abutment by the bridge superstructure framed into it. Furthermore, neither the structural nor geotechnical SLS is considered. In summary, the only thing that differs between a self-yielding IAB abutment and non-yielding conventional bridge abutment is simply the magnitude and distribution of the assumed lateral earth pressures used for structural design.

In general, to date the design lateral pressure diagrams for IAB abutments have been based solely on some assumed typical summer season when seasonal thermal expansion of the bridge superstructure is at a maximum so that the abutments have been jacked laterally outward from the superstructure and into the retained soil the maximum amount possible for that particular structure. Furthermore, it is implied, even if generally not assumed or stated explicitly, that this pressure diagram applies for the design life of the IAB which could be of the order of 100 years or more.

The assumed lateral earth pressure distribution acting on an IAB abutment is generally relatively complex geometrically compared to the simplistic equivalent-fluid-pressure triangular distribution usually used for both non-yielding and yielding RERSs as it is typically assumed that the passive earth pressure state will be substantially, if not fully, mobilized at least over some upper portion of the abutment where lateral displacements into the retained soil are maximum[33]. The specifics of at what depth the assumed lateral pressures transition to a state less than passive and what lateral pressures correspond to that lesser state varies from method to method.

One caution concerning this approach is that research noted previously (England 1994, England and Dunstan 1994, England et al. 1995) suggests that due to the inherent lifetime cyclic expansion-and-contraction nature of self-yielding RERSs such as IABs, the maximum summer-seasonal pressures do not remain constant with time. Due to the previously noted geomechanics phenomenon of ratcheting that is related to the non-linear, inelastic behavior of soils under cyclic loading, the summer pressures tend to increase over

---

[33] As noted previously, this is because lateral displacement of an IAB abutment tends to be a composite of the rotation-about-bottom plus translation modes, with the former much more dominant and thus significant than the latter.

time in a monotonic, permanent fashion. This behavior was observed in research by the writer (Horvath 2000) that was limited to numerical modeling even though a relatively simplistic constitutive model was used for the retained soil and only four complete annual cycles of temperature change were simulated. The underlying cause of this ratcheting, at least in the case of IABs, is explained subsequently.

As England and co-authors illustrated in the above-cited publications, the increase in lateral earth pressures due to ratcheting can, in theory, be significant and is believed in some actual cases to be the fundamental underlying cause of a structural ULS of a self-yielding RERSs developing at some time after construction when there had been no physical changes imposed on the RERS. Therefore, it is clear that ratcheting behavior should be studied further and incorporated into future design lateral pressure diagrams used for designing against the structural ULS of all types of self-yielding RERSs, including IAB abutments.

However, the larger and arguably more important technical issue is that research (Horvath 2000, 2004c, 2005) has clearly shown that the above-described design approach of considering only the structural ULS for IAB abutments (and possibly self-yielding RERSs in general) is problematic. The reason is that this design approach essentially considers IAB abutments in a behavioral vacuum as if they were isolated structural elements.

In reality, the abutments of an IAB are but one component of a complex *soil-structure interaction* (SSI) system that includes not only the IAB itself but also the approach roadways and ground (i.e. retained soil adjacent to the IAB abutments) underlying these roadways. So, while the summer condition may indeed be the most critical in terms of the structural ULS of IAB abutments, there are other conditions that can be more critical for the overall IAB-roadway system.

Specifically, the winter condition, when the superstructure contracts and pulls the abutments away from the retained soil the maximum amount, will tend to result in the geotechnical ULS being reached within at least some, if not all, of the retained soil behind each abutment. This is because of the well-known fact that the relative magnitude of lateral displacement of a soil mass necessary to mobilize the active state is approximately one order of magnitude less than that required to mobilize the passive state. So, while the summer displacements of an IAB abutment are generally insufficient to mobilize the passive state within the entire retained-soil mass, the winter displacements are typically more than sufficient to mobilize the active state within the entire retained-soil mass.

The consequence of developing an active-state failure within the retained soil behind an IAB abutment each summer season is that a soil wedge will tend to form within the retained soil and displace both laterally inward toward the abutment as well as vertically downward. This has two consequences:

- The <u>lateral component of retained-soil displacement</u> allows the retained soil to follow an abutment and slump against its inside face as the season moves from summer to winter, with the net result of the retained soil becoming wedged to a certain extent against the inside face of the abutment at the peak of the winter contraction of the bridge superstructure. When the season then moves from winter to summer and warmer weather causes the abutments to once again move outward and into the retained soil, two things happen. First, the soil that slumped against the inside face of the abutment will not return fully to its original position from the preceding summer due to the inherent non-linear, inelastic nature of soil behavior. Second, because the abutment displacement is starting from a position with the soil already wedged against it, the full summer displacement of the abutment will tend to mobilize more of the passive earth pressure state within the retained soil than was mobilized the preceding summer, leading to an incremental increase in lateral earth pressures against the inside face of the abutment compared to the preceding summer. This overall behavior of ever-increasing (with time) peak summer lateral earth pressures constitutes the ratcheting phenomenon noted and defined by England and co-workers.

- The <u>vertical component of retained-soil displacement</u> creates a depression in the surface of the retained soil adjacent to the inside face of the abutment. Assuming that the road-pavement system is placed directly on the surface of the retained soil, this means there will be a depression in the roadway surface as well. This depression will extend for some distance away from the IAB abutment in a direction parallel to the roadway alignment creating a three-dimensional settlement bowl. This settlement bowl will not only be permanent (because of the above-described non-recoverable nature of soil displacements) but may also increase in size (depth and/or lateral extent) over time. This settlement bowl is thus one mechanism for producing the well-known 'bump at the end of the bridge' condition for vehicles. Note that as an alternative to simply paving on the surface of the retained soil, a structural slab (called an *approach slab*) can be used as a transition from the abutment to the on-grade roadway surface some distance back from the abutment that is assumed to be unaffected by settlement. The primary purpose of an approach slab is to bridge over any settlement bowl that develops over time and eliminate the rideability issues that result from the bump-at-the-end-of-the-bridge roadway condition. However, this approach slab must be designed to span over the full extent of the surface depression without subgrade support which can be a demanding structural requirement. Research has shown that approach slabs are not always properly designed for this occurrence as structural ULS failures of them have been documented in the literature (Horvath 2000, 2004c, 2005).

In summary, it should be recognized that proper design of an IAB-approach roadway system requires not only that the abutments be designed for the structural ULS (taking long-term increases in lateral earth pressures due to ratcheting into account as appropriate) but that the abutment-to-roadway transition area be designed with appropriate, explicit consideration of both the geotechnical SLS and ULS within the retained soil as a consequence of seasonal abutment displacements. This latter aspect is something that is generally missing from designs at present as noted in Horvath (2000, 2004c, 2005).

The need to consider the effect of IAB-abutment displacements and the effect they have on the adjacent retained soil that underlies the approach roadway should not come as a surprise. The simple physics of all bridges is that bridge superstructures will always displace laterally with seasonal temperature variations. On the other hand, the ground adjacent to bridges is essentially unaffected displacement-wise by these seasonal temperature changes. Therefore, the interface between constantly displacing superstructure and non-displacing ground must be accommodated in some fashion. There is simply no way to avoid this and it cannot be 'wished away' as it appears that some IAB design professionals apparently think.

With traditional jointed bridges, the interface between the displacing and non-displacing components of the overall bridge-ground system was purely structural and consisted of expansion joints and bearings. This clearly and unambiguously defined the interface between the constantly displacing bridge superstructure and never-displacing adjacent ground as occurring essentially at the top of the abutments, with the important result that the abutments were, behaviorally at least, an extension or part of the ground. This was because the abutments were non-displacing like the ground.

While this combination of expansion joints and bearings worked well enough in principle, they were found to be perpetual maintenance issues. While the IAB concept neatly eliminated the structural maintenance of expansion joints and bearings, it did not and could not do anything, of course, to eliminate the basic physics of the problem. Essentially, all the IAB concept did was to transfer a structural issue (maintenance of expansion joints and bearings) to a newly created geotechnical issue (the displacement of the soil retained by the abutments as a result of seasonal abutment displacements). This is because the IAB concept essentially made the abutments part of the displacing superstructure as opposed to being part of the non-displacing ground as they were historically. In the process, design professionals forgot that the ground cannot be non-displacing on its own (it previously relied on the abutments to provide that function). Absent the support of the abutments, the ground displaces as well.

The logical conclusion is that to eliminate the bump-at-the-end-of-the-bridge and approach slab problems discussed above, as a minimum the ground adjacent to IABs needs to be rendered non-displacing in some manner as the abutments can no longer be counted on to provide this function as with conventional jointed bridges. Furthermore, as part of this there should also be proactive, explicit design attention paid to the interface between IAB abutments and adjacent ground. This is because with IABs the relative displacement between superstructure and ground is transferred from being accommodated within the superstructure to being accommodated within the ground as the abutments are no longer part of the non-displacing ground (as they are with conventional bridges) but part of the displacing bridge superstructure. Stated another way, the concept of an expansion joint is still required with IABs but now the 'expansion joint' is the contact zone between the displacing inside face of the IAB abutment and adjacent, non-displacing retained soil.

As discussed in detail in Chapter 6, the remedy for the geotechnical issues inherent in IABs is that a combined expansion-joint-plus-bearing system is needed as existed with traditional bridges. However, with IABs the 'expansion joints' and 'bearings' take on a completely different design concept as these must be geotechnical, not structural, elements. Essentially, on each end of an IAB there must be a 'bearing system' to hold up the ground year 'round and keep it more or less immobile and there must be an 'expansion joint' between the laterally displacing abutment and the adjacent non-displacing stabilized ground. However, the good news is that, if properly designed, this new geotechnical expansion-joint-and-bearing system can be maintenance-free, something that was never possible with the structural expansion joints and bearings of traditional bridges. Furthermore, as illustrated in Chapter 6, if properly designed this new geotechnical expansion-joint-and-bearing detail can substantially reduce the lateral earth pressures acting on IAB abutments, even during the critical summer-season timeframe.

3.2.3 Flexible Earth-Retaining Structures

FERSs are typically designed considering both the geotechnical and structural ULS. Whether or not the SLS is considered depends on the specific type and application of the FERSs. For example, although sheet-pile bulkheads are relatively deformable and the lateral displacements under service-load conditions may be of the order of tens of millimetres (several inches), serviceability is rarely a consideration. On the other hand, the design of a support-of-excavation (SOE) wall for a braced excavation might actually be governed by the geotechnical SLS in order to limit settlement and concomitant physical damage adjacent to the excavation.

3.2.4 Geofoam Walls

Despite their name, geofoam walls are really an extension of earthwork, not earth-retaining structure, concepts. Furthermore, other than a protective, architectural covering on the vertical exposed face of the geofoam, there is no structural component to a geofoam wall (Horvath 2003a). Consequently, the relevant design approach follows that for earthworks which typically involves a consideration of only the geotechnical ULS.

3.3 DESIGN OF GEOSYNTHETIC COMPONENTS

3.3.1 Introduction

This section illustrates and explains the basic differences between the two fundamentally different design concepts for using geofoam as the sole or at least primary geosynthetic component of a system to reduce lateral earth pressures on ERSs as well as corrects some misunderstandings about these functional applications that have evolved in practice to date. As will be seen, each function uses the geofoam product in a very different physical manner and with an assumed behavioral mechanism that is not always intuitive.

The simple generic problem shown in Figure 3.1 will be used for the purpose of this explanatory discussion. It also illustrates the baseline no-geofoam case that should always be investigated whether designing a new ERS or analyzing an existing ERS to provide the necessary technical and cost reference point against which various geofoam alternatives can be compared. Note that the use of a gravity retaining wall in this and subsequent related figures is in a qualitative context to represent any type of ERS. None of the concepts to be discussed is limited to this type of ERS.

To begin with, the lightweight-fill function is what is called a *small-strain function* of EPS-block geofoam. This means that all compressive normal stresses and strains[34] must be kept within the nominally linear-elastic initial portion of the EPS stress-strain curve (Horvath 1995b, Stark et al. 2004a, Horvath 2010b). This is typically defined as normal compressive strains not exceeding 1% (0.01) in magnitude.

On the other hand, compressible inclusion is a *large-strain function* based around substantial (typically of the order of several or even tens of percent strain) compression of the geofoam product under service-load conditions (Horvath 1995b, 1996c, 1997b, 1998a, 1998b, 2010b). The difference is important as geofoam materials tend to have very different stiffness under different strain levels, something that is, unfortunately, not always appreciated and understood in practice, even by academic researchers (Horvath 2010b).

---

[34] Unless noted otherwise, all normal stresses used or referenced in this document are *engineering* (a.k.a. *Cauchy*) *stresses* and all normal strains are *engineering strains*, i.e. both stresses and strains are based on the initial, undeformed geometry (cross-sectional area for normal stress and length for normal strain) of a test specimen as is typical for most civil engineering material testing. While this is adequate for small-strain geofoam functions such as lightweight fill, an argument can be made that large-strain applications such as compressible inclusion discussed subsequently might be more accurately portrayed using *true* (a.k.a. *Hencky*, *logarithmic*, or *natural*) *stress* and *true* strain. A detailed discussion of the difference between engineering and true stress/strain is presented in Appendix D. However, explicit determination or even estimation of true stress (as is sometimes done in soil testing) would add considerable complexity to geofoam testing protocols as the change in cross-sectional geometry of test specimens would need to be measured or at least estimated. This is neither easy nor obvious as many polymeric geofoam materials exhibit localized necking, which implies a negative Poisson's ratio, at large compressive normal strains.

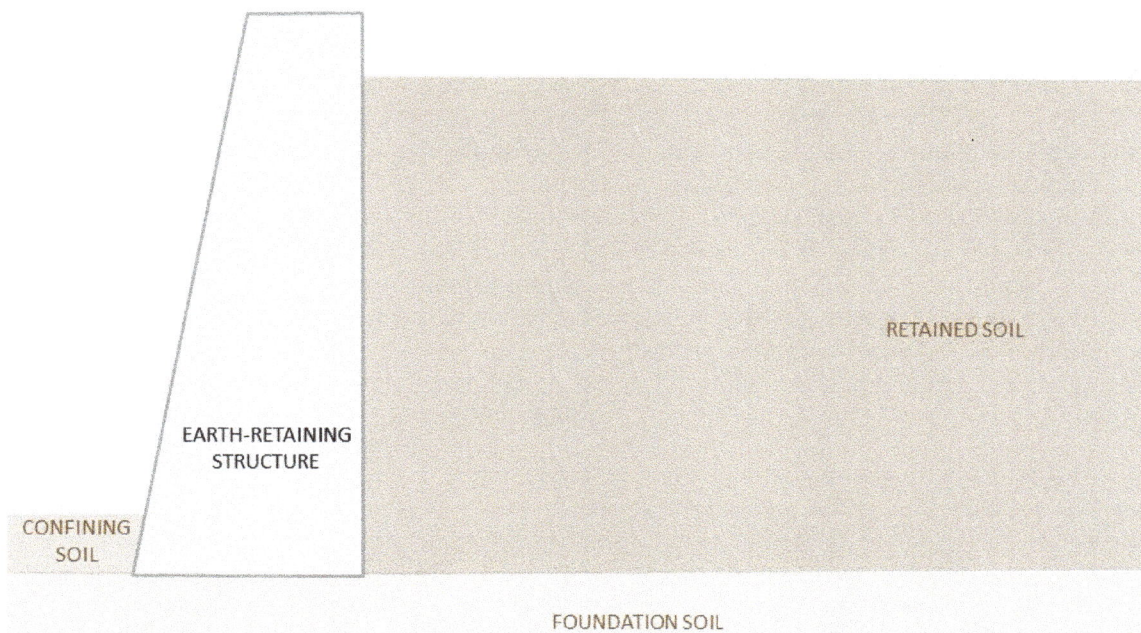

Figure 3.1. Generic ERS Problem Used to Illustrate Geofoam Functional Usage.

Regardless of which of these two primary functions is used as the basis for design, in virtually all applications the geofoam material(s) and product used can be designed to beneficially provide one or more additional, secondary geosynthetic functions. As noted in Chapter 2, this multifunctionality, which is a signature characteristic of EPS- and PS-based geofoams in particular, can be used to technical and economic benefit. These secondary functions were noted earlier and are discussed further in Chapter 7. They should be taken into account when selecting a specific geofoam product (discussed further in Chapter 8) and performing a project-specific economic assessment in order to optimize the cost balance between geosynthetic use and the ERS.

3.3.2 The Lightweight-Fill Function

3.3.2.1 Basic Concept and History of Use

The lightweight-fill function of geofoam in general, and EPS-block in particular, is arguably intuitive because of the inherent very low density of geofoam materials compared to normal, naturally occurring earth materials (although this intuitive aspect can be misleading in ERS applications as will be seen). For example, EPS blocks (a.k.a. *billets* as they were often referred to in the past and still are by some in the EPS industry) used for geofoam applications most commonly have a density, ρ, in the range of 16 to 40 kg/m³ (unit weight, γ, in the range of 1.0 to 2.5 lb/ft³) although a somewhat greater range of material densities/unit weights (both lower and higher) is achievable (Horvath (2011) contains a detailed presentation of the EPS block-molding process, including a discussion of achievable material densities). This is only about 1% to 2% of the density or unit weight of soil[35].

The lightweight-fill function was the second geosynthetic function recognized for geofoam (circa 1970, which was several years before the term 'geofoam' was used as a U.S.-registered trademark for an EPS product sold only in the State of Alaska and more than 20 years before 'geofoam' came into use in its current generic definition), preceded only by the function of thermal insulation circa 1960 (Horvath 1995b)[36]. Initially, both EPS and XPS were tried as lightweight fill as both had been both technically successful as well as cost effective when used in thermal-insulation geofoam applications throughout the 1960s. In fact, two known U.S. patents issued in the early 1970s timeframe covering the lightweight-fill function of polymeric foams, including for use with ERSs to reduce lateral earth pressures, were explicitly built around the use of XPS in general and Dow Chemical's well-known, blue-colored[37] *Styrofoam*™-brand product[38] in particular (Horvath 1995b). At least

---

[35] Strictly speaking, *density* and *unit weight* are not synonymous. Density is defined dimensionally as *mass per unit volume* and in SI units is generally expressed for geotechnical engineering usage in *kilograms per cubic metre* (kg/m³) although regional variations exist. For example, in Japan (and possibly elsewhere), *grams per cubic centimetre* (g/cm³ or g/cc) appears to be preferred in the geotechnical engineering literature even though use of centimetre as a SI unit of measurement is deprecated. In Imperial units, density has units of *slugs per cubic foot*. However, civil engineers, at least in the U.S., do not routinely use the unit of slug for mass. The unit of *pound-mass* (lbm) is used as an alternative. However, the writer has a profound dislike for this term, despite its official acceptance by ASTM and other groups, as historically a pound was solely a force-based, not mass-based, unit. Unit weight is defined dimensionally as *force per unit volume* and in SI units is generally expressed for geotechnical engineering usage as either *kilonewtons per cubic metre* (kN/m³) or *Newtons per cubic metre* (N/m³) depending on the magnitude and the user's desire to avoid use of fractional numbers. In Imperial units, *pounds per cubic foot* (lb/ft³ or pcf) is used. Note that the unit of *pound-force* (lbf) is considered by some to be an acceptable synonym and alternative term to pound used alone. However, the use of pound-force is equally disliked by the writer and thus eschewed in this monograph based on the argument that labeling a 'pound' a 'force' is redundant. In any event, to ensure clarity and correctness throughout this monograph as to what is being expressed, the parameters of density and unit weight will be used only in their rigorous and correct manner. However, for the convenience of readers and users of this monograph, whenever the density of a geofoam material or product is stated in kg/m³ the stated Imperial-unit equivalent will be always unit weight, not density, in lb/ft³ for the simple above-stated reason that density in Imperial units (slugs/ft³) is rarely, if ever, used in U.S. civil-engineering practice.

[36] Most polymeric geofoam materials as well as non-polymeric geofoam materials such as *cellular glass* were invented or developed specifically for use as thermal insulation although not specifically or initially for geotechnical applications. This is certainly true of EPS which was invented circa 1950 and *extruded polystyrene* (XPS) which was invented approximately a decade before that. The initial uses of both EPS and XPS, which are collectively referred to as *rigid cellular polystyrene* (RCPS) as noted previously, were all non-geotechnical in nature. However, geotechnical applications involving thermal insulation were a natural and intuitive evolutionary outgrowth for these materials so it is no surprise that they were used in what we would now call geofoam applications by at least circa 1960 (there are relevant U.S. patents that were issued in the mid-1960s which implies research and initial use had to occur some years earlier).

[37] Both EPS and XPS are inherently and naturally white in color although XPS that has had a flame-retardant chemical added will tend to be either a light tan or yellowish-brown in color. Both EPS and XPS can also be artificially colored and this is typically only done solely for product identification and marketing purposes. Historically, non-white EPS has been relatively uncommon but has been used in some countries (e.g. Canada and the U.K.) but not the U.S. until recently when an orange-colored product was introduced. On the other hand, colored XPS has been the norm for a long time, at least in the U.S. In addition to the aforementioned **[continued on following page]**

one test installation of XPS as a lightweight-fill material for an earthwork on soft ground is believed to have occurred during the early 1970s in the vicinity of Dow Chemical's corporate headquarter in Midland, Michigan. Whether or not XPS was ever actually used to reduce lateral earth pressures on an ERS is not known.

However, it was quickly realized in practice worldwide that using EPS-block in lightweight-fill applications was much more cost effective than using the relatively thin planks or panels of XPS (the standard XPS product form which is constrained by the XPS extrusion-based manufacturing process), even though EPS blocks of the era were somewhat smaller in all dimensions, especially length, compared to what is typically made at present (Horvath 2011, 2012)[39]. The fact that EPS is typically less expensive than XPS on a unit-volume basis was certainly another factor in tilting the lightweight-fill geofoam market completely in the favor of EPS-block. These facts coupled with the universally positive experience in Norway using EPS-block geofoam for lightweight fills beginning in 1972 established EPS-block as the worldwide geofoam material and product of choice for any and all lightweight-fill applications, something that continues globally to the present.

3.3.2.2 Full-Depth Placement Alternative

Despite the fact that the use of geofoam in general, and EPS-block geofoam in particular, as lightweight backfill and fill behind ERSs to reduce lateral pressures is perhaps intuitive, the actual mechanism by which this pressure reduction is currently hypothesized to occur is surprisingly non-intuitive and somewhat complex in nature. Experience to date indicates that there is a persistent, significant misunderstanding as to the correct physical and theoretical mechanisms by which lateral pressure reduction is achieved, at least based on the current state of knowledge (Horvath 2008a, 2010b).

Figure 3.2 is a generic depiction of what can, based on the current state of knowledge, be considered the basic, preferred use of EPS blocks for the lightweight-fill functional application behind any type of ERS (again, not necessarily limited to the gravity retaining wall shown in this figure). This is referred to as the *full-depth placement alternative* meaning that the EPS blocks are placed so that the bottom of the first level or course of blocks is at the same elevation as the heel of the ERS. Note, however, that even in this ideal case the EPS block placement would rarely be for the full geotechnical height, *H*, of the ERS (as defined in Figure 2.1). Even if the ground surface above the assemblage of EPS blocks were not paved or built on for some reason, it would not be considered good practice to leave the uppermost surface of the EPS blocks exposed permanently to the atmosphere as physical degradation and possibly vandalism might occur. Consequently, some nominal thickness of soil backfill/fill is should always be placed on top of the uppermost course of EPS blocks.

---

blue-colored XPS product, green, pink, and yellow have been used for XPS products in the U.S. by XPS manufacturers (referred to in the industry as *extruders*).

[38] Note that contrary to popular belief and colloquial usage in the U.S. at least, 'styrofoam' is not and has never been a generic term for any and all types of polymeric foam. It has always been and remains a trademark of a particular brand of XPS manufactured by Dow Chemical.

[39] This dimensional handicap of XPS compared to EPS-block in lightweight-fill applications does not exist in thermal-insulation applications where required product thicknesses are always substantially less.

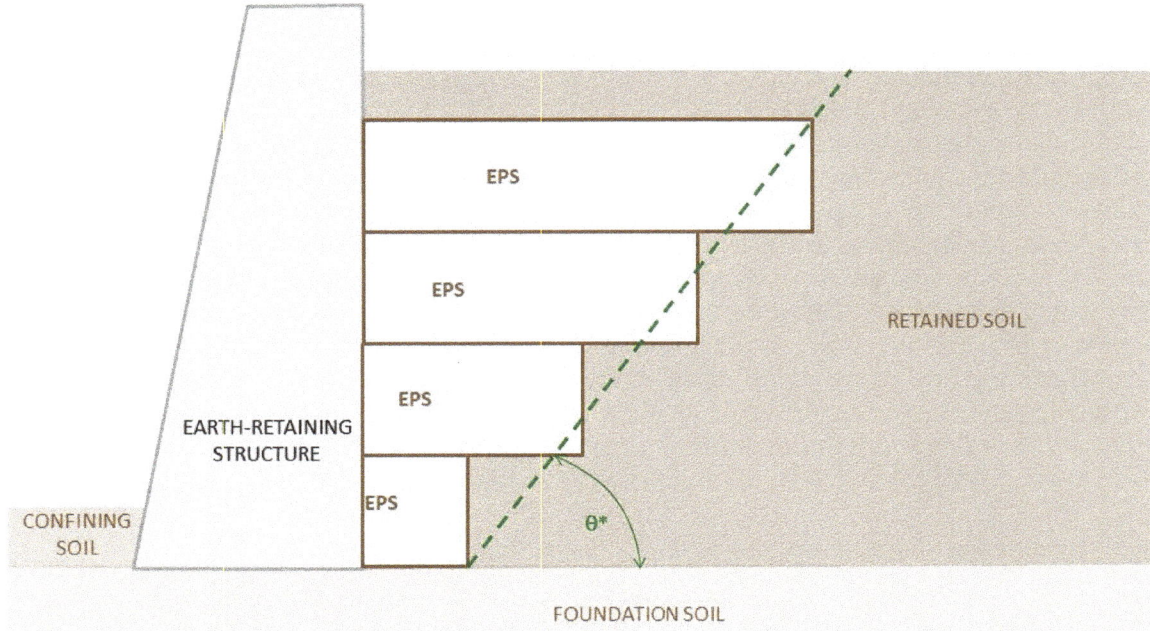

Figure 3.2. Lightweight-Fill Function - Basic Full-Depth Placement Alternative.

Based on the current state of knowledge as explained in detail subsequently, the single most important geometric detail of the full-depth placement alternative is that the EPS blocks be placed in a stair-step pattern against the retained soil as shown qualitatively in Figure 3.2 and defined by the angle θ^*. Note that this design detail was not appreciated in early applications of this concept which completely misunderstood the behavioral mechanism involved in full-depth placement. This explains why there are numerous photos in the published literature (such as one on the back cover of this monograph) showing full-depth placement applications where the assemblage of EPS blocks has a vertical face against the retained soil (i.e. $\theta^* = 90°$).

This full-depth use of EPS blocks is considered the norm for all new construction (with the only design variable being the value of θ^* used, with $0° < \theta^* < 90°$) except in cases where this would require substantial permanent submergence of the EPS blocks below water, e.g. a bulkhead in a marine application. Although EPS blocks are tolerant of permanent water submergence in terms of long-term material durability and load-bearing performance (Horvath 1995b, Stark et al. 2004a), the inherent, excellent buoyancy of EPS blocks requires that sufficient downward vertical force be applied to the assemblage of EPS blocks to counter the net uplift force from buoyancy. This typically requires a sufficient dead load in the form of soil cover above the EPS blocks although other design strategies such as anchorage systems have been used. However, this counter-buoyancy soil cover cannot be excessive otherwise the EPS may be overstressed which will results in excessive settlement of the ground surface, usually to the point of a SLS failure. Available information is that some design professionals, at least in the U.S., have neglected to consider this overstressing issue with the result that SLS failures have occurred (Horvath 2010a).

Note that there are geofoam products such as *EPS buoyancy blocks* (a proprietary EPS-shape product) as well as other geosynthetics such as *geocombs* that can be used where buoyancy is a controlling design issue (Horvath 2004d). However, neither is widely available in the U.S. at this time and will not be considered further in this monograph.

Full-depth placement is also considered the desired goal when retrofitting existing ERSs. However, when dealing with an existing ERS, it is not always practicable to excavate to the full depth behind the structure in order to place the EPS blocks. An alternative strategy called *partial-depth placement* is used in that case and is discussed in the following section.

As noted previously, although the use of EPS-block geofoam for lateral pressure reduction on ERSs is intuitive, the analytical methodology to use based on the current state of knowledge is somewhat non-intuitive. The common assumption made in U.S. practice early in the use of this geotechnology...and, unfortunately, still often to the present...is that for both non-yielding and yielding ERSs a classical active earth pressure 'wedge' somehow forms within the assemblage of EPS blocks as defined by the:

- planar (but not necessarily vertical) inside face of the ERS;

- planar (but not necessarily horizontal) ground surface; and

- an EPS-on-EPS failure surface that is, for simplicity, assumed to be planar and is depicted qualitatively by the red dashed line in Figure 3.3.

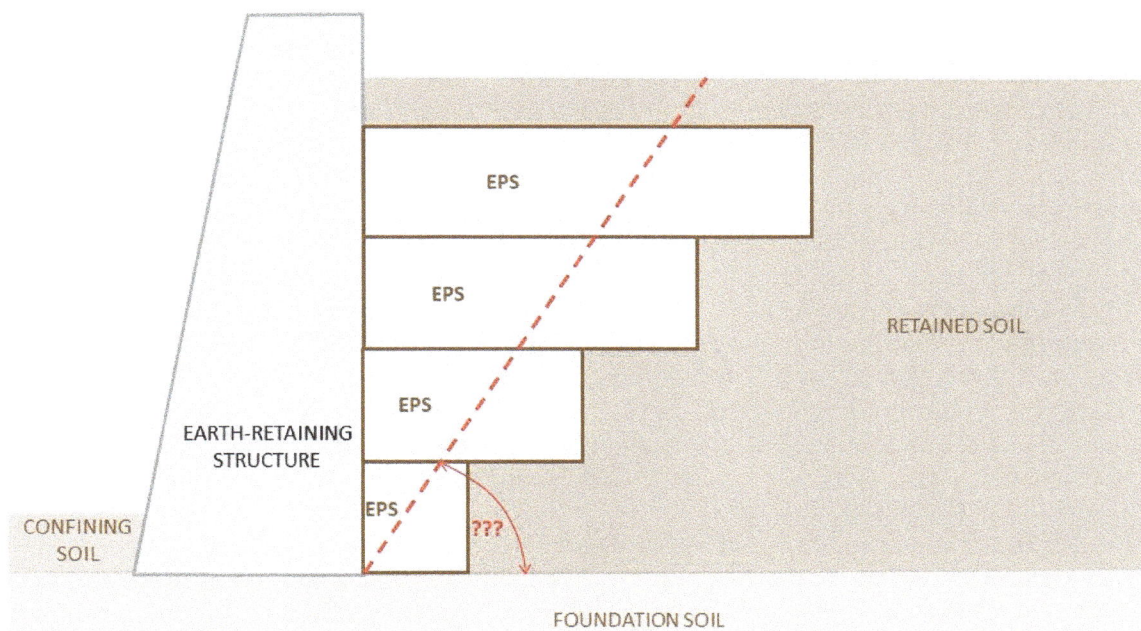

Figure 3.3. Lightweight-Fill Function - Traditional (But Incorrect) Assumed Failure Plane and Failure Mechanism for Full-Depth Placement Alternative.

Consistent with this assumed failure wedge, the reduction in lateral pressure on the ERS is presumed to occur because the unit weight of the EPS is used in the traditional equivalent-fluid-pressure formula for lateral 'earth' pressure as opposed to the unit weight

48

for soil. Because the difference in unit weights is typically of the order of 100, there is a drastic reduction in calculated lateral pressure acting on the ERS.

While this approach is intuitively and intellectually satisfying as well as attractive in its simplicity, it is, however, completely incorrect based on the current state of knowledge. This is a classic example of what has been done all too often in geotechnical engineering, i.e. hypothesizing a physical behavior (ULS mechanism in this case) that may appear to be eminently logical without verifying through physical or numerical modeling that the mechanism can or will actually occur. Stated another way, it is acceptable to hypothesize a behavioral mechanism but it is always a good idea to check with nature and see how nature will actually behave before proceeding further.

In this case, both research and experience indicate clearly that the hypothesized behavior depicted in Figure 3.3 simply does not occur for the simple reason that the hypothesized shear plane that physically cuts through the assemblage of EPS blocks will never develop (Horvath 1995b). In reality, EPS-block is sufficiently strong so that an assemblage of EPS blocks would be self-supporting (which is the basis of the geofoam-wall concept noted previously).

Based on the current state of knowledge, the correct behavioral mechanism for the problem shown in Figure 3.2 is that the overall assemblage of EPS blocks acts not only as a monolithic mass but as a de facto extension of the actual ERS as well. As a result, the green dashed line in Figure 3.2 that defines the average, overall geometry of the stair-stepped interface between EPS blocks and retained soil is assumed to act (with sufficient accuracy for routine analytical purposes) as the planar failure surface between the combined ERS + EPS and retained soil. Thus, the green dashed line shown in Figure 3.2 defines the new inside face of what will hereinafter be referred to as the pseudo-ERS that consists of the actual ERS plus assemblage of EPS blocks.

Furthermore, the retained soil to the right of the green dashed line in Figure 3.2 is assumed to be in an active earth pressure state even if the actual ERS is a non-yielding RERS. This is because the horizontal compression of the EPS blocks that occurs between the retained soil and actual ERS is assumed to be sufficient to mobilize the active state even if the ERS itself does not displace laterally.

For reasons discussed in a later chapter, it is important to note that this hypothesized behavioral mechanism is based solely on research conducted in Japan between the late 1980s and early 1990s. This work was conducted and coordinated by the Expanded Polystyrol[40] Construction Method Development Organization (EDO) which is a consortium of stakeholders involved in the manufacture and commercial use of both EPS and XPS[41]. Unfortunately, the English-language outcomes of this EDO-sponsored R&D were presented in the form of a design guide with no supporting documentation or verification of the behavioral mechanism described above.

The assumed benefit, then, of using the lightweight-fill function in its basic full-depth application is due to the fact that the lateral earth pressure from the retained soil,

---

[40] Polystyrol is an alternative term for polystyrene that is used in some countries, including Japan.

[41] As discussed in Horvath (1995b), in the past and in some regions of the world, the term 'expanded polystyrene' and acronym 'EPS' were used to refer collectively to both versions of closed-cell foams produced using polystyrene, i.e. what would nowadays in the U.S. at least be referred to as rigid cellular polystyrene (RCPS). The two forms of RCPS were called *molded expanded polystyrene* (MEPS)...what would nowadays be called simply expanded polystyrene (EPS)...and *extruded polystyrene* (XEPS)...what would nowadays be called simply extruded polystyrene (XPS). Thus, reading older literature or literature from other countries where this MEPS/XEPS terminology persisted longer than in the U.S. can be confusing at times.

Lateral Pressure Reduction on Earth-Retaining Structures Using Geofoam
John S. Horvath, Ph.D., P.E., Life Member.ASCE

which is assumed to be transmitted through the assemblage of EPS blocks to the ERS under both gravity and seismic conditions, is reduced compared to the no-geofoam baseline case as the magnitude of the angle θ* shown in Figure 3.2 is reduced from whatever it would be for the baseline no-geofoam case of the ERS alone (90° in most cases such as for the generic ERS depicted in Figure 3.1). In essence, the active wedge of retained soil is reduced in volume and thus magnitude as illustrated in Figure 3.4, with a comparison to the traditional assumption shown in Figure 3.5.

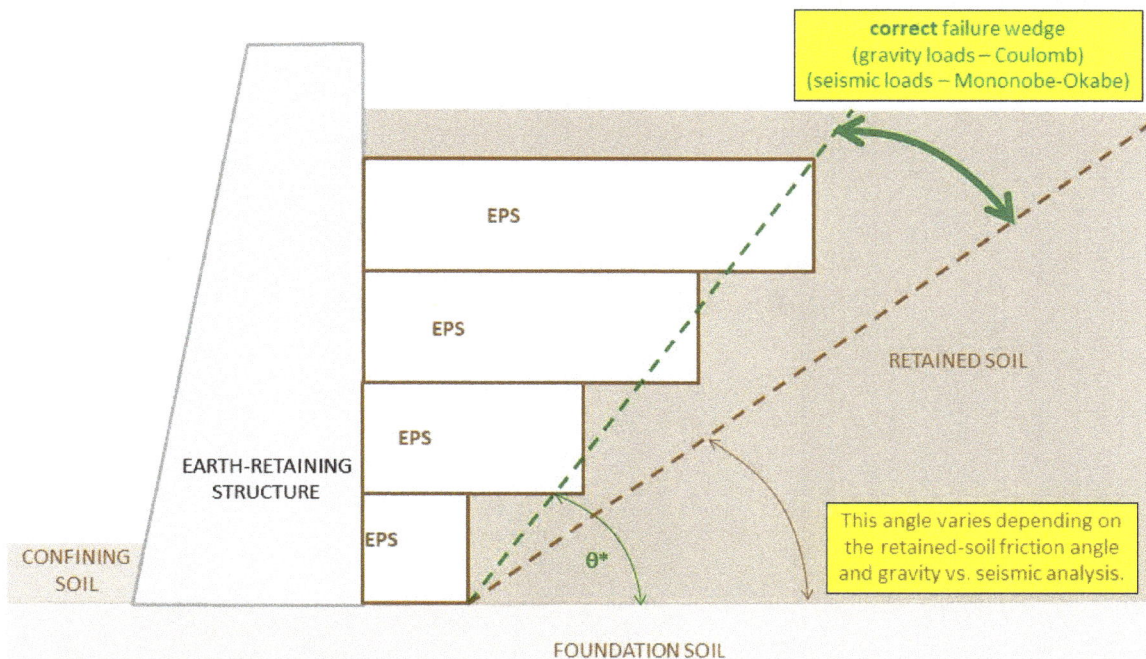

Figure 3.4. Lightweight-Fill Function - State of knowledge (and Assumed Correct) Failure Plane and Mechanism Assumptions for Full-Depth Placement Alternative.

Note that in Figures 3.4 and 3.5 the brown dashed line is the soil-on-soil failure plane that would develop within the retained soil independent of the presence of the geofoam (EPS blocks in this case). As noted in these figures, the angular orientation of this failure plane is a function of the retained soil's Mohr-Coulomb friction angle, φ, and whether gravity or seismic loading conditions are assumed.

Again, the benefit of using the lightweight-fill function is visually apparent in Figure 3.4 as the effect of the retained soil on the ERS is assumed to be limited to the soil between the green and brown dashed lines as indicated by the green curved arrow. Without the geofoam (EPS blocks), the soil wedge would extend from the brown dashed line to the inside face of the actual ERS. Furthermore, the critical importance of the stair-stepped geometry of the assemblage of EPS blocks is equally apparent from this figure. Again, this feature was not appreciated in early (circa 1970s and 1980s) project applications in many countries, including the U.S.

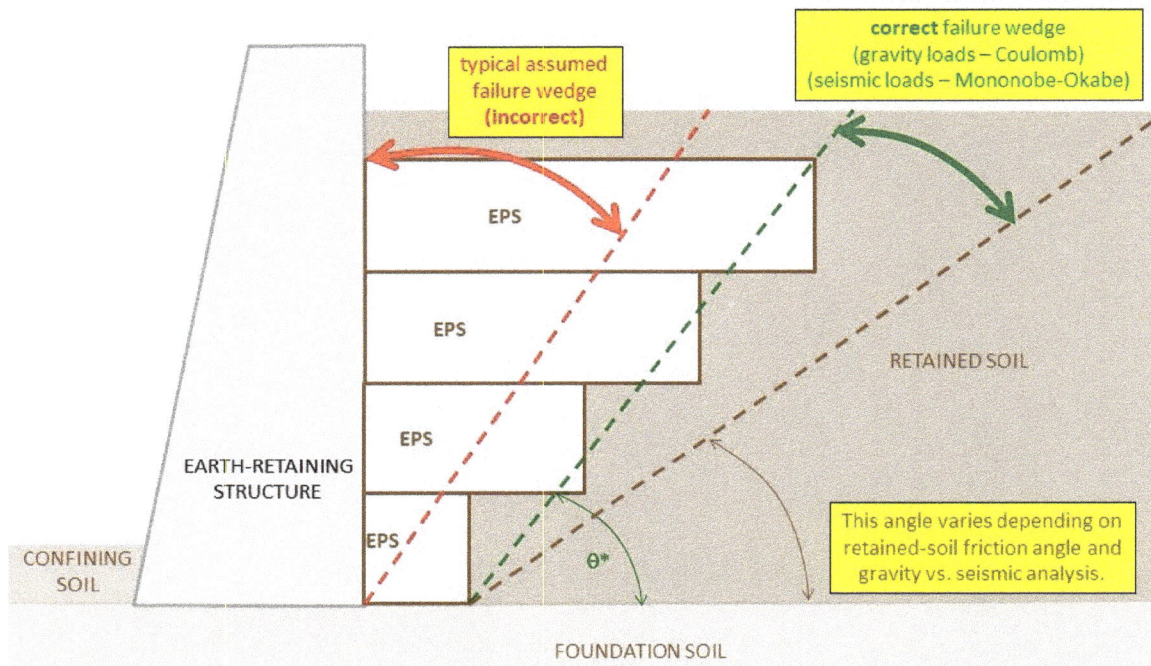

Figure 3.5. Lightweight-Fill Function - Comparison of Failure Plane and Mechanism Assumptions for Full-Depth Placement Alternative.

It should be obvious from Figure 3.4 that the reduced lateral earth pressure on the ERS that is the result of using geofoam can become zero, if desired, if a sufficiently small value of θ* is chosen such that the green and brown dashed lines become coincident. Based on the present state of knowledge, the solutions used to define the relationship between θ* and lateral earth pressure from the retained soil on the pseudo-ERS are Coulomb for gravity loads and Mononobe-Okabe for seismic loads. Thus, for gravity loads only the Mohr-Coulomb friction angle, φ, of the retained soil determines what this zero-earth-pressure value of θ* is while for seismic loads the seismic acceleration coefficient, predominantly in the horizontal direction, k_H, will have an influence as well.

Before proceeding further, it is worth reiterating that this understanding of the presumed (based on the current state of knowledge) 'correct' way in which the full-depth alternative of the lightweight-fill function works was something that evolved over time and did not become known in the U.S. until the mid-1990s initially (to the writer as a result of direct contact with EDO in Japan) and more than a decade later on a more-widespread basis. Therefore, as noted previously it is quite common when reviewing older case history applications of this concept to see photographs showing the EPS blocks with a vertical face against the retained soil, i.e. θ* = 90°. This should be viewed in the context of reflecting the design methodology of the time which is now presumed to be incorrect.

Continuing on, note that independent of lateral earth pressures, or lack thereof, from the retained soil acting on the pseudo-ERS, the assemblage of EPS blocks and soil overlying them will, in theory at least, always impart a lateral load on the actual ERS although in most cases the contribution of this component will be relatively small. This contribution is assumed to be a combined result of:

- lateral earth pressures (assumed to be at-rest state for a non-yielding RERS and active state for a yielding ERS) within the zone of soil overlying the EPS blocks plus the lateral stresses from any surface surcharge and

- the Poisson effect within the EPS blocks due to the vertical stress imposed on them by the overlying soil plus any surface surcharge (the effect of the self-weight of the EPS is typically so small as to be negligible but can be included if desired).

Note that where appropriate, both of these components would be increased for seismic loading.

In summary, perhaps the most notable and potentially dramatic positive aspect of utilizing the basic full-depth alternative is that, depending on the specific geometry of the assemblage of EPS blocks as defined by the angle θ* in Figure 3.4, lateral pressures on an ERS can be reduced from the baseline at-rest (for non-yielding RERSs) or active (for yielding ERSs) earth pressure state that would exist if no geofoam were used to close to zero under both gravity and seismic loading conditions or anything in between these two limiting cases. This can result in dramatic reductions in the ERS cost for new construction (or completely eliminate the need for an ERS) or increases in safety for existing ERSs.

However, this benefit must always be weighed against the biggest negative aspect of using geofoam in this manner which is that a relatively large volume of EPS is required, especially to achieve relatively large pressure reductions under seismic loading. The volume of EPS required typically represents a not-insignificant expense as EPS-block of sufficient load-bearing quality for geofoam applications costs more than soil on a per-unit-volume cost basis. In addition, depending on the location of adjacent property lines, rights-of-way, structures, pavements, utility lines, and other surface and below-ground obstructions there can be difficulties finding sufficient room to place a desired geometry of EPS blocks.

Nevertheless, experience in recent years indicates that the lightweight-fill function can be cost effective overall when the savings in ERS cost (for new construction) or increase in ERS safety (for existing structures) combined with ease and speed of implementation and other contributing factors are considered properly. The single most important factor to keep in mind is that this functional application is not an all-or-nothing proposition. This is because the geometry (as reflected in the angle θ* in Figure 3.4) of the assemblage of EPS blocks can be varied within easily predefined limits to analytically determine the optimum angle (from an overall cost-saving perspective) for a specific project that will also meet any physical constraints for that project site. As will be shown in Chapter 4 that deals in detail with analyzing and designing for the lightweight-fill function, all that needs to be done on a project-specific basis is to minimize the combined cost reduction of the ERS with increasing cost of EPS that occurs with decreasing θ* to find the optimal angle θ* for that project, subject to project-specific physical constraints.

3.3.2.3 Partial-Depth Placement Alternative

As noted previously, there are times with both new construction and, especially, existing ERSs when the basic full-depth placement alternative is not practicable for one reason or another. In such cases, *partial-depth placement* is an alternative as shown conceptually in Figure 3.6. Note that other than the fact that the first course of EPS blocks is placed at an elevation above the heel of the ERS, all other basic details as shown previously in Figure 3.2 for the full-depth placement alternative remain the same. Specifically, the blocks should be stair-stepped into the retained soil as with the full-depth case and there is typically at least some nominal thickness of soil placed above the assemblage of EPS blocks.

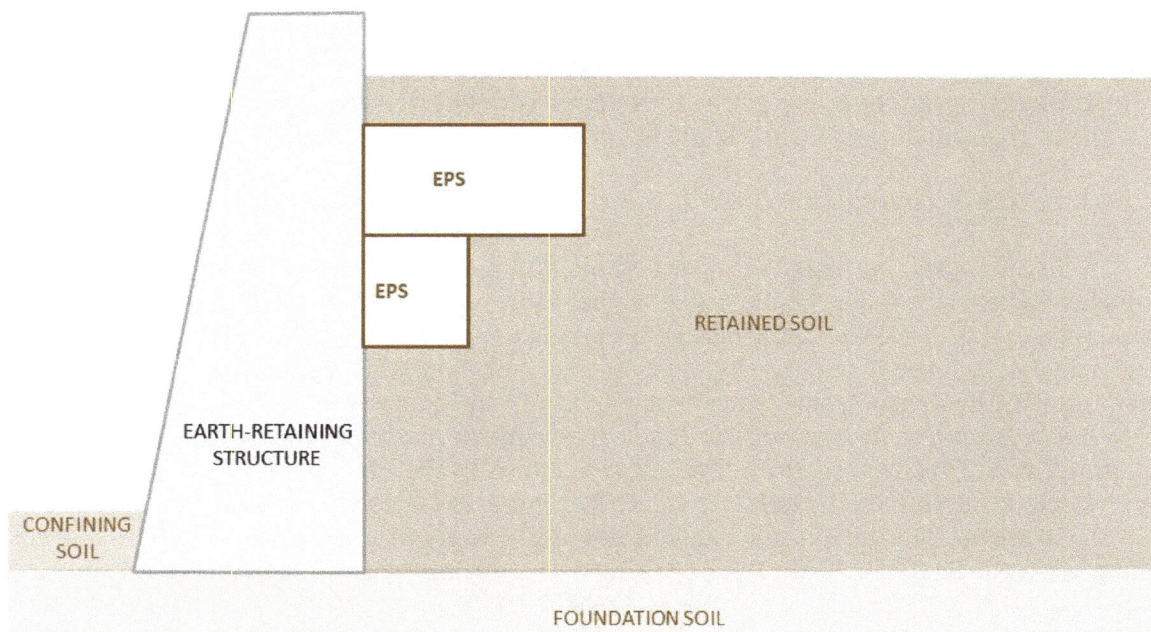

Figure 3.6. Lightweight-Fill Function - Partial-Depth Placement Alternative.

For reasons discussed in the following section, it is essential that any partial-depth placement design places the EPS blocks as shown in Figure 3.6, i.e. with the lowest course of EPS blocks placed above the heel of the ERS and the minimum necessary soil cover on top of the uppermost course of EPS blocks.

Available information suggests that the partial-depth placement alternative has been used relatively infrequently to date, with all known applications being with existing FERSs (anchored bulkheads to be specific). More importantly, there does not appear to have been any published fundamental research into the generic behavior of the partial-depth alternative in terms of defining the overall ULS mechanism for what appears to be a complex system involving the lower portion of the ERS where they are no EPS blocks and the upper portion of the ERS where there must be some as-yet-undefined interaction between the retained soil and EPS blocks. Therefore, some preliminary thoughts and concepts are presented in Chapter 4 for a logical analytical methodology for the partial-depth placement alternative. The primary purpose of doing so is to establish at least a working hypothesis that can be used as a starting point and basis of comparison for future research into this alternative.

3.3.2.4 Compressive Stresses on EPS Blocks

The lightweight-fill function as used with ERSs is a logical conceptual outgrowth of its original use as the primary component of earthworks where reduction of vertical stresses on a soil subgrade is the primary design goal. However, as part of this 'scope creep' from embankments to ERSs it appears that some potentially significant technical issues have not been given the consideration they deserve.

The most important of these issues is compressive stresses on and the stress system within the assemblage of EPS blocks. From the very beginning (early 1970s) of using EPS-block geofoam in earthworks, it was recognized that vertical compressive stresses from the overlying pavement system and vehicle live loads needed to be kept within limits that are a complex function of not just EPS density alone (as most still erroneously assume) but numerous manufacturing details as well (Horvath 2011, 2012). The reason for this limitation on stresses is so that both short- and long-term compressive strains within the EPS blocks stay within acceptable limits so that excessive settlement of the overlying pavement surface, that could result in a premature SLS failure (rideability issues) of the pavement system, do not develop. This is why in the partial-depth alternative for ERSs shown in Figure 3.6 it is essential that the EPS blocks be placed as close to the ground surface as practicable.

The most technically significant difference between earthworks and ERSs that utilize EPS-block geofoam as part of the overall system is that in most earthworks only vertical stresses on and within the EPS blocks need be considered, and only simple one-dimensional (1-D) stress-strain analyses need be performed as part of the design process. This is certainly true in stand-alone embankments where lateral pressures on the EPS blocks are essentially non-existent. This is also true in most single-sided (side-hill) embankments although perhaps less so in slope stabilization where lateral earth pressures from the soil on either or both sides of the assemblage of EPS blocks will result in a biaxial stress condition.

In any event, the point being made here is that because of more than 45 years of experience using EPS-block geofoam in earthworks, design professionals are conditioned to performing only a uniaxial stress analysis using vertical stresses when designing applications where EPS-block geofoam is used a lightweight-fill material. The fact that all of routine manufacturing quality control and assurance (MQC/MQA) stress-strain testing performed on EPS-block test specimens is and has always been uniaxial in nature adds to this mental conditioning.

However, it is clear from Figures 3.2 (full-depth alternative) and 3.6 (partial-depth alternative) that when the lightweight-fill function is used to reduce lateral pressures on ERSs there is always the potential that lateral earth pressures on the assemblage of EPS blocks are significant in addition to the vertical stresses. In fact, it is possible that in some cases the lateral earth pressures, especially toward the bottom of an ERS, might exceed the vertical stresses on the EPS blocks from the overlying soil cover and pavement system (if any). This clearly suggests that a biaxial stress analysis is a necessity for design.

Unfortunately, the biaxial stress-strain behavior of EPS-block has been little studied[42] and thus is poorly understood at the present time. Thus, the necessary guidelines for performing a correct biaxial-stress analysis do not exist at the present time.

3.3.3 The Compressible-Inclusion Function (Boundary-Yielding Concept)

3.3.3.1 Introduction

The compressible-inclusion function is conceptually and theoretically more complex than the lightweight-fill function. It utilizes the principle of *soil arching* in the horizontal direction as the physical mechanism by which lateral earth pressures are reduced. Because of the essential role that arching plays in understanding the basic principles of the compressible-inclusion function, a primer on this physical mechanism will be presented before discussing this function.

3.3.3.2 A Primer on Positive (Simple) Arching in Soil

The phenomenon of *positive* (sometimes called *simple*) *arching* within both natural and manufactured particulate (granular) materials in general, and soil in particular, is a logical extension of the architectural and structural engineering concept of using an arch composed of some solid material (natural stone, manufactured masonry, PCC, or metal) as the fundamental load-bearing/load-transfer mechanism for some structure such as a bridge or viaduct. In particular, the image of the classic solid-material arch is a useful visualization of the arch that can form within a mass of soil which, of course, is not visible from the surface[43]. In any event, positive arching is generally referred to simply as 'arching' in the geotechnical literature and will be so referenced throughout this monograph.

---

[42] Note that this should not be confused with the fact that it is well-established that EPS-block is an *isotropic* material which means that its uniaxial mechanical (stress-strain-time-temperature) properties are the same regardless of the orientation of the loading direction. This is not a trivial statement as other polymeric foams that have seen use in geofoam applications such as XPS are significantly *cross-anisotropic* as a result of their particular manufacturing process.

[43] One terminological difference of note is that in architecture and structural engineering the term 'arch' is reserved for a structural element whose load-bearing behavior is essentially planar or two-dimensional (2-D) in nature, with the term 'dome' being used for a behaviorally similar structural element that bears load three-dimensionally. This is along the same lines as using the terms 'beam' and 'plate' to define flexural elements acting in two and three dimensions respectively. On the other hand, in the geotechnical context as discussed here the term 'arch' is applied to all problems involving particulate materials, regardless of whether the behavior is nominally planar (2-D) in nature, as in virtually all cases involving ERSs, **[continued on following page]**

Other than the obvious material difference, the most significant difference between a solid-material arch and arching in soil is that the former is pre-existing and thus inherently structurally active and available as a load-bearing member from the start whereas the latter is not pre-existing but needs to develop within some mass of individual particles that is initially in a more-or-less homogeneous stress state.

Because an arch transmits forces, this means that a soil arch involves stresses, more specifically, some relatively complex two- or three-dimensional shear-stress state within a relatively localized region that differs significantly from the overall homogeneous stress state within the soil mass. The change in mobilized shear stresses within a soil mass necessary to produce a soil arch always requires soil strains that result from soil-particle displacements, at least within the localized portions of the soil mass where the arch develops. Thus, arching in soil always implies displacement within the soil mass, even if the magnitudes of this displacement are not calculated explicitly, as there can never be arching in soil absent inter-particle displacement.

Thus, the fundamental causative mechanism for creation of a soil arch is inter-particle displacement within some relatively localized two- or three-dimensional region. These localized particle displacements do not occur on their own but are induced by some localized initiating or trigger displacement (which can be natural or the result of some human activity) within the overall soil mass, a process that is called *induced yielding*. These displacements produce the strains that allow the necessary shear-strength mobilization but again, only in the immediate vicinity of the displacements.

As a result of these localized displacements, normal stresses <u>decrease</u> in the direction of the initiating or triggering displacement as the soil forms what amounts to a natural arch through inter-particle friction around the localized region of displacement. It is important to note that because the overall system must remain in force equilibrium, there must be corresponding normal-stress <u>increases</u> in adjacent regions that either do not displace or displace less and thus essentially act as the support reactions of the arch (just as an arch of solid material produces reactions at its supports). Thus, soil arching does not, and cannot, 'get rid' of stresses as overall force equilibrium must be maintained. Soil arching simply redistributes stresses locally within some portion of the overall soil mass while maintaining overall force equilibrium. It is important to note that depending on the problem or application, this stress redistribution can either be a benefit or a detriment.

An interesting outcome of this stress redistribution is that at some distance away from the localized region of displacement and concomitant stress redistribution, the overall soil mass does not 'feel' the effect of the localized arching so there is no displacement/strain and the stresses are unchanged. Thus, in a very broad sense arching within a mass of particulate material such as soil is conceptually similar to Saint-Venant's principle applied to solids. For example, for a hole in a steel plate under in-plane loading, the stresses and strains in the immediate vicinity of the hole would reflect the presence of the hole. But at some distance away from the hole the plate would not 'know' that the hole even existed.

Another aspect of arching is that it can, in principle, occur in any direction within a soil mass. However, for practical reasons the arching problems that arise in geotechnical engineering generally involve either the vertical or horizontal direction, with the latter being a key element of this monograph.

Soil arching in the vertical direction was the first to be recognized and is sometimes referred to as the *trapdoor problem* or *silo problem* (the latter referring to the behavior of grain within the confined space of a storage silo). Consequently, vertical arching was one of

---

or truly three-dimensional (3-D) in nature as in the classic 'trapdoor problem' (in which case an argument could be made for calling this 'soil doming').

the earliest granular-material phenomena identified and studied when modern soil mechanics theory began to evolve in the early decades of the 20th century (Terzaghi 1943). Vertical arching continues to be studied to the present (Costa et al. 2009).

However, the recognition, study, and practical use of vertical soil arching actually predates its formal treatment in modern soil mechanics. For example, vertical soil arching had been recognized as the primary physical mechanism for both increasing and reducing vertical loads on underground conduits such as pipes and culverts in a wide variety of applications well before Terzaghi's seminal works were published (Spangler and Handy 1982). Vertical arching of particulate materials in general was also the basis for Janssen's theory for conditions that exist within silos storing granular materials such as grain, also decades before Terzaghi's work.

As an aside, as noted previously, a brief introduction to the use of geofoam and the compressible-inclusion function to beneficially induce vertical arching over underground conduits as well as in other applications where reduction of vertical earth loads is the priority is presented in Chapter 10. Geofoam-related research into this application dates back to at least the 1980s.

In any event, despite research related to vertical soil arching in particulate materials that dates back to the 19th century, significant research into horizontal arching in soil only occurred in the latter decades of the 20th century even though Terzaghi had noted its occurrence with ERSs decades earlier. At around the same time that horizontal soil arching was being researched and written about as a geomechanics mechanism to better explain the long-observed non-hydrostatic lateral earth pressure distributions behind RERSs[44] (Handy 1985, Harrop-Williams 1989), it was being recognized as the underlying mechanism in the use of compressible inclusions (not limited to geofoam as other materials, both natural and manufactured, had been tried in earlier years) to reduce lateral pressures on ERSs. As noted in the Preface to this monograph, it was the desire to find a practical, behaviorally predictable, environmentally inert, manufactured material to use as a compressible inclusion with ERSs that led the writer to begin investigating what we now call geofoam in 1987. The writer's first published work on the subject was in 1990 (Horvath 1990b).

As a final comment on the subject of arching, it should be noted that there is a geotechnical behavioral phenomenon referred to in the literature as *negative arching*. This term usually comes up with regard to the vertical stresses on underground conduits and is discussed further in Chapter 10.

3.3.3.3 Basic Concept and History of Use

In simple terms, the basic, fundamental objective of the compressible-inclusion function when used to reduce lateral pressures on ERSs is to induce horizontal arching within the retained-soil mass adjacent to an ERS, especially in cases where the ERS is relatively rigid and thus non-deforming and possibly non-yielding as well. The specifics of how this is done and the benefits in terms of reduced lateral pressures on the ERS that derive from this are illustrated subsequently.

As discussed in the preceding section, an arch within a soil mass is never pre-existing but must always be 'constructed' by inducing relatively localized displacement and concomitant shear stresses and shear-strength mobilization within an otherwise

---

[44] More recently, horizontal soil arching has been used to explain the long-observed behavior of certain types of FERSs such as anchored bulkheads (Horvath 2014).

homogeneous stress state. Thus, there must always be some trigger mechanism to induce soil arching to develop.

Consequently, the design objective of the compressible-inclusion function is fundamentally completely different than that of the lightweight-fill function. The goal of a compressible inclusion is not to fill some relatively large volume with a solid material whose primary material property is density much lower than that of earth materials but to use a material that will <u>efficiently</u> induce the magnitude of displacement required for soil arching to develop. Note that 'efficiently' is the operative term here and means using as little (in terms of thickness) of a relatively compressible geofoam material as possible to induce the desired displacement. The density per se of the geofoam material is irrelevant. In addition, the geofoam product used as the compressible inclusion must be placed in a configuration, orientation, and location where it will induce horizontal displacement of the retained soil so that it will have maximum load-reduction benefit for the ERS.

The fact that *stiffness (compressibility)* and not density/unit weight per se is the single most important material property of the geofoam product in compressible-inclusion applications cannot be overemphasized. Experience to date indicates that this is not always completely and, more importantly, <u>correctly</u> understood and appreciated by design professionals in both practice and research (Horvath 2010b). More to the specific point, the stiffness properties of the geofoam product used in compressible-inclusion applications <u>must</u> be consistent with the operational (actual) stress levels of a specific application if the geofoam product is to function as intended. Again, this cannot be overemphasized as the mechanical (stress-strain-time-temperature) properties of geofoam materials are, in general, very complex as they include both non-linear and inelastic (non-recoverable or plastic) behavior components that are also significantly time dependent (creep and relaxation).

As in the situation noted previously with the lightweight-fill function wherein many early project applications did not stair-step the EPS blocks into the retained soil because the importance of this was not recognized at the time, many project and research applications of geofoam as a compressible inclusion that can be found in the published literature did not recognize, or properly recognize, the importance of using or matching geofoam material and product stiffness under the operational stress levels relevant to the specific application. This includes applications for both types of arching, i.e. vertical arching to reduce vertical loads on underground conduits as well as horizontal arching to reduce lateral pressures on ERSs. Unfortunately, unlike lightweight-fill applications where it appears most design professionals have 'gotten the message' about the need to stair-step EPS-block layouts, it appears that many who are either researching or using geofoams as a compressible-inclusion have not gotten the message appropriate to that functional application (Horvath 2010b).

Therefore, it is important to understand that there is a high probability that a publication discussing the results of either research or a project application of a geofoam product as a compressible inclusion will have either used an inefficient or even completely inappropriate geofoam material or product and/or incorrectly interpreted the observed behavior in some way. This is more than a little unfortunate as it means that there is a relatively substantial and still-increasing body of published literature that contains misinformation of one kind or another that will inevitably be passed on and used by others who will assume that the author(s) of the published work knew what they were doing.

3.3.3.4 Application Basics

Research to date has identified two different concepts for using the compressible-inclusion function to reduce lateral pressures on an ERS:

- the *Reduced Earth Pressure Wall* (REP-Wall) concept and

- the *Zero Earth Pressure Wall* (ZEP-Wall) concept

with each potentially applicable in both full- and partial-depth placement as with the lightweight-fill function. These terms were coined by the writer in the late 1980s.

There are common elements among all four variations of the compressible-inclusion application. The most significant is that the geofoam product[45] typically used has the form of a relatively thin (of the order of tens of millimetres or several inches thick) panel that is placed in a 'chimney' orientation along the inside face of the ERS (as will be seen, the ERS is most commonly a non-yielding RERS). Specific geometries and other basic design details are described and illustrated subsequently.

The geofoam product in both REP- and ZEP-Wall applications is intentionally designed[46] to be significantly more compressible than the EPS-block used for lightweight fill. This is so the geofoam product is the most compressible component of the overall ERS-geofoam-retained soil system under a wide operational stress range and thus compresses readily compared to the other two components under the lateral pressures imposed by the retained soil, even when these pressures are of relatively small magnitude. This fulfills the mandate noted previously that the stiffness of the geofoam product is the single most important design variable in this functional application.

This compression of the geofoam product, which is essentially one-dimensional in nature in the direction of the thickness dimension of the product, allows the retained soil to yield (displace) in a direction perpendicular to the geofoam panel (the phenomenon of induced yielding noted previously) even though the ERS itself may be otherwise rigid and non-yielding. Essentially, the geofoam acts as a vertically oriented trapdoor to allow the retained soil adjacent to it to yield laterally and in the process mobilize its inherent shear strength and arch around the geofoam panel, thereby reducing the lateral pressures on the ERS. This, then, is a classic case of soil arching as described and discussed previously but in the to-date-less-common horizontal direction. In this case, it is the geofoam panel acting as a compressible inclusion that provides the necessary trigger mechanism to induce the yielding necessary for the soil arch to form.

The unique aspect of the compressible-inclusion function in general when implemented using geofoam[47] is that the magnitude of this yielding is always quantifiable during the design process, at least in principle, because the geofoam product used is composed of engineered material(s) with readily measurable mechanical (stress-strain-time-temperature) properties. This is notably unlike the early applications of vertical

[45] As discussed in Chapter 8, the preferred geosynthetic product for most compressible-inclusion functional applications is actually a geofoam-based geocomposite, with R-EPS being the primary component and primary source of the desired product compressibility.
[46] As alluded to earlier and as will be seen subsequently in detail, there are geofoam materials that have been crafted specifically for compressible-inclusion applications.
[47] This function is, in principle, valid with any compressible material and a wide variety of natural and synthetic, organic and inorganic materials have been tried over the years, especially for inducing the vertical arching over underground conduits that was noted previously.

arching above underground conduits that used bales of hay and other organic materials whose compressibility was unknown and unpredictable. Thus, the use of a geofoam compressible inclusion results in a condition called *controlled induced yielding* or, more simply and commonly, *controlled yielding* with the retained soil mass. The term *boundary yielding* as used by researchers in the U.K. is a synonym for controlled yielding.

As noted previously, compressible inclusion is one of the relatively few geofoam functions that is large-strain in nature meaning that relatively large compressive normal strains of the geofoam product are actually desirable, if not necessary, for the function to work. It is worth repeating that 'large strain' in this context is typically several percent or even tens of percent compressive strain based on the original thickness of the geofoam panel. This contrasts to most geofoam functions, including lightweight fill, that are small-strain in nature wherein compressive strains need to be kept to relatively low levels (typically 1% or less for EPS-block) to avoid potential serviceability issues related to creep.

Note that this means that a geofoam material and product that is optimal for lightweight-fill applications will <u>never</u> be optimal for compressible-inclusion applications. This point has, historically, not been appreciated by either researchers or practitioners who have routinely adopted a 'one size fits all' mindset when using geofoam materials and products and think that one material and product (and in extreme cases, such as in Norway in the 1970s and 1980s, one <u>density</u> of one material/product) can universally be used for all geofoam functions. This is what has led to inefficient use, misuse, and behavioral misinterpretation of geofoam materials and products as compressible inclusions as discussed previously.

As will be seen, a significant benefit of the compressible-inclusion function is that, assuming the most efficient material and product have been specified for a given project, substantially less geofoam material is required compared to the lightweight-fill function to achieve significant lateral pressure reductions on ERSs. As noted previously, a correctly designed compressible inclusion may only be several tens of millimetres (several inches) thick which is substantially less geofoam material than required when the lightweight-fill function is used. In addition, the geofoam product used can easily be designed to be multifunctional to also provide insulated drainage for the ERS which is desirable in most applications. Overall this makes the compressible-inclusion function very attractive for use with ERSs from a cost perspective.

The primary downside of the compressible-inclusion function (the writer does not consider it to be a 'negative' as the problem does not lie with either the concept or the geofoam material or product) is that it is based on the geomechanics concept of arching that is much more complex to deal with analytically and, as a result, much less intuitive and thus much less familiar to design professionals, compared to the lightweight-fill function. Consequently, even though both research and actual project experience dating back to the 1980s as summarized in numerous publications by the writer (Horvath 1990b, 1991a, 1991b, 1991c, 1991d, 1995b, 1995c, 1996a, 1996b, 1996c, 1997b, 1998a, 1998b, 2001a, 2003b, 2003c, 2004e; 2010b) have clearly demonstrated the viability of the compressible-inclusion function, experience to date has shown that even experienced geotechnical engineers have been slow to recognize its cost effectiveness in project applications. As a result, practitioners have either not made use of geofoams at all to reduce lateral pressures on ERSs or used the lightweight-fill function when it offered no advantages[48] over and cost more than the compressible-inclusion function.

---

[48] For the sake of completeness and objectivity, it should be noted that there are certain project-specific applications where the lightweight-fill function offers clear technical advantages over the compressible-inclusion function that would justify **[continued on following page]**

In addition, there has been relatively little fundamental research by others to develop and publicize analytical methodologies suitable for use in routine design practice beyond material in the writer's publications noted above. However, this has started to change in recent years, at least for full-depth applications involving the REP-Wall concept and seismic loading, what has been coined the *seismic-buffer* concept by others. Nevertheless, all evidence indicates that geofoam compressible inclusions have the potential to revolutionize the way that ERSs are designed and built (Horvath 2003c, 2004e).

3.3.3.5 The Reduced Earth Pressure Wall (REP-Wall) Concept

3.3.3.5.1 Full-Depth Placement Alternative

The most basic application of the compressible-inclusion function is shown in Figure 3.7 and is referred to as the *full-depth placement alternative* of the REP-Wall concept. In this application, the geofoam panel is placed along the entire inside face of the ERS.

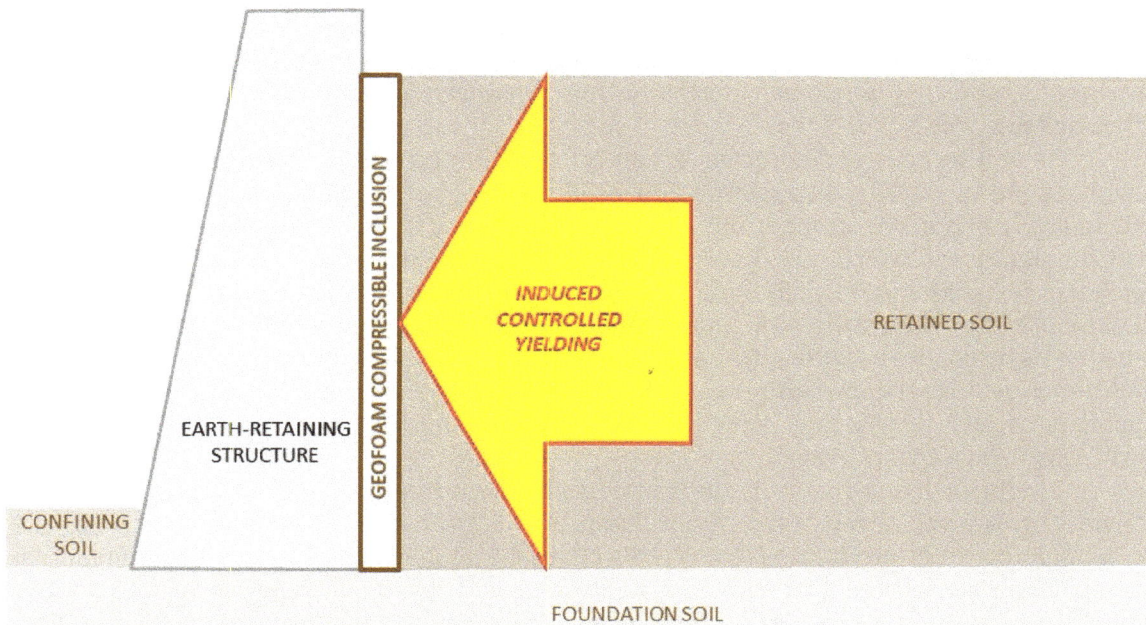

Figure 3.7. Compressible-Inclusion Function - REP-Wall/Full-Depth Placement Alternative.

The required thickness of the geofoam panel is always a project-specific assessment. However, for visualization purposes, this panel is typically of the order of 150 to 300 millimetres (6 to 12 in) thick, assuming that the <u>appropriate</u> geofoam material is used. Note that this includes the thickness of the sheet-drain component of the geocomposite product

---

the greater cost of the former compared to the latter. The most common situation is when the foundation soil underlying the retained soil adjacent to the ERS is highly compressible and settlements of the retained soil adjacent to the ERS are to be minimized. Some of these applications are discussed further in Chapters 6 and 9.

that would be used in most REP-Wall applications as providing for positive drainage of the retained soil is desirable in virtually all applications.

As in all compressible-inclusion applications, the design goal here is for the geofoam to compresses easily and significantly in the horizontal direction under the lateral earth pressures from the retained soil placed against it. This, in turn, induces the retained soil to yield (displace) in the same direction as depicted in Figure 3.7 by the large arrow, even if the ERS is rigid and non-yielding (as is typically the case). Because the geofoam product is composed of an engineered material with known, predictable mechanical (stress-strain-time-temperature) properties, the horizontal yielding within the retained soil occurs in a controlled, predictable fashion, hence the use of the term 'controlled yielding' noted previously.

For 'normal' soils that are the focus of this monograph, the consequence of this controlled yielding is that the soil mobilizes its inherent shear strength sufficiently to arch laterally around the geofoam panel. The minimum resultant lateral force that can be achieved is typically equal to or somewhat less than that corresponding to the active earth pressure state calculated using an appropriate solution (Coulomb or an exact solution such as Caquot-Kerisel but <u>not</u> Rankine for the reasons discussed in detail in Appendix A).

As will be seen in Chapter 5, this horizontal arching produces distributions of lateral earth pressure that are approximately parabolic in shape, with a peak pressure located at a depth that is well above the heel of the ERS. The triangular equivalent-fluid-pressure distribution with peak pressure at the heel as normally assumed in routine practice for the active earth pressure state does not develop although some simplified analytical methodologies for the REP-Wall alternative that have been developed by others assume such a distribution for simplicity.

The implication of this overall behavior is that the REP-Wall concept would appear to be:

- primarily useful with non-yielding RERSs as this would allow them to be designed reliably for the active earth pressure state instead of the at-rest state as would normally be appropriate. Furthermore, the presence of the compressible inclusion would also reduce the compaction-induced lateral pressures that typically develop and persist behind non-yielding RERSs (Reeves and Filz 2000); and

- not inherently useful with yielding RERSs or FERSs retaining normal soils as such ERSs are typically assumed to be capable of mobilizing the active earth pressure state within their retained soil on their own.

However, with regard to the latter item, an argument can be made that the REP-Wall concept would still be useful with yielding RERSs to the extent that it would <u>ensure</u> that the active earth pressure state was achieved within the retained soil under service-load conditions without having to rely on happenstance (which might be called *uncontrolled yielding* in contrast to the predictable, well-defined controlled yielding that is the result of using the REP-Wall concept). As discussed in Chapter 2, research has shown that even though a yielding RERS may be designed geotechnically and structurally for the active earth pressure state within the retained soil, very often the active state is not achieved or at least not achieved completely because of the excess geotechnical capacity designed and built into the system as a safety margin. As also discussed previously, this inability to achieve the assumed active state implies that the structural safety margin is always less than that assumed during design. Thus, as a minimum, the use of the REP-Wall concept even with yielding RERSs would have a benefit for the structural safety margin of the ERS as it would guarantee that active earth pressures would always exist on the ERS, even under service-load conditions.

An additional argument for using the REP-Wall concept even with yielding RERSs is that a positive drainage system would normally be desirable for such ERSs in virtually all applications. Such a drainage system would have as its primary component a chimney drain along the inside face of the ERS, similar to the orientation of the geofoam compressible inclusion shown in Figure 3.7.

As discussed in Chapters 7 and 8, there are a number of compelling reasons to use not only a geosynthetic product for this chimney drain (as opposed to the traditional natural aggregate) but to use a geofoam-based geosynthetic product in order to provide insulated drainage. Such products tend to act naturally as a compressible inclusion by virtue of the geofoam material that is typically used (GPS-PB) so to go one step further and use a composite geofoam product to simultaneously provide the functions of insulated drainage and compressible inclusion in a single, multifunctional geocomposite makes economic sense.

Regardless of the type of ERS, the maximum lateral pressure reduction that can be achieved using the REP-Wall concept is always less than the maximum that can be achieved using the lightweight-fill function. As noted previously, lateral pressures can, in theory at least, be reduced to practically zero if desired using the lightweight-fill function. However, also as noted previously, these impressive lateral pressure reductions come at relatively significant construction cost because of the relatively large volume of EPS-block required that may not be cost-effective for the additional benefit (lateral earth pressure reduction) obtained.

On the other hand, the geofoam-product cost of the REP-Wall concept is always significantly less than that associated with the lightweight-fill function. In addition, the REP-Wall concept has the added benefits of being able to be used when space and/or access behind the ERS is at a premium (often an important consideration when dealing with existing ERSs) as well as being able to provide the additional desirable function of insulated drainage as noted above.

One final item of note is that all discussion of the REP-Wall concept up to this point has involved normal soils. The situation when expansive soils are involved is quite different. This is because the relationship between lateral earth pressures and ERS yielding is much more complex. Consequently, as will be seen subsequently, the REP-Wall concept has clear benefits even with yielding ERSs when expansive soils are involved.

3.3.3.5.2 Partial-Depth Placement Alternative

As with the lightweight-fill function, there can be REP-Wall applications, especially with either existing ERSs or FERSs such as anchored bulkheads where groundwater within the retained soil is relatively shallow, where full-depth placement of the geofoam panel as shown in Figure 3.7 is not practicable for one reason or another. In such cases, the *partial-depth placement alternative* as shown conceptually in Figure 3.8 can be used.

Note that unlike lightweight-fill applications involving partial-depth placement where the EPS blocks need to be placed near the top of ERS as shown in Figure 3.6 to avoid overstressing, when partial-depth placement is used with REP-Wall applications the geofoam panel can be placed anywhere along the inside face of the ERS depending on project-specific constraints and technical needs: top (likely to be most common in practice), bottom, or somewhere in between as shown in Figure 3.8.

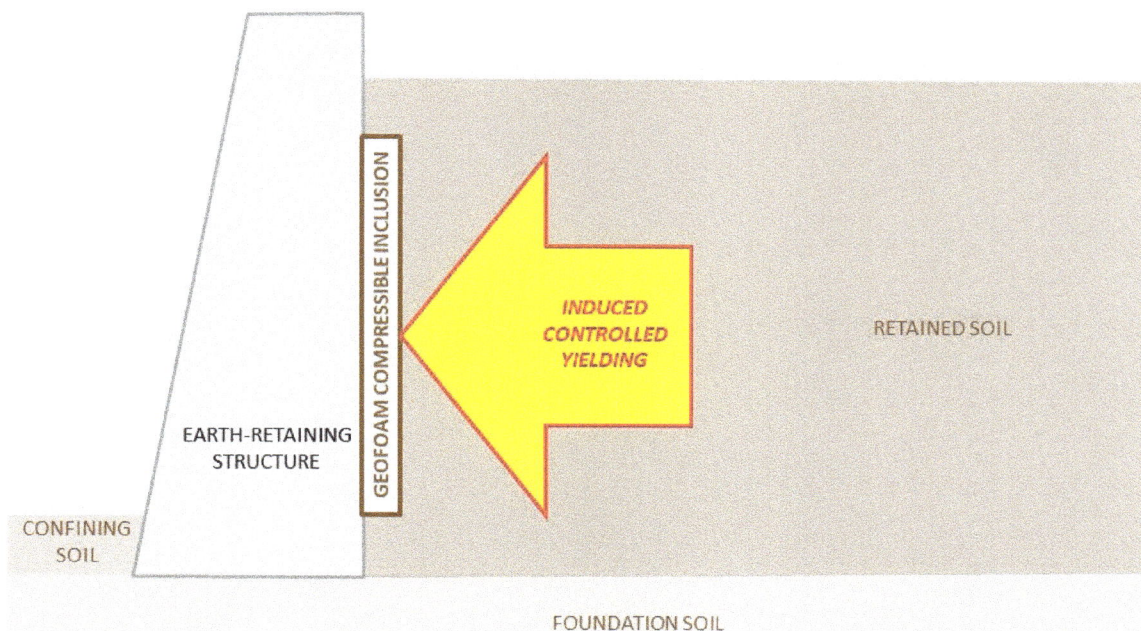

Figure 3.8. Compressible-Inclusion Function - REP-Wall/Partial-Depth Placement Alternative.

As with the lightweight-fill function, theoretical work performed to date has focused on the full-depth placement alternative. This is understandable as this is the basic REP-Wall case and the design-of-choice in most applications. Consequently, full-depth placement is the primary focus in this monograph and is discussed at length in Chapter 5. However, there will be some discussion of partial-depth installations as well.

Interestingly, one of the very first case history applications of the REP-Wall concept (Partos and Kazaniwsky 1987), and possibly the first to use a geofoam product in this application, used partial-depth placement over approximately two-thirds of the geotechnical height of a non-yielding RERS (the basement wall of a high-rise building in Philadelphia, Pennsylvania). The geofoam panel was positioned in a manner broadly similar to that shown in Figure 3.8. This case history is discussed at length in Chapter 5.

3.3.3.6 The Zero Earth Pressure Wall (ZEP-Wall) Concept

3.3.3.6.1 Full-Depth Placement Alternative

The second concept for implementing the compressible-inclusion function with ERSs is an evolutionary outgrowth of the REP-Wall concept and is illustrated conceptually in Figure 3.9. The use of a relatively thin panel of geofoam placed against the inside face of the ERS is carried over from the REP-Wall concept and remains the key element in the overall system. The new element is the addition of multiple horizontal layers of *geosynthetic tensile reinforcement* (*geogrids*, *geotextiles*, or metallic elements)[49] within the retained soil.

---

[49] As noted in Chapter 2, in most publications the term 'geosynthetic tensile reinforcement' is limited to polymeric-based products such as geogrids and geotextiles, and is not inclusive of metallic elements which nowadays would typically be either **[continued on following page]**

This effectively creates a MSE mass within the retained soil. Note that in the basic application of this concept both the geofoam and reinforcement are placed full-depth as shown in this figure.

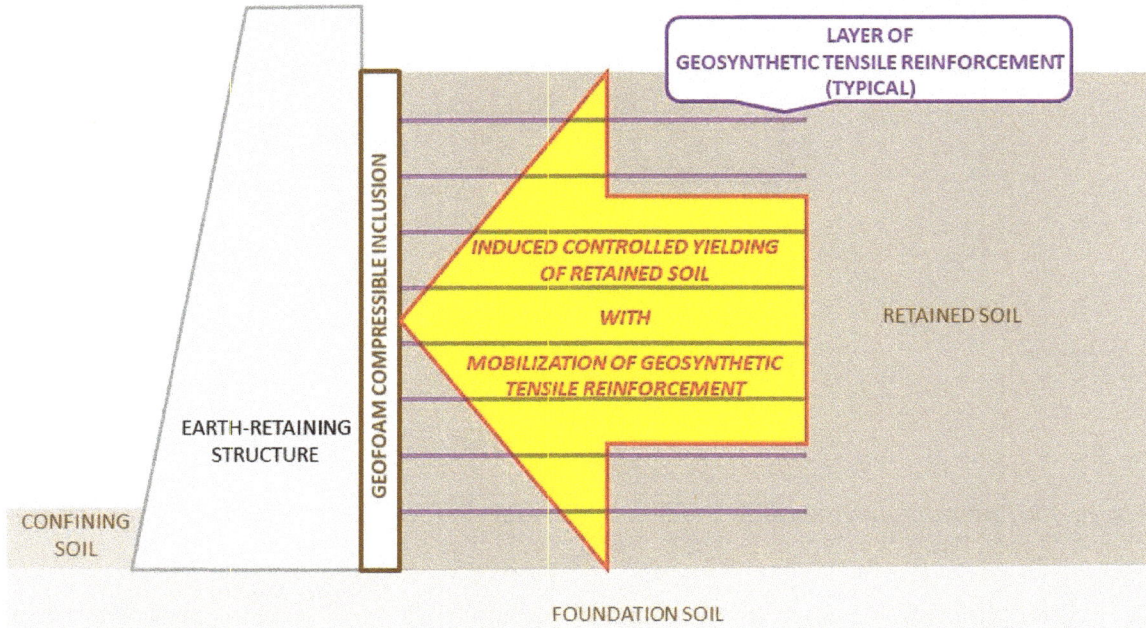

Figure 3.9. Compressible-Inclusion Function - ZEP-Wall/Full-Depth Placement Alternative.

The benefit of doing this is that as the geofoam compresses horizontally due to the lateral earth pressure exerted by the retained soil, the concomitant induced controlled yielding of the retained soil causes the development of tensile strains within the embedded geosynthetic tensile reinforcement in the standard manner for MSE. This is something that otherwise cannot occur when a MSE mass is placed adjacent to an ERS unless the ERS yields sufficiently to activate the reinforcement in tension.

Research indicates that by choosing appropriate absolute and relative stiffnesses for the primary system components (geofoam panel and geosynthetic tensile reinforcement), the resulting lateral pressures on the ERS can be reduced to well below the nominally active earth pressure lower-bound of the REP-Wall application to practically zero. Hence the

galvanized steel strips as used in the original proprietary *Reinforced Earth®* system or galvanized steel mesh as used in other tradenamed systems. This curiously and arguably illogical narrow definition of geosynthetic tensile reinforcement, which, in the writer's opinion, appears to have business-political origins dating back to the earliest days of geosynthetics, has never made sense to the writer. By definition, any synthetic material, whether polymeric or metallic in composition, that is artificially and intentionally placed on or in the ground should logically be considered a 'geosynthetic'. Therefore, the term 'geosynthetic tensile reinforcement' is broadened in this monograph to include all such materials and products, both polymeric and metallic.

generic name for the application of the compressible-inclusion function shown in Figure 3.9 is the *Zero Earth Pressure Wall* (*ZEP-Wall*) concept[50].

In the writer's opinion, the ZEP-Wall concept is extremely attractive in practice. From a technical perspective, it has all the benefits of using the lightweight-fill function in terms of the relative magnitude of lateral earth pressure reduction that can be achieved in principle, i.e. to near zero if desired under both gravity and seismic loads. Furthermore, the ZEP-Wall concept has unambiguous potential benefit for both non-yielding and yielding RERSs that are either new or existing. From the perspective of economics, it requires much less geofoam material than the lightweight-fill function to achieve its technical goals although the cost of the geosynthetic tensile reinforcement and its placement must obviously be considered and factored into the overall cost analysis.

However, despite the fact that the ZEP-Wall concept was formally identified, defined, and publicized beginning in the early 1990s by the writer (and late 1980s if the published work of McGown, Andrawes, et al. is included), it remains too new in terms of technical development and use in actual practice to be able to draw conclusions as to how it compares in terms of overall costs in actual projects to the use of the lightweight-fill function for lateral pressure reduction on ERSs. Nevertheless, the ZEP-Wall concept has the potential to displace use of the lightweight-fill function in the vast majority of applications. In addition, as will be seen, research to date indicates that the ZEP-Wall concept also appears to provide the optimal design alternative when self-yielding ERSs such as IABs are involved.

3.3.3.6.2 Partial-Depth Placement Alternative

As with the REP-Wall concept, there can be applications where full-depth implementation of the ZEP-Wall concept is not practical, especially with existing ERSs. In such cases, partial-depth installation can be used. Note that the geosynthetics can be placed at the top (as shown in Figure 3.10, which is likely to be the most common situation in practice), bottom, or somewhere in between as with the REP-Wall partial-depth placement alternative. However, regardless of where and to what extent geosynthetics are used, the vertical extent of the geofoam panel and layers of geosynthetic tensile reinforcement must always be approximately the same if they are to work together synergistically.

Also, as with the REP-Wall concept, all theoretical work performed to date for the ZEP-Wall concept has focused on full-depth placement as this is the design-of-choice in most cases. Consequently, full-depth placement is the primary focus in this monograph in Chapter 5. However, there will be some discussion of partial-depth placement as well.

---

[50] In the late 1980s to early 1990s published works of A. McGown, K. Z. Andrawes, and colleagues in the U.K. (McGown et al. 1987, 1988; Andrawes et al. 1991b; Ahmad 1991; Andrawes et al. 1991a; Andrawes 1991; Andrawes et al. 1992, 1993) that dealt with the compressible-inclusion function (although they never used that term), there was no terminological differentiation made between an unreinforced vs. reinforced retained-soil mass, i.e. the REP- and ZEP-Wall concepts as defined by the writer. Both applications were collectively referred to by the U.K. researchers using the term 'boundary-yielding concept'.

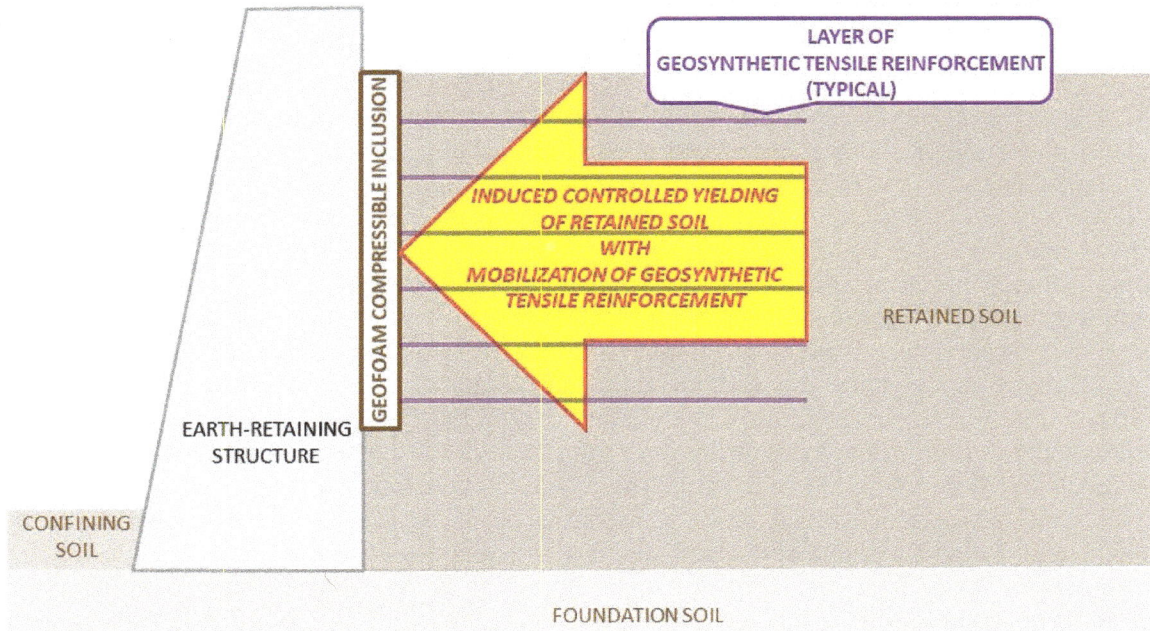

Figure 3.10. Compressible-Inclusion Function - ZEP-Wall/Partial-Depth Placement Alternative.

3.4 LOAD-REDUCTION ASSESSMENT

Because the primary, if not sole, reason for using geofoam with ERSs is to reduce lateral earth pressures relative to some baseline no-geofoam design alternative (for new construction) or current condition (for an existing ERS), it is logical that there should be some appropriate metric for quantifying this reduction so that the technical benefit of different alternatives can be made simply and rationally. One way of doing this is to define a parameter, preferably dimensionless, that expresses the reduced lateral pressure relative to some baseline value.

There have been at least two efforts to date, including one by the writer, to define just such a parameter for cases involving gravity loading and both efforts are discussed in detail in Appendix B. For reasons explained in Appendix B, the writer's methodology is used throughout this monograph.

Also discussed in Appendix B is extending these concepts to seismic loading, an issue that has not been adequately addressed to date in the published literature in the writer's opinion.

3.5 OTHER ASSESSMENTS

3.5.1 Cost Estimating

Cost estimating is always a key component of the civil engineering design process, whether for new construction or when modifying an existing structure, and is not a unique feature or aspect of the technologies addressed in this monograph. However, experience indicates that cost estimating when geofoam is used for any functional application, not just to reduce lateral pressures on ERSs, is often done incompletely and thus incorrectly. Consequently, it is essential that cost estimating be done properly whenever any type of geofoam product is used for any functional application with ERSs.

One key aspect of cost estimating that is relevant to this monograph is that EPS or EPS-related geofoam materials and products are always the materials and products of choice for load-reduction on ERSs. *Petroleum* (*crude oil*) is the primary base component of the raw material (called *expandable polystyrene* or, more colloquially, *beads* or *resin*) used to make EPS in all its forms and derivatives (Horvath 2011). In the U.S. at least, typically more than half the final cost of an EPS product delivered to a project site reflects the cost of raw material alone, with manufacturing and shipping costs making up the balance. So, the cost of EPS and related geofoam products are directly affected by both the absolute cost of petroleum as well as the temporal volatility of petroleum costs.

This volatility can be especially problematic on larger projects where several months and sometimes even years can elapse between when an EPS molder submits unit prices to a contractor to supply EPS products and those products are actually produced and delivered to the project site. This fact has had a variety of negative cost and technical impacts on projects in the past (Horvath 2012). Only relatively recently has it been proposed that project specifications for EPS-geofoam products provide for some objective assessment of and compensation for temporal cost volatility along the lines of what is done for other petroleum-based construction materials such as asphalt concrete (Arellano at al. 2011a, 2011b; Horvath 2012; Arellano at al. 2013a, 2013b).

Even neglecting the issue of price volatility, another key aspect related to cost estimating is the simple fact that petroleum costs are always a significant component of EPS costs. The net result is that EPS products are almost always more expensive on a unit-volume basis compared to the cost of earth materials. In the writer's experience, the single most common and consistent error made in practice is that only the unit cost (usually on a volumetric basis) of the geofoam product as-delivered to the project site is considered and compared to the unit cost of traditional materials that the geofoam is replacing. This is particularly problematic when the lightweight-fill function is used as relatively large volumes of EPS-block are involved compared to the compressible-inclusion function.

In reality, a correct, complete cost estimate for any project application using geofoams in any application should always consider the following general components:

- Total capital construction cost, not just of the geofoam material alone but including all aspects of shipping and on-site handling and placement. The reason these latter issues are so important is that geofoam products are inherently relatively light in weight and can, therefore, be handled with much less effort and much faster compared to traditional construction materials. In addition, with few exceptions geofoam products can typically be placed in any weather which means that work can be done in cold and/or wet weather that would typically inhibit other construction materials and processes, especially traditional earthwork using soil (see the photo on the back cover as an example). Overall, the placement of geofoam products typically improves construction productivity overall compared to other alternatives.

- Accelerated construction which is actually the outcome of the aforementioned construction efficiency and productivity but is listed as a separate item intentionally to highlight its specific importance by the U.S. Federal Highway Administration (FHWA) and others. First of all, this is a relatively new consideration in cost estimating, at least in a formal, explicit sense. Only in recent years has formal credit been given to technologies that may cost more than other alternatives but can be implemented faster. This is based on the recognition that in certain construction scenarios (road construction being the one identified first and most prominently to date by the FHWA), the ability to minimize inconvenience and the concomitant delays to the traveling public has a tangible and potentially significant monetary value associated with it. This monetary value often exceeds the premium in capital construction cost of a technology that may cost more but can be implemented faster. Therefore, in any application of geofoam to reduce lateral pressures on an ERS where the benefits of accelerated construction can be clearly identified they should be included explicitly in the overall cost assessment.

- <u>Life-cycle costs</u> for the completed structure which includes all operation (including energy- and 'green'- related issues which are discussed separately in the following section) and maintenance (O&M) costs for the design life of the structure. This is a cost center that is often overlooked or neglected during design for various reasons, especially if the owner-builder of a structure during its construction will not be the owner-operator of the structure in the long term. However, life-cycle costs are particularly important to consider whenever geofoam is used as it can often lower life-cycle costs in a wide variety of ways compared to other design alternatives.

In addition to these broad areas of consideration that apply to every project that involves geofoam (not just for load reduction on ERSs), there are several additional considerations that are specific to the focus of this monograph:

- EPS-based geofoam products used to reduce lateral pressures on ERSs, whether as lightweight-fill or compressible-inclusion, produce results that are either difficult or impossible to achieve any other way so provide the design professional with truly unique design tools and alternatives. This is certainly true of the ZEP-Wall concept that synergistically combines two different categories of geosynthetics for a net result that neither one could provide if used alone, at least when non-yielding RERSs are involved. Furthermore, as noted at the beginning of this chapter, Equation 3.1 must always be satisfied, whether for new construction or with an existing structure. Design professionals have historically only had the option of increasing system capacity (resistance) when dealing with ERSs, something that can be difficult and expensive to achieve for existing structures in particular. Using geofoam allows Equation 3.1 to be satisfied alternatively by reducing the demand (load) side of the equation, something that is again particularly attractive when dealing with existing ERSs.

- As noted already and as will be discussed further in Chapters 7 and 8, on most, if not all, designs utilizing the compressible-inclusion function one or more additional functions can be beneficially provided by the geofoam product used so that one product does the work of two or more individual products or materials. This enhances the overall cost effectiveness of the geofoam product.

- For designs utilizing the lightweight-fill function, there is never a unique technical solution to a problem. With reference to Figure 3.2, there are numerous solutions as defined by the angle θ^* within a range of well-defined limits. How to develop a project-specific value for θ^* that optimizes cost vs. benefit is discussed in Chapter 4.

3.5.2 Energy Issues

In recent years, various aspects of energy usage and environmental impact of all aspects and stages of engineered construction in the form of the *carbon footprint* of some product or process have become increasingly important considerations during the design phase of a project. It is likely that such considerations will only become even more important in the future and in the process both guide and shape technical decisions by design professionals.

Note that this new, significantly broader aspect of energy usage is an extension of the issue of energy usage that was noted in the preceding section. Historically, energy usage and associated costs were generally limited to a consideration of operation only of the completed structure as part of the life-cycle O&M costs. It was this historical aspect of energy usage and concomitant costs what was noted in the preceding section.

The new, more inclusive consideration of energy extends to material and product manufacturing and shipping to a project site as well as construction. So, for example, whereas in the past thermal insulation of a house was viewed solely from the perspective of optimizing insulation costs with energy savings over time that accrued to the completed structure, nowadays consideration would also be given to the energy-related aspects of manufacturing the insulation and placing it in the structure.

The 'green' aspects of design as they relate to projects where geofoam is used to reduce lateral loads on an ERS are not explored in any detail in this monograph. This issue is simply mentioned here for the sake of completeness and to note that there are both positive and negative aspects of geofoam usage that should be considered in any all-inclusive energy assessment and carbon-footprint assessment. As with cost estimates, there is always the danger that only one or some aspects will be considered as part of some energy assessment that could produce misleading results.

There is no doubt that the single biggest environmental negative for the geotechnologies discussed in this monograph is that the EPS and EPS-related products are heavily dependent on petroleum hydrocarbons at many steps in their manufacturing process, beginning with the petroleum used to make the basic *styrene monomer* (*styrene*)[51] that is the underlying chemical compound for all PS-based products. Petroleum is also the source of the *blowing agent* (a gas, usually a blend of *pentanes* which are *volatile organic carbons* or VOCs) that is a key element incorporated into *styrene polymer* to make the raw material (the aforementioned expandable polystyrene a.k.a. beads or resin) that is used to make EPS (Horvath 2011). Furthermore, it takes a substantial amount of steam to expand the expandable polystyrene into the *prepuff* that is then fused together using more steam into the final EPS product (prismatic block or custom shape). Of course, it takes energy generated using some type of fuel to make all this steam.

However, downside of the carbon footprint of EPS can be offset to some extent by the fact that it is possible to use recycled EPS in the EPS production process. However, as discussed in Horvath (2011), the use of recycled EPS, called *regrind* or simply *grind* by the North American EPS industry, can compromise the small-strain stiffness of the final EPS product. While this can be problematic for the lightweight-fill function, it can actually be beneficial for products used for the compressible-inclusion function where low stiffness is a desirable trait.

---

[51] Although styrene is a naturally occurring organic chemical compound found in some trees, plants, and even food products (e.g. coffee beans and peanuts), for all practical purposes all the styrene used to produce polystyrene (***poly**merized **styrene***) is produced synthetically from petroleum.

Also, on the positive side, it is important when making any kind of energy/carbon-footprint assessment to consider that geofoam can provide, either directly or indirectly, energy reductions during both construction and in service for the design life of the structure. To begin with, the fact that the geofoam is replacing other materials that would be used for a non-geofoam design alternative means that there is an energy saving from not using those materials. Next, the inherent low density and concomitant relatively light weight of geofoam products that was noted previously with regard to cost estimates reduces the demand for equipment to haul and place the geofoam. Finally, the life-cycle O&M savings also noted previously under cost estimating typically involve considerable energy savings either directly or indirectly.

In summary, when making any sort of energy/carbon-footprint assessment for a design alternative incorporating geofoam it is important, as with cost estimating, not to focus solely on the geofoam product(s) and, in this case, the energy issues associated with their manufacture. It is essential that the entire life-cycle energy issues of any design alternative be considered as every alternative has positive and negative elements with some net overall impact. What varies from one design alternative to the next is that the positive and negative aspects occur in different parts of the overall assessment. So, to only focus on one part or element of an overall behavior will not produce an accurate absolute assessment that will allow an accurate relevant assessment between or among design alternatives.

Chapter 4

Analysis and Design Based on the Lightweight-Fill Function

4.1 INTRODUCTION

In the current states of knowledge and practice, the overall design process for the lightweight-fill function follows the same basic procedure for all ERSs except for self-yielding RERSs that are dealt with separately in Chapter 6. There are only relatively slight differences in the assumed system of resultant forces between the various applications, i.e. non-yielding vs. yielding, RERS vs. FERS. Therefore, this chapter is structured to first present the overall design process that is used regardless of specific application. Then the specifics of each application are addressed in separate sections.

As indicated in Chapter 3, the contents of this chapter focus on the basic full-depth placement application shown in Figure 3.2. Not only would this be the choice in most applications, it is the only condition that has received extensive research and development of design concepts to date. However, partial-depth placement as shown in Figure 3.6 will receive limited presentation and discussion so that the current state of knowledge is defined for use in future research and development of this concept.

4.2 ANALYSIS AND DESIGN BASICS

A design for the lightweight-fill function should always have the basic cross-sectional geometry shown in Figure 3.2. As noted in Chapter 2, stair-stepping the EPS blocks into the retained soil as shown qualitatively in this figure is now taken to be the single most important design detail for this functional application. As will be seen, the angle θ^* that defines the linear approximation of the overall geometry of the stepped interface between EPS blocks and retained soil relative to the horizontal is the primary design variable for this functional application.

The criticality of this stepped geometry is emphasized here yet again because on many early projects that were constructed before there was the proper understanding of how an assemblage of EPS blocks behaves in this functional application this stepped geometry was often absent. Unfortunately, photographs of some of these early projects where the EPS blocks were simply stacked vertically against the retained soil have appeared in print and other venues over the years and continue to appear in presentations and publications (one of the most popular ones is intentionally reproduced on the back cover of this monograph to make this point), including on magazine covers where they are highly visible and thus memorable. As a result, this now-deprecated design detail continues to be seen by potential users[52].

---

[52] Other now-deprecated design and construction details that are applicable to all lightweight-fill applications of EPS-block geofoam (earthworks, slope stabilization, ERS load reduction), such as failing to properly orient blocks to minimize the continuity of vertical inter-block joints by rotating successive layers of blocks 90° relative to the layer below, also appear in these older, well-publicized photos.

72

4.3 OVERVIEW OF THE DESIGN PROCESS

As noted in Chapter 3, in any project-specific application of this function there is never a single 'correct' answer in terms of the required value of the angle θ* and concomitant lateral earth pressure reduction. There is always a range of potential outcomes that is a function of the angle θ* at which the assemblage of EPS blocks is stepped. This means design is always a project-specific iterative process in which economics combined with any physical constraints of the project site play primary roles in the design process.

As a minimum, the design process for implementing the lightweight-fill function consists of performing a preliminary design considering gravity loads. Where and when appropriate, a separate preliminary design is also performed for seismic loads. The results from the two preliminary designs are then compared to see which one governs the overall final design for the project.

Note that although seismic loads are always larger in magnitude than gravity loads, they may not govern the final design because different safety margins (whether ASD safety factors or LRFD load and resistance factors) are often used for these different load cases. So, it is not known a priori whether gravity loads with a larger required margin of safety or seismic loads with a lower required margin of safety produces the more critical design.

Ideally, each design should consist of a suite of analyses for several different cases:

- A baseline case assuming no geofoam usage and using the traditional at-rest or active earth pressure state as appropriate depending on whether the ERS is non-yielding or yielding. This will provide a technical-performance and cost reference against which designs using geofoam can be compared.

- Unless there are pre-defined restrictions on the layout geometry of the EPS blocks that are dictated by project-specific considerations such as a surface or subsurface obstructions or constraints (e.g. another structure, underground utility line(s), property line, etc.), a design based on maximum lateral pressure reduction should be performed. This requires an analysis (the details of which are presented in this chapter) based on the angle θ* of the stair-stepped EPS blocks at which the resultant earth force from the retained soil will just be zero[53]. Design for this limiting case will provide a benchmark for the maximum level of load reduction and the cost associated with this. Note, however, that this will not necessarily be the most economical design overall as it always requires the largest volume of both EPS blocks and associated excavation.

- One or more additional applications of the analysis procedure discussed in this chapter for various arbitrary, assumed geometries (i.e. angle θ*) of the assemblage of EPS blocks between the above-described limiting conditions of no geofoam/no earth-load reduction (which will effectively correspond to θ* = 90⁰ in most cases) and maximum geofoam/maximum earth-load reduction. The purpose of these additional analyses is to generate a sufficient number of data points to be able to develop project-specific design curves relating the costs of the EPS and ERS as a function of the volume of EPS used (which is related to θ*) as shown in Figure 4.1[54]. As noted previously, as the volume and concomitant cost of EPS blocks increases, the cost related to the ERS (to build if new, to

---

[53] As noted previously, there will still be some relatively small load on the ERS from the assemblage of EPS blocks even if the loads from the retained soil are zero.
[54] The shapes of all curves shown in this figure are descriptive and intended to indicate broad trends only. The actual shapes of the curves will vary from project to project.

upgrade if existing) decreases because less structural material is required to support the reduced loads. Thus, the overall project cost is represented by a curve defining the combined costs of the EPS and ERS (also shown in Figure 4.1) and it is this overall cost that is compared to the baseline cost of using no geosynthetics to decide what will be the final design. Note that it is not necessary to choose the lowest point on this combined curve as multiple points on the curve represent a net cost saving. In addition, there will always be situations, especially when dealing with the retrofit, repair, or rehabilitation of an existing ERS, where considerations other than cost, e.g. achieving a desired safety margin or limitations of physical access (which may put a limit on the angle θ*), may dictate the final design.

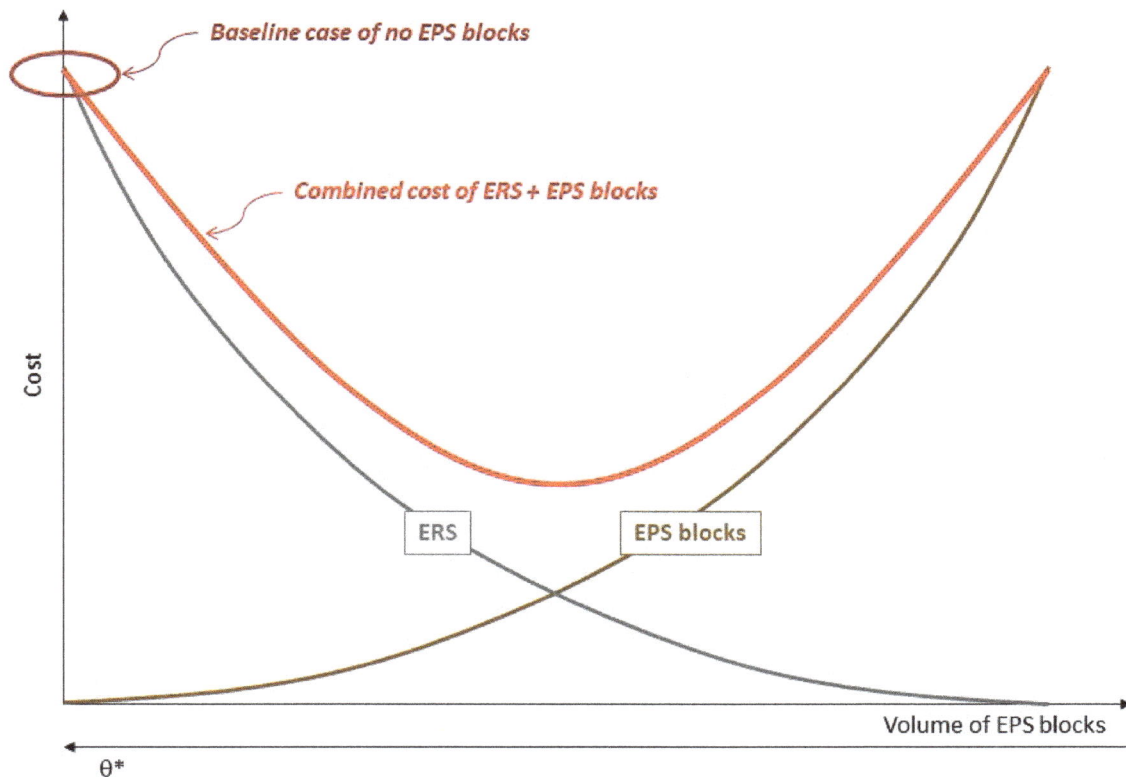

Figure 4.1. Lightweight-Fill Function - Generic Design-Cost Optimization Concept.

Because the results depicted conceptually in Figure 4.1 reflect an optimization process, the analytical steps outlined above would appear to lend themselves to automation using computer software. In particular, the *Solver* function of *Excel* might be useful for this purpose. The writer has used this function for a variety of research applications unrelated to this monograph and Arellano et al. (2013a, 2013b) used it to optimize both the volume and placement of EPS-block geofoam used as lightweight-fill in slope stabilization. So clearly, with some thought and effort to develop an automated procedure the process reflected in Figure 4.1 could be made less onerous than it might appear.

4.4 ANALYTICAL MODELS AND METHODS FOR RIGID EARTH-RETAINING STRUCTURES

4.4.1 Full-Depth Placement Alternative

4.4.1.1 Basic Assumptions

Whether dealing with a non-yielding or yielding RERS, the current state of knowledge analytical procedure developed in Japan models the overall three-component (RERS-geofoam-retained soil) system shown in Figure 4.2[55] as illustrated in Figures 4.3a (qualitative visualization) and 4.3b (quantitative visualization). There are some additional elements related to seismic loading that are discussed separately later in this chapter.

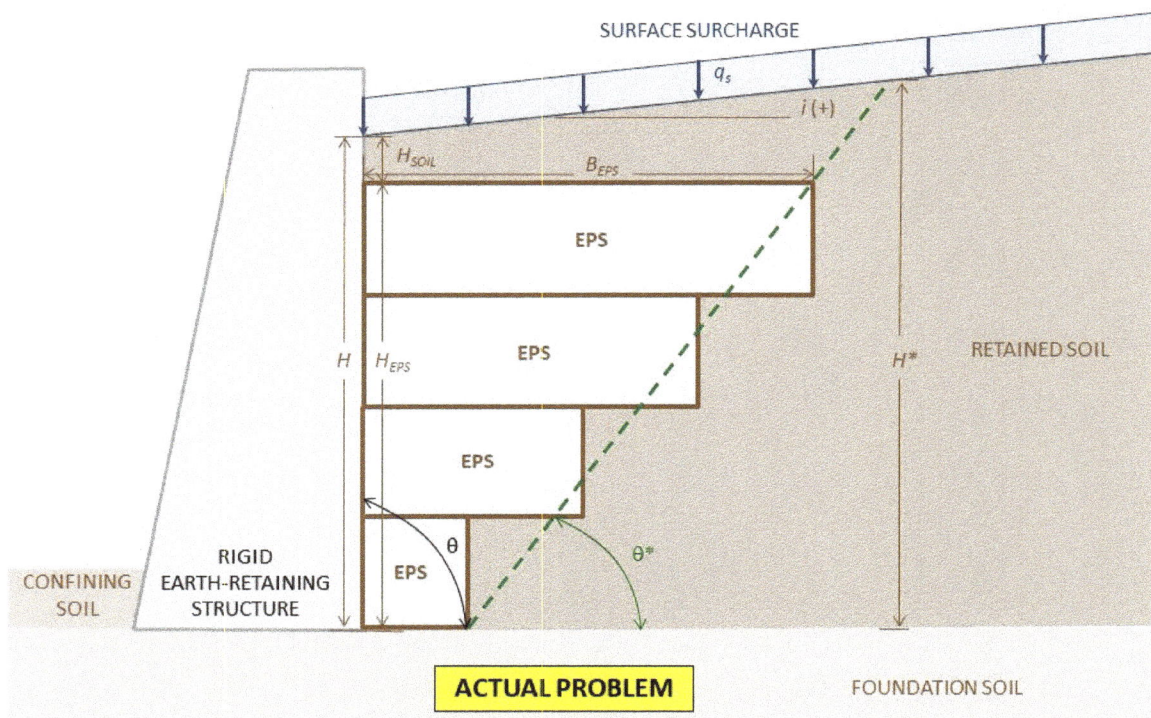

**Figure 4.2. Lightweight-Fill Function - Full-Depth Placement Alternative
Actual Problem Components.**

Omitted from Figure 4.2 and subsequent related figures simply for clarity is that good practice would call for a geosynthetic sheet drain to be placed between the inside face of the actual RERS and the assemblage of EPS blocks. While any number of products could be used for this purpose, the preferred product in this case would be a geofoam-based geocomposite that uses GPS-PB for its core. This product is discussed in Chapter 8.

---

[55] This is essentially the same as the generic depiction shown previously in Figure 3.2 but with some additional annotations and a sloped retained-soil surface with surcharge included for generality.

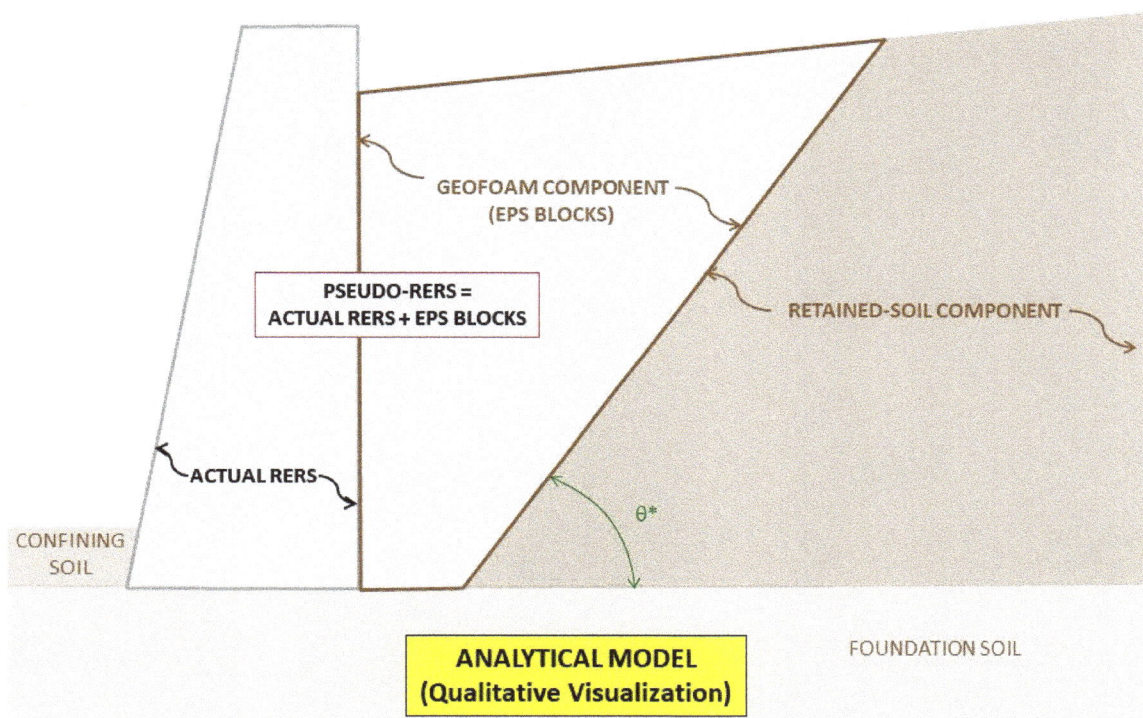

Figure 4.3a. Lightweight-Fill Function - Full-Depth Placement Alternative Qualitative Visualization of Basic Analytical Model.

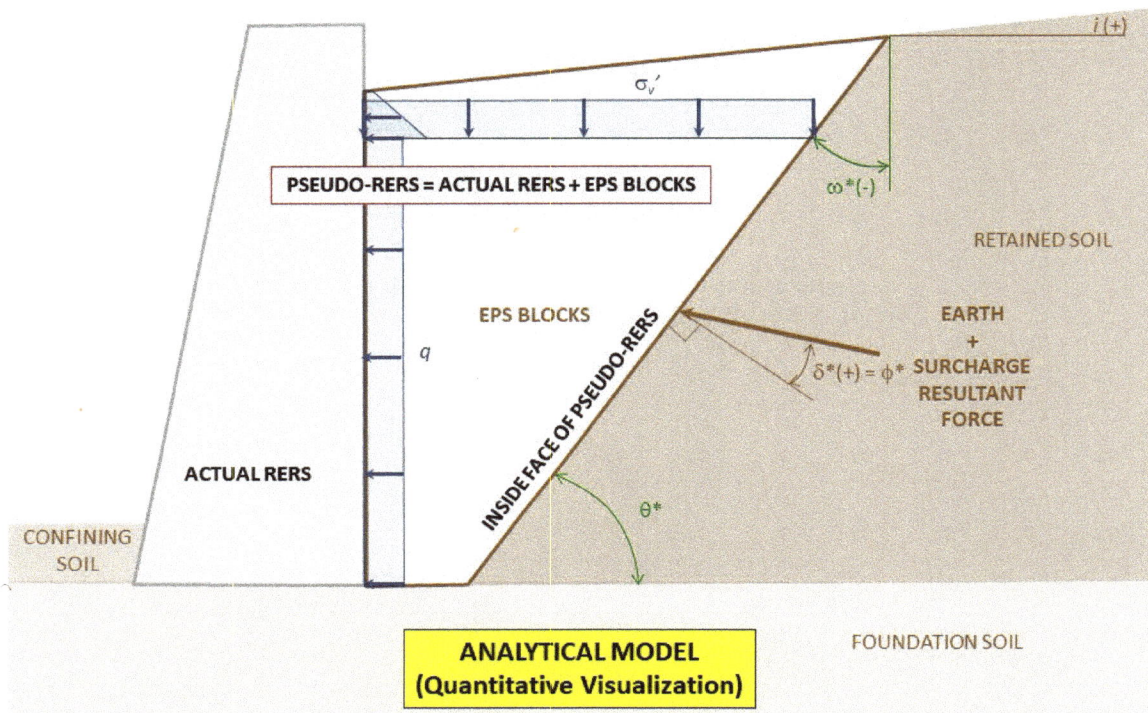

Figure 4.3b. Lightweight-Fill Function - Full-Depth Placement Alternative Quantitative Visualization of Basic Analytical Model.

With regard to these figures, note that specific details as to the actual type and geometry of the RERS and retained soil are not important for the present discussion. Thus, as in Chapter 3, the gravity retaining wall depicted in these figures is simply a generic stand-in for any type of RERS, yielding or non-yielding. In addition, the ground surface of the retained soil is shown as sloping upward[56] from the inside face of the actual RERS and with a surface surcharge of uniform magnitude, q_s, simply for generality in formulating an analytical model.

The key element of the model depicted in these figures is that the assemblage of EPS blocks and any materials (soil, pavement) directly overlying the blocks are assumed to act collectively as a physical part and extension of the actual RERS to create a larger, combined pseudo-RERS. The assumed equivalent planar surface defined by the stair-step geometry of the assemblage of EPS blocks in contact with the retained soil (as represented by the green dashed line at an angle θ^* with respect to the horizontal in Figure 4.2) defines, at least for analytical purposes, the inside face of the pseudo-RERS as shown in Figure 4.3b, with the lateral pressures and concomitant resultant forces from both the retained soil and any surface surcharge above the retained soil acting on this planar surface. Note that if the ground surface behind the actual RERS is not horizontal ($i \neq 0°$), the geotechnical height of the combined pseudo-ERS will be H^*, not H which is the geotechnical height of the actual ERS. This is illustrated in Figure 4.2. As will be seen, this distinction is important.

Based on this model, the design values of the lateral pressures acting on the inside face of the actual RERS are assumed to come from two distinct sources, each of which is discussed in detail below, that are assumed to be uncoupled analytically for computational simplicity and efficiency, i.e. the calculation of one source of lateral pressure on the actual RERS does not depend on calculations for the other. In addition, as will be seen later in this chapter, for one of these load sources it is useful to break it into multiple components (the exact number depends on the specifics of the types of loads involved) for ease of calculation. As a result, the lateral pressure diagram applied to the actual RERS for analysis and design purposes shown later in this chapter will always consist of two or more separate components that are collectively superimposed on the actual RERS.

The first of the two load sources assumed to act on the actual RERS is related to the retained soil acting on the inside face of the pseudo-RERS, i.e. all retained soil and surface surcharge (if any) to the right of the imaginary plane defined by the green dashed line and angle θ^* in Figure 4.2 which is the same as the labeled "inside face of pseudo-RERS" in Figure 4.3b. This is the load that can, in theory, be zero if an appropriate value of θ^* is chosen as was discussed in Chapter 3.

The active earth pressure state is always assumed to develop within the retained soil regardless of whether the actual RERS is yielding or non-yielding. The rationale behind this is that even if the RERS itself is non-yielding there is always a combination of several factors[57] that, although difficult to quantify either individually or collectively in terms of

---

[56] Note that the sign of this slope is not trivial and a positive sign for this angle, i, is as shown in this figure. A zero or positive slope angle is most common by far in practice. A negative slope angle is typically found only with some specialized RERSs such as floodwalls (a.k.a. structural levees).

[57] The major factors include: the inherent compressibility of the EPS blocks; the cumulative horizontal 'slop' between EPS blocks arising from gaps at vertical joints between blocks; the fact that the area adjacent to the actual RERS is completely open initially and is then progressively backfilled or filled with EPS blocks and soil; and compressibility of the chimney drain that is typically placed between the EPS blocks and inside face of the actual RERS, especially if a geofoam-based geocomposite product is used. Also considered is the fact that only relatively small lateral displacements are required to mobilize the active earth pressure state.

78

their exact magnitude, are assumed to act cumulatively and synergistically to provide sufficient lateral compressibility and resulting net lateral displacement within the overall assemblage of EPS blocks to allow the active state to develop within the retained soil regardless of the specific type and state of yielding of the actual RERS. Consequently, for a given problem the angle θ^* will be the primary variable governing the load magnitude from this source.

As shown in Figure 4.3b, the loads from the retained soil plus any surface surcharge on the retained soil are assumed to act at an orientation that properly accounts for both the problem geometry as defined by the angle θ^* and the assumed interface friction along the inside face of the pseudo-ERS. Note that although these loads are shown in this figure as a single resultant force with a single point of application, this is solely for illustrative purposes at this point. As will be seen later in this chapter, for structural and geotechnical engineering purposes these earth and surcharge loads are calculated and shown as separate lateral pressure components, each having a different point of application for their respective resultant force.

With further regard to the spatial orientation of the resultant force shown by the brown arrow in Figure 4.3b, to begin with, note that the interface friction angle, δ^*, acting along the imaginary plane that defines the interface between the inside face of the pseudo-RERS and retained soil is assumed to be equal in magnitude to the Mohr-Coulomb friction angle of the retained soil, ϕ^*, and positive in sign with the sense shown in this figure. This sign implies that the active failure wedge that forms within the retained soil displaces <u>downward</u> relative to the pseudo-RERS. This assumed behavior is not only the more common one in reality but is judged to be particular appropriate here given that the retained soil has a density approximately 100 times that of the EPS blocks that comprise a significant portion of the pseudo-ERS. However, as discussed in detail in Chapter 2, this assumption concerning the sign of the interface friction is not trivial in its potential implications regarding the magnitude of earth loads acting on the pseudo-RERS.

From geometry, it can be seen that this resultant force acts at an angle equal to (θ^* - 90° + δ^*) with respect to the horizontal. Noting that $\delta^* = \phi^*$ here, this becomes (θ^* - 90° + ϕ^*). Consistent with the geometry and angle definitions shown in Figure A.1 in Appendix A, (θ^* - 90°) is equal to the angle ω^* as defined in Figure 4.3b which means the resultant force acts at an angle equal to (ω^* + ϕ^*) with respect to the horizontal. Note that angle ω^* is always negative in sign in this case because θ^* will always be less than 90°. This, in turn, means that (ω^* + ϕ^*) will often be negative in sign as ω^* is usually greater than ϕ^*. In turn, this implies that the line of action of the resultant force will below the horizontal as shown qualitatively in Figure 4.3b. Note that in such cases the vertical component of this force will act upward. This is atypical for ERSs so is noted here to avoid confusion in practice.

These loads from the retained soil are assumed to be transmitted through the assemblage of EPS blocks to the back of the actual RERS without any reduction in magnitude (an exception for yielding RERSs under seismic loading is discussed subsequently). It is further assumed that these transmitted loads act on the actual RERS at the same angle and points of application at which they act on the fictitious planar surface that defines the inside face of the combined pseudo-RERS.

A subtle, but important, outcome of these assumptions is that the loads from the retained soil and any surcharge overlying the retained soil are calculated using the H^* dimension shown in Figure 4.2 but then applied using the H dimension for the design of the actual RERS. As will be seen, this means that for situations where the surface of the retained soil is sloped so that H and H^* are unequal (typically the slope angle, i, is positive in sign so H^* will be greater than H), the pressure distribution applied to the vertical dimension H

along the actual RERS must be adjusted so that the resultant force delivered to the actual RERS is the same as that applied to the vertical dimension $H*$ defined by the fictitious planar surface of the equivalent pseudo-RERS.

Moving on to the second distinct load source for which the actual RERS must be designed, this is due to the material (soil, pavement, etc.) plus surface surcharge (if any) directly overlying the assemblage of EPS blocks. Note that this load component can never be zero, regardless of what value is used for $\theta*$, unless the EPS blocks extend all the way to the ground surface which is not good practice so it unlikely in any actual application.

For a given overall problem geometry, this load source is assumed to create a vertical effective stress (shown as σ_v' in Figure 4.3b) that is assumed, for simplicity, to be uniform in magnitude across the top of the EPS blocks. It is further assumed that this vertical stress creates a lateral stress, q, also of uniform magnitude, on the back of the actual RERS due to the Poisson effect within the assemblage of EPS blocks. In addition to this stress q, within the relatively thin zone of material overlying the EPS blocks there will be a lateral earth pressure in either the active or at-rest state depending on whether the actual RERS is yielding or non-yielding respectively.

As noted above and illustrated in detail in the following section, it is assumed that the contribution to the overall design lateral pressure diagram for the actual RERS from each of these two load sources can be calculated in an uncoupled fashion, i.e. independently of the other source. Thus, the net effect of all loads combined for which the actual RERS is analyzed or designed is then assumed to be the simple superposition of each of the independently obtained pressure diagram or diagrams for each source of load.

4.4.1.2 Loading Conditions

4.4.1.2.1 Overview

The necessary lateral earth pressure coefficients for the assumed active earth pressure state within the retained soil are calculated using Equation A.4 for the gravity (Coulomb) case and whichever analytical methodology for the seismic case (Section A.2.3.2) is desired using the problem geometry defined in Figure 4.3b. Note again that the angle $\omega*$ is always negative in sign in this application due to the atypical 'overhang' geometry of the inside face of the pseudo-ERS compared to the more-typical ERS geometry shown throughout Appendix A where the comparable parameter ω is always positive in sign. Also, as noted previously, the interface friction angle, $\delta*$, which in this case is the friction angle along the assumed planar interface between the EPS blocks and retained soil, is taken to be the operative value of the Mohr-Coulomb friction angle, $\phi*$, of the retained soil.

4.4.1.2.2 Gravity Conditions

In the most general case for gravity loads with a surface surcharge, the two load sources acting on the inside face of the actual RERS that were described previously (from the retained soil and Poisson effects from the EPS blocks) produce a combined total of three components of lateral pressure. How to calculate each of these components will now be illustrated in detail.

The first and primary component is an assumed triangular distribution of lateral earth pressure from the retained soil. As noted previously, this is always assumed to correspond to the active earth pressure state, independent of the yielding condition of the

RERS. Because of the previously noted situation wherein this pressure component physically acts on the geotechnical height $H*$ of the combined pseudo-RERS but is applied to the geotechnical height H of the actual RERS (see Figure 4.2), Equations A.7a and A.7b in Appendix A that define the pressure magnitudes of this triangle have to be modified empirically to account for this.

The underlying logic used to make this empirical modification is that the resultant force of the actual pressure must be the same whether it is applied to the H or $H*$ height. So, the magnitudes of the pressures applied to the actual RERS for analysis and design purposes are adjusted to accomplish this.

With this as a guideline, the modified active earth pressures, p_a, applied to the actual RERS have magnitudes:

$$= 0 \text{ at the ground surface} \tag{4.1a}$$

$$= \left[K_a^* \cdot \gamma_{eff}^* \cdot H* \cdot \left(\frac{H*}{H} \right)^2 \right] \text{ at the heel} \tag{4.1b}$$

where K_a^* is the Coulomb coefficient of active earth pressure for the retained soil and γ_{eff}^* is the *effective unit weight* of the retained soil[58].

Note that, in general, this earth pressure triangle acting on the actual RERS will be skewed relative to the horizontal at an angle equal to $(\omega* + \phi*)$ as derived previously and shown in Figure 4.3b so will produce both horizontal and vertical components with the latter usually acting upward as noted previously. Whether or not this vertical component is used for analyzing or designing the RERS generally depends on whether the RERS is non-yielding or yielding in nature. This is because the vertical component of earth pressures acting on a RERS is typically used only for assessing the SLS (specifically, the magnitude and distribution of bearing stresses along the base of the RERS to check if uplift will occur at the heel of the RERS) as well as ULS margins of safety against the geotechnical modes of failure that, by definition, are only relevant for yielding RERSs: translation (sliding), rotation (overturning), and bearing capacity.

The second potential component of lateral pressure acting on the actual RERS is a rectangular distribution of surcharge pressures (if any) from the retained soil. Because of the same $H*$ vs. H issue, Equation A.55a in Appendix A has to be modified to account for this. The surcharge pressure applied to the actual RERS is:

$$K_a^* \cdot q_s \cdot \left(\frac{H*}{H} \right). \tag{4.2}$$

Again, this rectangle will, in general, be skewed relative to this horizontal at the same angle as the aforementioned earth pressure triangle and thus produce both lateral and (usually upward) vertical components.

Before discussing the third component of pressure used to design the RERS for gravity loads, it is of interest to digress and discuss one aspect of these first two

---

[58] This is defined as the unit weight that will produce an effective stress in a given application. Above the groundwater table, as will be typical in most applications, this would be the *total* (a.k.a. *damp, moist, wet*) *unit weight*, γ_t. Below the groundwater table, this would be the *buoyant unit weight*, γ_b (sometimes written as γ').

components in greater detail. As noted previously, on most projects, one of the limiting cases to be examined as part of the design process is maximum usage of EPS to reduce lateral pressures from the retained soil under gravity loading to zero (at least in theory). An examination of Coulomb's solution for the active earth pressure state indicates that both the earth force and surcharge (if any) from the retained soil under gravity loading will be zero if K_a^* is zero. This can be achieved if the layout of the EPS blocks shown in Figure 4.2 is configured so that:

$$\theta^* = \phi^*. \tag{4.3}$$

This result is intuitively understandable as this simply means that the retained soil is self-stable (albeit with a safety factor exactly equal to one) if inclined at precisely its *angle of repose*. In such a case, the RERS would have to be designed only for the loads that result directly from the EPS blocks and overlying materials in direct contact with the structure as discussed next.

Continuing on, the third and final component of lateral pressure for which a RERS is analyzed or designed is unique to the use of the lightweight-fill function. The primary portion of this stress, shown as q in Figure 4.3b and assumed to always have a purely horizontal orientation, is the direct result of the vertical stress, shown as σ'_v in the same figure, caused by the soil, pavement, etc. plus any surface surcharge directly overlying the assemblage of EPS blocks. For simplicity and ease-of-use, an average value of σ'_v is used, even when the ground surface is sloped above the EPS blocks as is shown qualitatively in Figure 4.3b.

It is worth nothing that in any practical application, this vertical stress σ'_v and the lateral pressure q it creates on the inside face of the actual ERS can never be eliminated as there will always be some overburden placed on top of the EPS blocks. This is simply because the EPS blocks should never extend up to the ground surface and be left permanently exposed to the atmosphere (a long-term durability issue) and potential vandalism in the long term.

Based on observation of actual structures and consistent with the relatively small magnitude of Poisson's ratio for EPS within its nominally elastic small-strain range (Horvath 1995b), it is typically assumed for simplicity that:

$$q = 0.1 \cdot \sigma'_v. \tag{4.4}$$

Note that Equation 4.4 implies that the self-weight stresses from the EPS blocks are neglected. This is reasonable given the relatively very low density of EPS and the insignificant magnitude the EPS self-weight stresses have compared to typical values of σ'_v in practice.

With reference to Figure 4.3b, note also that q only extends over the height of the EPS blocks where they are in direct contact with the inside face of the actual RERS (the dimension H_{EPS} in Figure 4.2). This leaves open the question of how to deal with the lateral pressures on the RERS caused by the soil, etc. overlying the EPS blocks that is retained by the ERS and acting over the dimension H_{SOIL} in Figure 4.2.

This pressure component is sometimes neglected in practice based on the argument that it is already considered as part of the pressures from the retained soil because this material and surcharge (if any) overlying the EPS blocks is generally physically continuous in a horizontal direction from the inside face of the actual RERS to the retained soil as shown in Figures 4.2 and 4.3b. However, the position argued in this monograph is that this

82

lateral pressure should always be considered as a separate entity although this is probably somewhat conservative as noted subsequently. The rationale is based on the limiting case defined by Equation 4.3 where EPS is placed so that the lateral pressures on the actual RERS from the retained soil are zero. There would still be a lateral pressure on the actual ERS from the soil, pavement (if any), and surface surcharge (if any) above the EPS blocks.

If the lateral pressure from the soil, etc. overlying the EPS blocks is considered (as it is in this monograph), logic dictates that the active earth pressure state be assumed for yielding RERSs and the at-rest earth-pressure state for non-yielding RERSs unless a layer of relatively compressible material (e.g. a geofoam-based geocomposite sheet drain as discussed in Chapters 7 and 8) is placed between the RERS and soil. In such a case, the active state would be appropriate as well.

As noted previously, the lateral pressures on the actual RERS produced by the EPS blocks and overlying material are assumed to develop and act independently of those from the retained soil and remain constant in magnitude regardless of what angle θ^* is chosen for the interface between the EPS blocks and retained soil. Stated another way, the lateral pressures produced by the EPS blocks and overlying material is assumed to remain constant no matter how much or how little geofoam is used.

This is acknowledged as being a conservative approximation of reality. It should be intuitively obvious that as θ^* approaches θ for the actual RERS the volume of EPS approaches zero, i.e. it approaches the baseline case of no geofoam, just retained soil against the inside face of the actual RERS. Therefore, as this baseline case is approached the lateral pressures from the EPS blocks and overlying material must approach zero as well. Nevertheless, the assumption that these lateral pressures remain constant is judged to be reasonable because it greatly simplifies the solution while yielding a conservative outcome for all values of θ^*.

These three load components on the actual RERS are illustrated conceptually and qualitatively in Figure 4.4. The assumptions made when developing this figure (which do not detract from the generality of the methodology for applications deviating from these assumptions) are:

- The actual RERS has a vertical inside face (i.e. $\theta = 90^0$ and $\omega = 0^0$) which is the predominant situation by far in practice for all types of RERSs, yielding and non-yielding.

- The ground surface slopes upward from the inside face of the actual RERS (i.e. the slope angle i is positive in sign) and is subjected to a uniform surface surcharge of magnitude q_s.

- Only the horizontal components of the two retained-soil load components are shown (as noted previously, the load component arising from loads directly on the EPS blocks is always assumed to be only horizontal) which accounts for the cosine ($\omega^* + \phi^*$) term in the equations shown in the figure. Also, as noted previously, only lateral pressures are assumed to affect the structural behavior of the RERS regardless of whether it is yielding or non-yielding. Vertical stresses and their resultant forces acting along the inside face of the actual RERS would only affect the geotechnical-stability assessment of a yielding RERS.

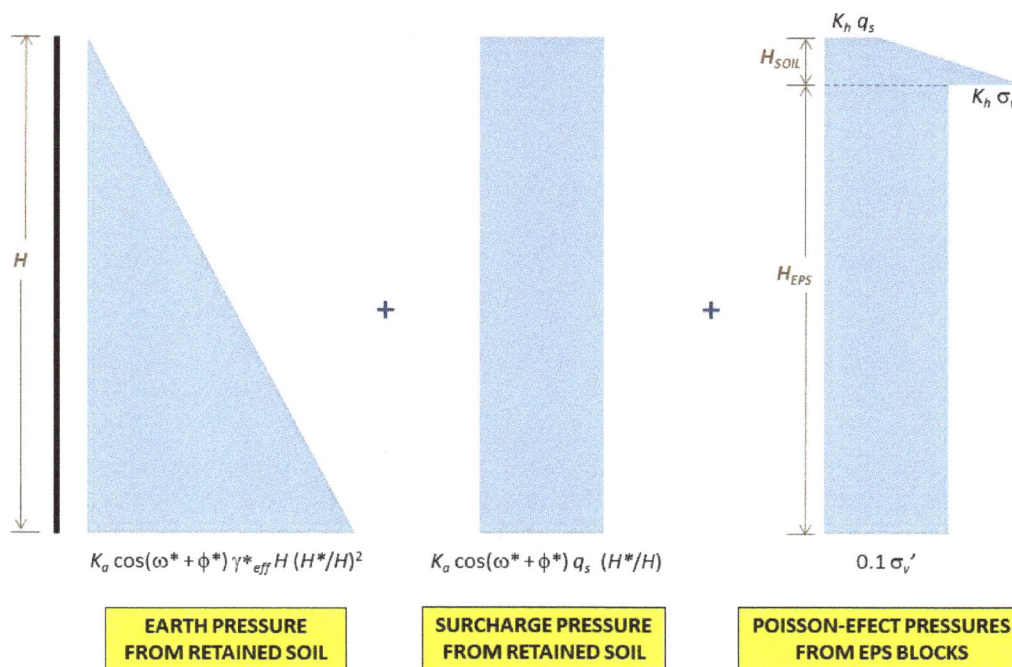

Figure 4.4. Lightweight-Fill Function - Full-Depth Placement -
Design Lateral Pressure Diagrams (Horizontal Components Only).

The problem-geometry and soil-property variables that appear in Figure 4.4 have all been defined previously but their definitions are summarized here for ready reference and ease of use:

- H is the geotechnical height of the actual RERS [shown in Figure 4.2].

- H_{SOIL} is the vertical distance from the ground surface to top of the EPS blocks adjacent to the inside face of the actual RERS [shown in Figure 4.2].

- H_{EPS} is the height of the EPS blocks adjacent to the inside face of the actual RERS [shown in Figure 4.2].

- H^* is the geotechnical height of the pseudo-RERS formed by the actual RERS plus assemblage of EPS blocks [shown in Figure 4.2].

- K_a^* is the Coulomb coefficient of active earth pressure for the retained soil and is calculated using Coulomb's solution in Appendix A and the angles i, δ^*, ϕ^*, and θ^*.

- K_h is the coefficient of lateral earth pressure for the operative earth pressure state for the material overlying the EPS blocks. If the actual RERS is yielding in nature or if a compressible geosynthetic sheet-drain is used between the actual RERS and material overlying the EPS blocks, then $K_h = (K_a \cdot \cos \delta)$ where K_a is the Coulomb coefficient of active earth pressure for the material overlying the EPS blocks and is calculated per Appendix A and the angles i, δ, ϕ, and θ. Otherwise, $K_h = K_o$ for the overlying material should be used. In the absence of more-definitive information, $K_o = (1 - \sin \phi)$ can be used.

- q_s is the magnitude of the uniform surface surcharge, if any [shown in Figure 4.2].

- δ is the interface friction angle between the actual RERS and material overlying the EPS blocks.

- δ^* is the interface friction angle between the pseudo-RERS and retained soil and is always taken to be equal to ϕ^*.

- ϕ is the operative value of the stress-dependent Mohr-Coulomb strength parameter (a.k.a. angle of internal friction or friction angle) for the material overlying the EPS blocks.

- ϕ^* is the operative value of the stress-dependent Mohr-Coulomb strength parameter (a.k.a. angle of internal friction or friction angle) for the retained soil.

- γ^*_{eff} is the effective unit weight of the retained soil.

- θ is the angle with respect to the horizontal defined by the assumed planar inside face of the actual ERS [shown in Figure 4.2].

- θ^* is the angle with respect to the horizontal defined by the assumed planar interface between the retained soil and inside face of the pseudo-ERS [shown in Figure 4.2].

- σ'_v is the average vertical effective stress acting on the top of the assemblage of EPS blocks and includes not only the weight of the material (soil, pavement, etc.) overlying the EPS blocks but any overlying surface surcharge, q_s, as well [shown in Figure 4.3b].

- ω ($= \theta - 90°$) is the angle with respect to the vertical defined of the assumed planar inside face of the actual RERS.

- ω^* ($= \theta^* - 90°$) is the angle with respect to the vertical defined by the assumed planar interface between the retained soil and inside face of the pseudo-RERS and is always negative in sign [shown in Figure 4.3b].

4.4.1.2.3 Seismic Conditions

The current state of practice for dealing with seismic loads when the lightweight-fill function is utilized is based almost entirely on research and methodology development conducted in Japan in the late 1980s to early 1990s. As will be seen, there was subsequently some further investigation and refinement in the U.S. of the basic Japanese methodology.

The lateral pressures acting on the inside face of the actual RERS under seismic loading conditions are assumed to come from the same three sources discussed in the preceding section for gravity loads:

- the retained soil acting on the assemblage of EPS blocks that is essentially a structural extension of the actual RERS;

- the surcharge, if any, on the surface of the retained soil; and

- the Poisson effect of the assemblage of EPS blocks and overlying material on the actual RERS.

The primary difference between the gravity and seismic load cases is, of course, that the latter includes inertial effects in addition to gravity effects. Note that these inertial effects are independent of and in addition to inertia effects on the actual RERS itself. The inertial effects of the actual RERS can be relatively significant in magnitude and are assumed to develop independently of loads from whatever material is retained by the RERS.

As will be seen, the specific way in which the seismic inertial effects from the three load sources that are external to the RERS are modeled in the current state of practice varies with each load source. In addition, there is a significant difference between how the seismic load case is handled between non-yielding and yielding RERSs (as seen in the preceding section, there is assumed to be virtually no difference for gravity loads). Although the lateral pressures from the three load components are the same regardless of the yielding potential of the RERS, for yielding RERSs only there is a resisting force that is assumed to develop that has the net effect of reducing lateral pressures transmitted through the EPS blocks from the retained soil and surface surcharge (if any) overlying the retained soil. Furthermore, this resisting force is assumed to be uniquely present under seismic loading only which is why it was not mentioned in the preceding discussion of gravity loads.

Beginning now the discussion of the three load sources on the actual RERS, the first one considered is the earth force from the retained soil acting on the assumed planar interface defined by the angle θ* that forms the inside face of the pseudo-RERS formed by the actual RERS plus assemblage of EPS blocks as defined in Figures 4.3a and 4.3b. Consistent with the current state of practice that is summarized in Appendix A, the seismic effects for this load component can be dealt with using one of the three methods presented in Section A.2.3.2 that is related to the original Mononobe-Okabe Method[59]. These methods vary in terms of:

- how the resultant earth force under seismic loading, P_{ae}, is evaluated, i.e. using the original Mononobe-Okabe equation or some arbitrary, empirical version of it; and

- where, i.e. at what height, is that resultant force placed on the inside face of the RERS.

Note that this implies that there are no phase variations between the base and top of the ERS (actual or pseudo) so that the retained soil moves as a single mass of material. This is a common assumption in the current state of practice for retaining walls, independent of the use of geofoam. It is also a common assumption for below-grade walls such as the basement walls of buildings. However, recent research as discussed in Appendix A suggests that for deep basements this assumption becomes increasingly approximate and inaccurate.

In any event, because there is no current consensus as to which combination of these two variables (i.e. the magnitude of the seismic resultant force and its point of

---

[59] Note that the Seed-Whitman Method cannot be used as this empirical methodology was developed only for a very simple problem geometry, i.e. a RERS with a vertical inside face (θ = 90°) and retained soil with a horizontal ground surface.

86

applications) is 'correct' or at least the 'best', choices are left to the design professional performing an application-specific assessment. However, there are some general comments that can be made:

- The calculations are based on the height of the pseudo-RERS, H^*, with the resulting lateral pressure distributions scaled-up in magnitude when applied to the inside face of the actual RERS that has a geotechnical height equal to H.

- The effects of vertical acceleration on the unit weight of the retained soil, γ_{eff}^*, can be included or neglected as desired.

- As with gravity loading, the calculated pressure diagram will be skewed relative to the horizontal at an angle = $(\omega^* + \phi^*)$ so will produce both horizontal and vertical (usually upward) pressure components on the actual RERS. Whether or not the vertical component is used in analyzing or designing the actual RERS usually depends on whether it is non-yielding or yielding in nature.

The second load component is a rectangular distribution of surcharge pressures (if any) from the retained soil. In this case, the gravity and seismic effects are lumped into one equation that is essentially the same as Equation 4.2 for the gravity-load case but with a larger lateral earth pressure coefficient, K_{ae}^*, that is based on one of the aforementioned methodologies presented in Appendix A:

$$K_{ae}^* \cdot q_s^* \cdot \left(\frac{H^*}{H}\right). \tag{4.5}$$

Note that:

- this equation is somewhat simplified and approximate in that it does not include all possible effects related to the vertical component of seismic acceleration. In order to be rigorous with regard to vertical acceleration, the q_s^* term should be multiplied by $(1 \pm k_v)$; and

- as with the previously described earth-load component, this rectangle is skewed relative to the horizontal at an angle equal to $(\omega^* + \phi^*)$ and thus produces both vertical (usually upward) and lateral components on the actual ERS.

Before discussing the third load component used to design the actual ERS for seismic loads, it is of interest to digress and discuss the first two components in greater detail. As with the gravity-load case, it is theoretically possible to reduce the lateral pressures on the actual RERS due to the retained soil and overlying surcharge (if any) under seismic loading to zero.

An approximate[60] way of achieving this is by selecting a geometry for the EPS blocks (as defined by the angle $\theta*$ in Figure 4.2) such that K_{ae}^* is equal to zero. On most projects, this limiting case would be investigated routinely as part of the design process to create the relationships shown in Figure 4.1 that would be used as part of the design-related decision-making process. The required geometry for this condition is developed as follows as it is not quite as obvious as for the gravity load case.

With reference to Equation A.13 in Appendix A for the Mononobe-Okabe Method and the parameters defined there, $K_{ae}^* = 0$ will occur when:

$$\cos(\phi* - \psi* - \omega*) = 0 \tag{4.6}$$

which is when:

$$\phi* - \psi* - \omega* = 90°. \tag{4.7}$$

Note that the influence of $\theta*$ on the problem is through its direct geometric relationship with $\omega*$ (the fact that $\omega*$ will always be negative in sign in this application as shown in Figure 4.3b is automatically accounted for in this case) as noted previously for the gravity-load case:

$$\theta* = \omega* + 90°. \tag{4.8}$$

Thus, the angle $\theta*$ required for $K_{ae}^* = 0$ can be found by substituting Equation 4.8 into Equation 4.7 and simplifying to obtain:

$$\phi* - \psi* - \theta* = 0° \tag{4.9}$$

or

$$\theta* = \phi* - \psi*. \tag{4.10}$$

Thus, the required angle $\theta*$ for zero earth and surcharge load under seismic loading is always smaller in magnitude than that required for gravity loading only ($\theta* = \phi*$) which is intuitive and so comes as no surprise.

As with the gravity-load case, even if the loads from the retained soil and its surcharge (if any) were designed to be zero by satisfying Equation 4.10, the lateral pressure on the actual RERS will not be zero under seismic loading. This is because there will still be lateral pressures related to the material and surcharge (if any) overlying the assemblage of EPS blocks that are in direct contact with the back of the actual ERS. As with the gravity-load case, there are assumed to be two distinct components of lateral pressure but the analytical model used is quite different for the seismic-load case compared to the gravity-load case discussed previously.

---

[60] The approximation exists only for yielding RERSs, not non-yielding ERSs, and is related to the aforementioned sliding resistance presumed to develop along the bottoms of the EPS blocks only under seismic loading conditions. This sliding resistance is discussed later in this chapter. The result of this approximation for yielding RERSs is that the angle $\theta*$ estimated here will be somewhat smaller and thus more conservative than necessary to achieve theoretical zero-net-pressure conditions.

In the seismic-load case, the first component of lateral pressure, from Poisson effects within the assemblage of EPS blocks that produce the lateral pressure q shown in Figure 4.3b, is assumed here and throughout the remainder of this monograph to be the same as in the gravity-load case (Equation 4.4) if the effect of the vertical component of seismic acceleration is neglected. Strictly speaking, vertical acceleration affects the σ'_v term in Equation 4.4 (it should be multiplied by $(1 \pm k_v)$) which implies that q should logically be multiplied by $(1 \pm k_v)$ as well.

However, it is the second component of lateral pressure (the lateral inertia force from the material plus surcharge (if any) overlying the EPS blocks that causes σ'_v) that is unique to the seismic case. The triangular assemblage of EPS blocks and overlying material (see Figure 4.2) form a top-heavy system overall because of the combination of:

- the density of the overlying material (approximately 100 times greater than that of the EPS blocks) and

- geometry (the triangle of EPS blocks is inverted, with its widest part at the top where all the denser material is located).

While this combination of factors has no particular significance under gravity loading, research in Japan in the late 1980s and early 1990s indicates it is quite important under seismic loading. Specifically, under horizontal shaking this triangular wedge of EPS with overlying soil, pavement, etc. does not displace laterally as a rigid block as is assumed for the wedge of retained soil in the Mononobe-Okabe Method (which was used to model seismic effects on the retained soil). Rather, the assemblage of EPS blocks plus overlying material exhibits characteristics of a flexible system and thus must be modeled taking its flexibility into account.

Consideration of system flexibility, when it occurs, is quite important in seismic applications in general because it often results in seismic amplification of the ground motion relative to the input base motion. This amplification produces lateral inertial forces greater than what would be calculated assuming rigid-body response based on ground-surface accelerations. In the applications considered in this monograph, these lateral inertial forces act against the inside face of the actual RERS so are important to consider.

As an aside, note that this means that seismic analysis and design for the lightweight-fill function is an unusual combination of seemingly conflicting assumptions in the same overall analytical model. Specifically, the inertial effects on the retained soil and surface surcharge above the retained soil are modeled as a rigid mass assuming no phase difference or seismic amplification exists while the adjacent assemblage of EPS blocks and overlying fill, pavement, etc. is modeled assuming a flexible single-degree-of-freedom (SDOF) system with seismic-amplification effects.

The simplest approach for dealing with system flexibility, and the one that represents the current state of practice for a wide variety of applications that involve the lightweight-fill function of EPS-block geofoam, is to model this triangular wedge of material in the actual problem (shown in Figure 4.5 as the red-shaded area) as a traditional SDOF system consisting of a lumped mass, axial spring, and axial damper (dashpot). In this particular application, the SDOF model is visualized as shown in Figure 4.6 where:

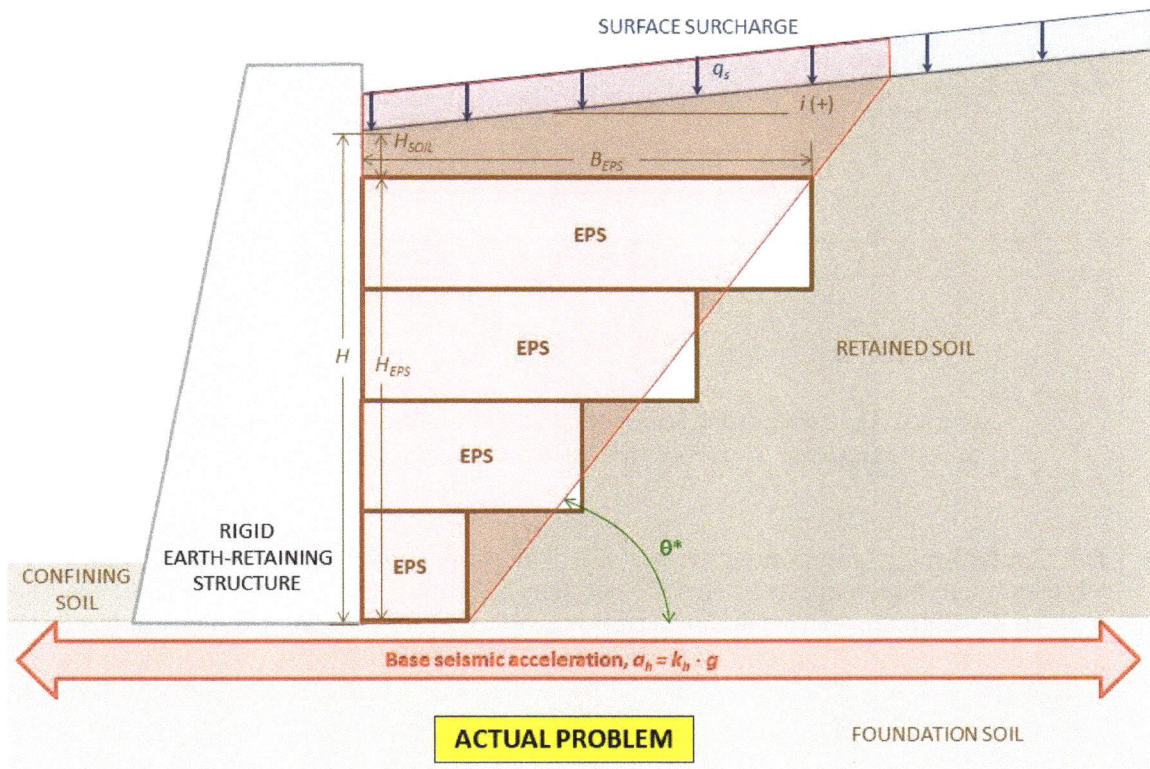

Figure 4.5. Lightweight-Fill Function - Full-Depth Placement Alternative Seismic Loading - Actual Problem Components.

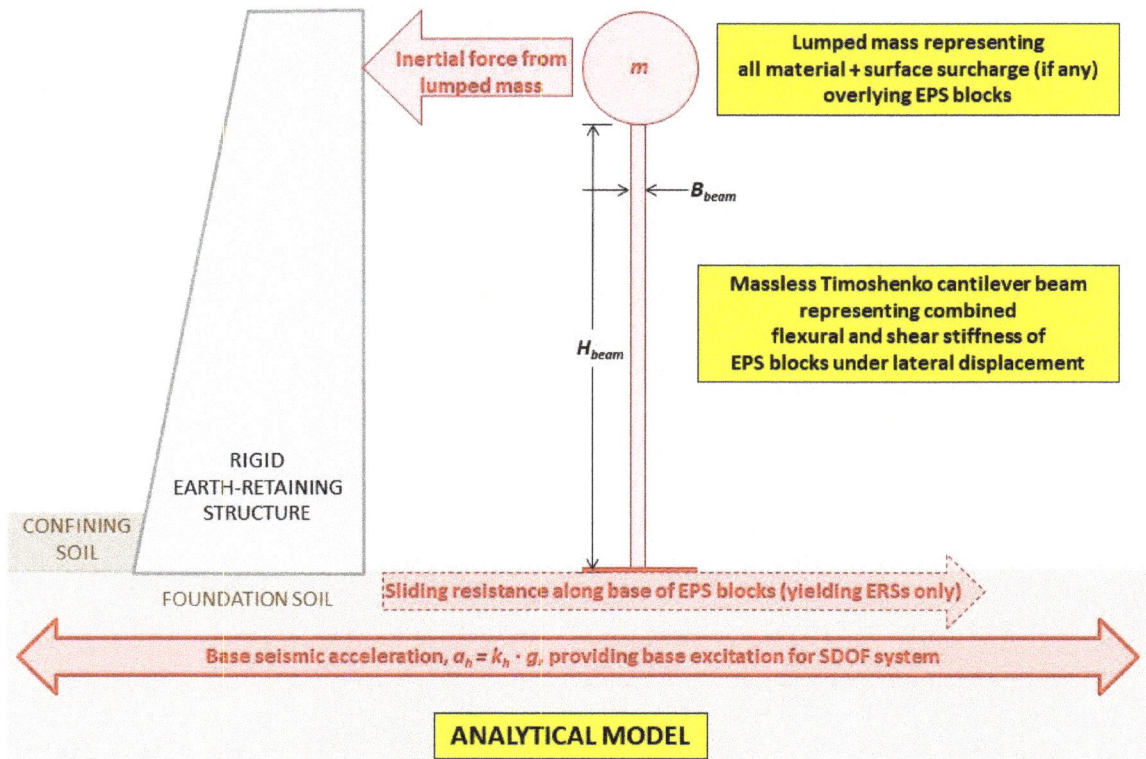

Figure 4.6. Lightweight-Fill Function - Full-Depth Placement Alternative Seismic Loading - Analytical Model (Geofoam Components Only).

- The <u>lumped-mass</u> component of the SDOF system comes solely from the mass of material (soil, pavement) and surface surcharge (if any) overlying the EPS blocks. The mass of the EPS blocks is considered to be relatively insignificant compared to the mass of this material and is ignored. The magnitude of the lumped mass can be multiplied by $(1 \pm k_v)$ if desired to account for the effects of the vertical component of seismic acceleration.

- The <u>spring component</u> of the SDOF is assumed to be provided solely by the lateral stiffness of the assemblage of EPS blocks. Specifically, the EPS blocks are visualized and modeled as a linear-elastic cantilever beam oriented vertically with its fixed-base at the bottom (defined as the assumed horizontal interface between the retained soil and foundation soil at the heel of the ERS). The ratio of lateral force to lateral displacement of the upper (free) end of this beam where the beam connects with the lumped mass is taken as the equivalent axial-spring constant of the SDOF system. Note that both flexure- and shear-related contributions to lateral displacement of this beam are included in the formulation. This is sometimes referred to in the literature as a *Timoshenko beam*.

- The <u>damping component</u> of the SDOF is assumed to be provided by a combination of material damping within each block of EPS plus sliding friction between EPS blocks along the horizontal joints between blocks.

Note that for both the actual problem (Figure 4.5) and SDOF model (Figure 4.6), all calculations are actually performed on a unit-width (metre, foot, etc.) basis for the dimension perpendicular to these figures. Therefore, problem variables and calculated quantities will always have force-per-unit-width dimensions.

As with the evaluation of any SDOF system under dynamic loading, calculating the fundamental period, T_o, of the system is a crucial first step in the process as it can affect the calculated acceleration and, ultimately, the magnitude of the calculated seismic inertial force of the system. The methodology for calculating T_o for an EPS-block geofoam fill was first illustrated and outlined in detail in Horvath (1995b). Since then, there have been two important additions to the state of knowledge in this area.

First, there can be some variation in calculated results for T_o depending on the specific equation used to quantify the equivalent spring stiffness of the cantilever beam used to model the EPS blocks. Three equations (two completely theoretical ones derived by the writer and one empirically-modified theory based on the aforementioned Japanese research) have been identified to date. A limited comparative assessment of these equations indicates that the variation in T_o and, ultimately, the calculated horizontal inertial force on the actual RERS from the assumed lumped mass, is relatively small but, nonetheless, noticeable to the point where a design professional might want to exercise professional judgment as to which equation to use in order to produce conservative results. A complete discussion of this issue can be found in Horvath (2004a).

The second and much more significant issue involves estimating the equivalent length, H_{beam}, and depth (thickness), B_{beam}, of the Timoshenko cantilever beam as shown in Figure 4.6 based on the actual dimensions, H_{EPS} and B_{EPS}, of the assemblage of EPS blocks as defined in Figure 4.5. The original formulation for this based on research in Japan was that B_{beam} should always equal B_{EPS} and then H_{beam} should always be chosen such that the area of the beam (= B_{beam} x H_{beam}) equaled the cross-sectional area of the assemblage of EPS blocks.

As an aside, it is important to understand that the original Japanese research that produced these guidelines for evaluating H_{beam} and B_{beam} was based on the use of EPS-block geofoam as a lightweight-fill material for traditional stand-alone embankments with either a trapezoidal or rectangular cross-section and not primarily with ERS applications in mind. For the basic case of a rectangular standalone embankment, the result is trivial and H_{beam} = H_{EPS}. However, for side-hill embankments or as extended to problems involving ERSs where the cross-sectional geometry of the EPS blocks is overall trapezoidal-trending-to-triangular in shape, the result is always that H_{beam} < H_{EPS}, sometimes significantly so (note that H_{beam} = $H_{EPS}/2$ for a perfect triangle). This difference between the actual problem and model can have a significant impact on the calculated value of T_o.

The technical basis or justification of the B_{beam} = B_{EPS} assumption in the Japanese methodology is not known to the writer and so has long been a source of uncertainty to the writer. In recent years, the impact of this assumption was evaluated by comparing values of T_o calculated using this assumption to values calculated using an alternative modeling approach postulated by the writer that H_{beam} = H_{EPS} always and it is the magnitude of B_{beam} that is varied so that the beam area (= B_{beam} x H_{beam}) is equal to the cross-sectional area of the assemblage of EPS blocks.

A detailed investigation into this alternative guideline was performed at the writer's suggestion at the University of Memphis (Tennessee) as part of National Cooperative Highway Research Board Project 24-11(02) that focused on using EPS-block geofoam for slope stabilization. Non-rectangular geometries of EPS blocks are very common in such applications as is seismic loading so it was relevant to investigate this issue as part of that study.

The results were published in Arellano et al. (2011a, 2011b, 2013b) and the conclusion was that the difference between values of T_o calculated using the two different methods (the original Japanese approach and the writer's alternative approach) could be quite significant. Furthermore, these significant differences could, in some cases, be carried over into the estimated seismic inertial force. Because neither of these two methodologies was found to consistently produce a result that is conservative relative to the other, the conditional (pending further research) recommendation by Arellano et al. was that in project applications a design professional should calculate T_o using both methods; see which value of T_o produces the greater, more-conservative estimate of seismic inertial force; and then use that greater value in analysis or design.

The writer concurs that this seems like a prudent approach to take (which admittedly is not an objective assessment given that the writer co-authored the Arellano et al. documents) unless and until future research clearly indicates which method better models actual behavior as defined by physical measurements on actual structures, centrifuge modeling, or advanced numerical modeling using computer software capable of explicitly simulating dynamic behavior. Alternatively, future research may indicate that some new, presently undefined methodology is superior to both.

In any event, once T_o has been calculated, there are several ways in which it can be used to calculate the lateral seismic inertial force acting on the actual RERS that is shown conceptually in Figure 4.6. The most theoretically rigorous approach requires development of a site-specific acceleration response spectrum for the SDOF base-input motion shown in Figure 4.6. A less-rigorous alternative would be to use a generic response spectrum such as might appear in a building or design code. In either case, when developing the input motion, appropriate consideration should be given to possible modification of the bedrock motion at the site by the soil column between the top of rock and ground surface on which the RERS is constructed.

Once the acceleration response spectrum has either been developed or selected, the calculated value of T_o is plotted on this spectrum to determine the lateral acceleration, $a_{h\text{-}SDOF}$, for the lumped mass and from this the dimensionless lateral acceleration coefficient, $k_{h\text{-}SDOF}$ ($= a_{h\text{-}SDOF}/g$). In most cases, $k_{h\text{-}SDOF}$ will be greater in magnitude than k_h for the SDOF base motion. This increase in lateral acceleration between the base and top of the EPS blocks is referred to as *seismic amplification* and is a typical occurrence for flexible systems subjected to seismic motion. This is because the dominant period of most seismic motion is usually relatively close in magnitude to the fundamental period of a wide range of constructed facilities.

An important technical issue to keep in mind is that response spectra are most commonly used for structural analyses where the system being analyzed has perhaps 1% or 2% of critical damping. Research to date indicates that EPS-block geofoam fills have significantly higher damping, in the range of 2.5% to 8% of critical (Horvath 1995b). This is due to the fact that the damping of an assemblage of EPS blocks comes from not only material damping within each EPS block but also friction between blocks along horizontal joints[61]. Therefore, when performing a seismic analysis as outlined above, a damping level that is appropriate to EPS-block geofoam fills should be used with a response spectrum.

---

[61] It is worth noting that it is now well proven that lateral displacement will occur along the horizontal planar joints between EPS blocks even if the well-known and widely used barbed connector plates are used to supplement inherent inter-block friction and provide additional resistance to lateral sliding. This is because the resistance provided by the current designs of connector plates require lateral displacement of the pointed barbs into the EPS in order for resistance to be mobilized.

This will usually produce horizontal accelerations at the top of the EPS blocks that is smaller in magnitude than for typical structures.

An alternative, simpler approach to using response spectra is to use some empirical equation that relates T_o to seismic acceleration directly. Such equations are often found in design-oriented codes such as the American Association of State Highway and Transportation Officials (AASHTO) code for road-related structures. While attractive in their simplicity, it should be kept in mind that such equations will usually produce conservative results (in the form of an overestimate of the inertial force) for the applications discussed here. This is because these equations have generally been developed and calibrated for normal structural damping levels that are smaller in magnitude than the above-noted damping levels observed with assemblages of EPS blocks. As a rule, seismic amplification of a flexible structure decreases with increasing damping, all other variables being equal.

4.4.1.3 Solution Assessment

The preceding analytical methodology for the basic full-depth placement alternative for the lightweight-fill function under both gravity and seismic loading reflects the outcome of research and development (R&D) that began more than 30 years ago in Japan and achieved a national consensus among all stakeholders there more than 20 years ago. Furthermore, all key components are based on a logical application of theoretical concepts and do not rely on empiricisms. Most importantly, it is an analytical methodology that has presumably not only seen extensive use but, given the frequency of significant seismic events in Japan, has likely been subjected to such events at some locations during this timeframe. As a result, it is tempting to fall in line with the groupthink on this issue and assume that the overall methodology has been properly vetted and verified and can be used worldwide with confidence.

That having been said, unpublished FE analyses (gravity loading only) by the writer suggest that a critical, zero-based review and assessment of this Japanese analytical methodology would not only be appropriate but is arguably necessary. This is because there are indications that the behavior of actual applications of this geotechnology may not be accurately captured by the current analytical methodology. Furthermore, as was noted in Chapter 3, there is the never-asked and thus to-date-unanswered/'elephant-in-the-room' question of the effect of a biaxial-stress system acting on the assemblage of EPS blocks. The long-term effect of biaxial loading on EPS-block in geofoam applications is simply not well-understood at the present time.

Note that this does not mean that the current Japanese analytical methodology should be scrapped. It may prove acceptable and conservative to use, much the same way that Coulomb's solution for yielding RERSs remains in use to the present even though it has been known for the better part of a century that it does not completely and correctly capture or replicate all aspects of actual behavior. This is part of the eternal tradeoff in civil engineering between analytical accuracy and ease-of-use in routine practice.

In any event, in the writer's opinion, it is essential that field measurements of full-scale installations be included in this proposed assessment and that it not rely solely on numerical analyses. This is because there are certain elements of the lightweight-fill function, especially the fact that individual EPS blocks are placed in the field, that are not easily or accurately replicated in a numerical analysis.

This assessment also needs to include a comprehensive, zero-based effort to research the effect of biaxial stresses on the performance of EPS-block that includes developing new laboratory-testing protocols for assessing the behavior of EPS-block under biaxial stresses. Only after such work is conducted can a correct decision be made as to whether or not the current practice of performing a simple 1-D stress analysis is adequate or if a true two-dimensional (2-D) analysis needs to be performed on a routine basis.

4.4.2 Partial-Depth Placement Alternative

4.4.2.1 Introduction

As noted previously, there is no technical reason why the lightweight-fill function cannot be implemented over only a portion of the geotechnical height of an ERS as shown conceptually in Figure 3.6[62] in what has been defined in this monograph as the partial-depth placement alternative[63]. In practice, this alternative to the basic full-depth placement would likely be considered and used primarily with existing RERSs for at least two reasons:

- Physical site constraints, excavation safety issues, and other pragmatic considerations can frequently limit the vertical and horizontal extent of excavation possible adjacent to the inside face of an existing RERS making full-depth placement simply unfeasible in such situations.

- Reasonably significant improvement in calculated performance of an existing RERS can often be achieved with surprisingly limited replacement of existing soil. Recall from lateral earth pressure theory that the resultant lateral earth force on a RERS varies with the square of the height (depth) of the retained soil. So, a 25% reduction in retained-soil height results in a 44% reduction in earth force; a 50% reduction in height yields a 75% reduction in force, etc. While these example calculations are theoretical and simplistic, they do make the intended point that in most cases when it is desired to reduce earth loads on an existing ERS to increase its calculated safety, 'a little goes a long way' when it comes to reducing the depth of retained soil behind an ERS by replacing it with EPS blocks.

4.4.2.2 Overview

Although partial-depth placement is a potentially useful design alternative in practice, the writer is not aware of any comprehensive published study of this alternative geared toward developing a logical analytical methodology along the lines of what was presented in the preceding sections for the basis full-depth placement alternative. Such a study would, of necessity, require both analytical work based on theoretical soil mechanics and/or numerical modeling combined with verification and calibration through physical modeling using instrumented 1-*g*, full-scale RERSs (or similar large-scale test facilities) and/or centrifuge testing. While it is beyond the resources of the writer and scope of this

---

[62] Although a horizontal ground surface of the retained soil is shown in this figure, the discussion in this section is completely general and applicable to cases where this surface slopes either upward or downward.
[63] For the reasons discussed previously, use of partial-depth placement should always be done in the general manner shown in Figure 3.6, i.e. with the EPS blocks placed as close to the surface as possible consistent with maintaining adequate cover, pavement system, etc. above the EPS blocks. This is to minimize the vertical compressive stresses to which the EPS blocks are subjected.

monograph to conduct and present such a comprehensive investigation, some preliminary thoughts on the subject are presented here that will hopefully provide at least a starting point for future research by others.

It is always a challenge to postulate a zero-based SLS or ULS mechanism for any new problem or application in geotechnical engineering. One must avoid the temptation and pitfall of simply extending or reusing an existing behavioral mechanism, no matter how logical, intuitive, or appealing it might be to do so, without verifying that the proposed mechanism is both kinematically feasible and consistent with what will actually occur in nature. This was noted in Chapter 3 and illustrated in Figure 3.5 for the full-depth alternative where the ULS mechanism that is currently considered to be correct for this problem (although itself possibly flawed) is quite different from what many have intuitively and initially thought the failure mechanism would or should be.

In summary then, academic researchers and design professionals alike always need to remember that just because they hypothesize a given SLS or ULS behavioral mechanism, there is no assurance that nature will oblige them and behave according to that mechanism. As a minimum, numerical modeling using FE or finite-difference (FD) methods should be used to verify a hypothesized behavioral mechanism, at least initially. This should eventually be followed up by verification using physical modeling using instrumented full-scale RERSs.

As will be seen in the following sections, unlike for the full-depth placement alternative where one basic analytical model is currently used for both non-yielding and yielding RERSs, the writer's initial, preliminary investigation of the partial-depth placement alternative indicates that non-yielding and yielding conditions need to be dealt with separately. The reason for this will be made clear by presenting the proposed analytical model and concomitant analytical methodology for yielding conditions first as this appears to be the easier problem to solve based on current knowledge.

4.4.2.3 Proposed Analytical Method for Yielding Conditions

4.4.2.3.1 Basic Concept

Study of Figure 3.6, which illustrates the generic application of the partial-depth placement alternative, suggests that for a yielding RERS a classical active earth pressure failure wedge can, depending on the vertical and horizontal extent of the assemblage of EPS blocks, develop naturally within the retained soil such that this wedge has the necessary soil-on-soil failure surface while completely encompassing the assemblage of EPS blocks. This is in sharp contrast to the full-depth placement alternative where a failure wedge is forced to develop separately from the EPS blocks and entirely within the retained soil because of interference from the EPS blocks. Thus, the hypothesized ULS mechanism for the partial-depth alternative would be similar to the conventional ULS mechanism for a yielding RERS retaining only soil (i.e. a nominally triangular failure wedge within the retained soil) except that the presence of the EPS blocks within the failure wedge would act as a weight-reducer for the overall failure wedge. Because the effective weight of the retained soil acting on the inside face of the RERS would be reduced, this means the resultant earth force acting on the RERS would be reduced as well and this would reflect the benefit of using the lightweight-fill function in this application.

As an aside, it is of interest to note that this hypothesized manner in which EPS-block geofoam provides a benefit in the partial-depth alternative is the same as the oft-imagined, but incorrect, manner (shown in Figure 3.3) in which it acts in the full-depth alternative. Note also that because the development of an active earth pressure failure wedge within the retained soil is assumed, this implies that this ULS mechanism is only applicable to yielding RERSs which is why a completely separate analytical methodology for the partial-depth alternative when used with non-yielding RERSs is required.

Logic dictates that there must be consonance between the full-depth and partial-depth alternatives because as the vertical and horizontal extent of a partial-depth application increases and, in the limit, becomes a full-depth application there must be a point at which the ULS mechanism must transition as well. However, it is beyond the scope of this monograph to explore what this transition looks like, i.e. is it an abrupt transition or is there some intermediate, hybrid ULS mechanism, and at what point it develops. The remainder of the present discussion will be limited to exploring the basic soil-wedge ULS mechanism for the partial-depth applications as proposed above.

4.4.2.3.2 Solution for Gravity Loading

The basic process proposed for solving this problem is the well-known *trial-wedge method* which is arguably the oldest, most basic concept in lateral earth pressure theory as it formed the underlying logic of Coulomb's solution in the 18th century (Huntington 1957). The essential aspect of this solution methodology is that there is, in principle, an infinite number of potential failure wedges defined geometrically in two dimensions by:

- the inside face of the actual RERS (assumed to be planar but not necessarily vertical in orientation);

- the ground surface of the retained soil (which can be irregular in geometry); and

- a soil-on-soil failure surface within the retained soil that is planar in shape, with zero thickness, and oriented at an angle θ^*_i with respect to the horizontal.

Note that although the basic physical model for the trial-wedge method is the same for gravity and seismic loading, the force systems are slightly different. Consequently, only gravity loading is discussed here, with presentation of seismic loading dealt with in the following section.

To begin with, Figure 4.7 shows a typical trial failure wedge (bordered and shaded in green) and associated resultant-force vectors for an intentionally simplistic problem geometry as the focus here is on the basic elements of problem formulation and solution, not the full breadth of generalities and possibilities of geometric boundary conditions. Note that applied loads on the surface of the retained soil are also permissible and can be completely general in nature, i.e. variable in magnitude and distribution in the plane of Figure 4.7. However, only that portion of any surface loading that actually lies within the limits of an assumed failure wedge is included in the free-body diagram of forces for that wedge. The resultant force from that portion of the surcharge is simply added vectorially to the weight of the wedge, W^*_i.

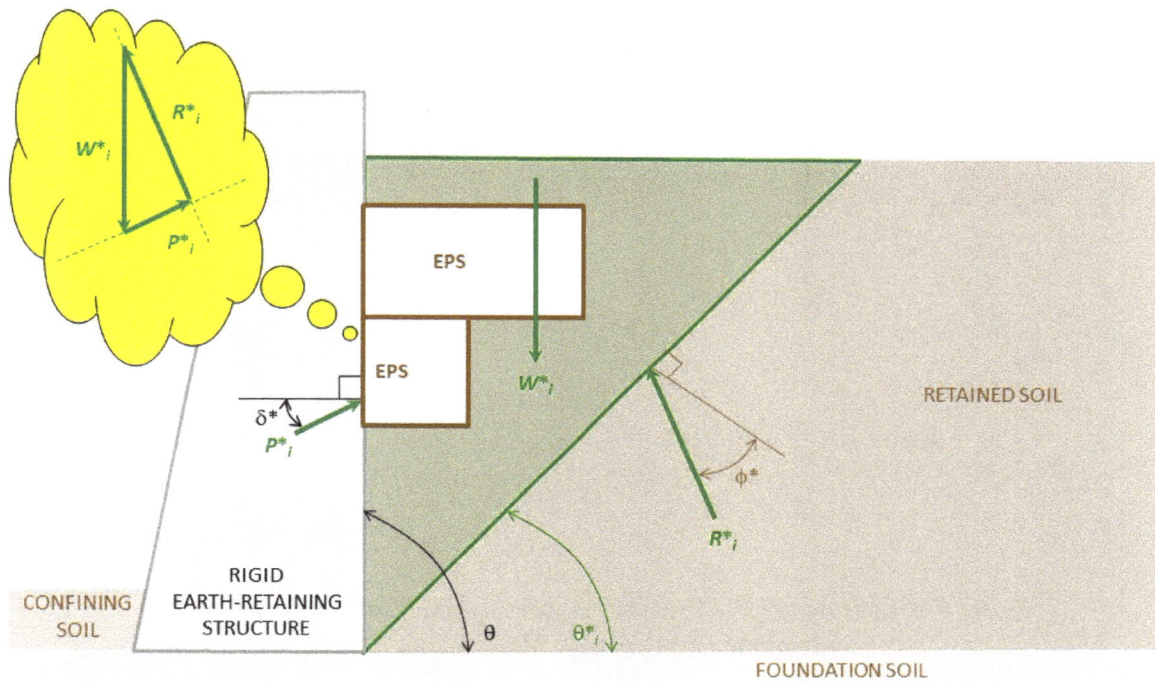

Figure 4.7. Lightweight-Fill Function - Partial-Depth Placement Alternative Gravity Loading - Analytical Model Based on Trial-Wedge Method.

Fortunately, there is always a unique solution to this problem and the solution process begins by recognizing that θ^*_i always has upper- and lower-bound values (even in the original all-soil, no-geofoam usage) defined as θ^*_{max} and θ^*_{min} respectively as shown conceptually in Figure 4.8. The former angle is assumed to be defined by a failure plane that just touches the corner of the EPS block that protrudes farthest into the retained soil (in the all-soil/no-geofoam problem, this angle would be defined by the angle θ defining the planar inside face of the RERS) and the latter angle is equal to ϕ^*, the friction angle of the retained soil which, in this situation, is sometimes referred to as the angle-of-repose as noted previously.

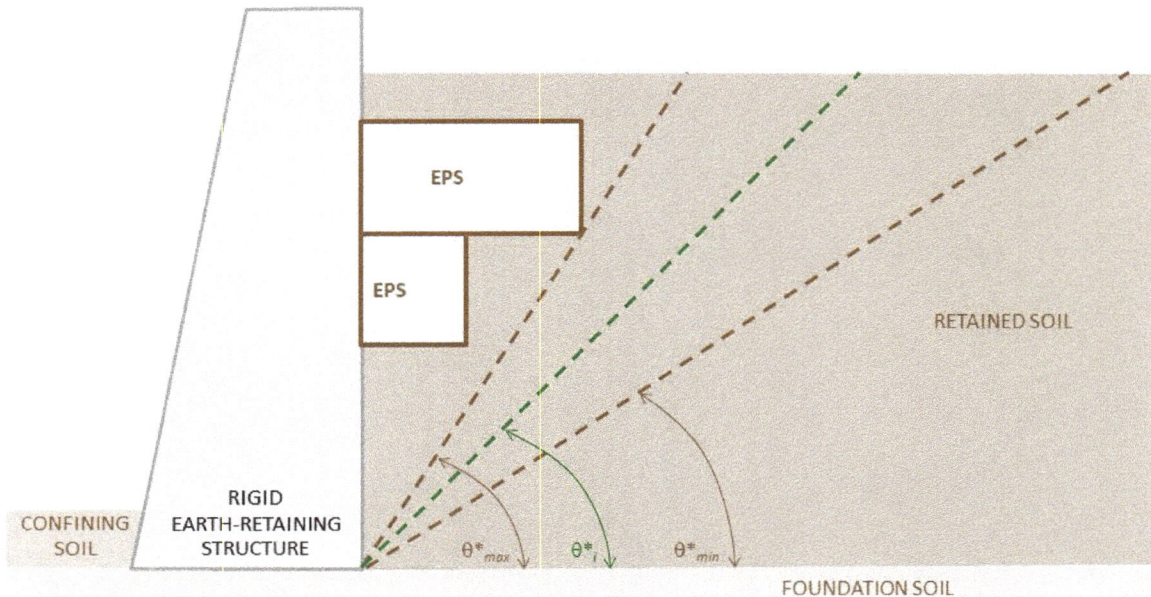

Figure 4.8. Lightweight-Fill Function - Partial-Depth Placement Alternative Gravity-Loading - Trial-Wedge Method Solution Limits.

Note that a failure plane at an angle greater than θ^*_{max} (but less than θ) is possible in principle. However, such a failure plane would have to physically cut through one or more EPS blocks which would be physically impossible. It would also violate the assumptions used to formulate this methodology and thus be inadmissible within the constraints of this analytical methodology.

An actual failure surface at an angle steeper than θ^*_{max} is likely be complex in geometry (i.e. not a single plane) and unique to that particular problem. It is most likely to occur when the assemblage of EPS blocks is extensive, i.e. approaching the basic full-depth placement alternative. As such, a complex failure surface would appear to be typical of the aforementioned transition zone from partial- to full-depth placement that is not being considered here and is left to future research by others.

It is worth noting that the solution methodology for this problem that is developed and presented subsequently would actually alert a user to the potential for a failure surface steeper than the input θ^*_{max} value for that problem. This would indicate an application that could not be solved using this solution methodology that would have to be analyzed either using an application-specific numerical analysis using the FEM or finite-difference method (FDM) or perhaps approximately analyzed as a full-depth application.

Continuing with the solution methodology, for each assumed angle θ^*_i and concomitant trial failure wedge, a system of three forces as shown in Figure 4.7 is defined as follows:

- W^*_i is the weight of the wedge plus any surface loading within the limits of the assumed failure wedge. Its magnitude is always known (by simple calculation) although it will change from one trial wedge to another. Its orientation (vertical) is also always known.

- R^*_i is the resultant force acting along the assumed soil-on-soil failure plane. Its magnitude is unknown initially but its orientation relative to the failure plane is always known (as shown in Figure 4.7, where ϕ^* is the friction angle of the retained soil). However, its absolute orientation, which is function of not only ϕ^* but θ^*_i as well, will change from one trial wedge to another.

- P^*_i is the resultant force acting on the failure plane that is defined by the inside face of the actual RERS. Its magnitude is also unknown initially but its orientation is always known (as shown in Figure 4.7, where δ^* is the friction angle of the RERS-retained soil interface and typically $0 < \delta^* < \phi^*$) and is assumed to be the same for every trial wedge in a given problem. Note that this is an approximation and simplification of reality as the material interfacing with the RERS is a combination of soil and block-molded EPS, each of which likely have different frictional characteristics with respect to the material comprising the RERS.

By definition, the solution to this problem is the active earth force, P_a^*. This is the largest value of P^*_i and the plane defined by the angle θ^*_i corresponding to this value of P^*_i would be the theoretical failure plane for the system and is defined here as $\theta^*_{critical}$.[64]

There are a number of ways in which $\theta^*_{critical}$ can be determined and the concomitant P^*_a obtained. Coulomb developed a solution to this problem, as summarized in Appendix A, mathematically using calculus but his solution is limited for use with problems that have or can reasonably be assumed to have relatively simple geometries (which, fortunately, includes the vast majority of applications in routine practice).

A more-general solution methodology is referred to simply as the 'trial-wedge method' in most text and reference books. This more-general methodology is useful for problems involving an irregular ground surface for the retained soil and/or a discontinuous or complex surface loading for which Coulomb's solution does not apply.

Historically, solution of the more-general trial-wedge method was done graphically. For each trial wedge, this involves drawing the three resultant-force vectors head-to-tail to create a triangle as shown in the yellow-highlighted 'cloud' in Figure 4.7. All drawing is done to scale with regard to both vector magnitude (initially only W^*_i is known) as well as the angular orientation of the lines-of-action of each vector (known a priori for all three) in two-dimensional space so this triangle will be unique for a given trial wedge. The

---

[64] It is recognized that unless the retained soil-RERS interface friction angle δ^* were zero this failure plane is only a theoretical approximation of the actual soil-on-soil failure surface which would be curved due to interface friction. The soil-on-soil failure surface is assumed here to be planar for the same reason Coulomb assumed it to be planar, i.e. to facilitate problem solution based on force equilibrium alone and ignoring moment equilibrium. This is because a curved soil-on-soil failure surface complicates determination of the orientation of the line of action of the R^*_i force component.

100

magnitude of P^*_i is then scaled off this triangle to yield the desired result. This process is then repeated for other trial wedges (how many is always at the discretion of the analyst) until the desired maximum value of P^*_i $(= P_a^*)$ is identified.

The traditional manner for performing this graphical construction, as shown in Huntington (1957) and many other older textbooks, is to do all work on a single plot (essentially W^*_i remains in the same place on a sheet of paper and simply gets physically longer or shorter for different wedges) and to draw a curve that connects the tips of all the P^*_i vectors (Huntington calls this the *pressure locus*) so that the maximum value of P^*_i becomes visually obvious.

This graphical-solution methodology of the trial-wedge method, especially for yielding RERSs such as rigid retaining walls, was once a staple of both civil engineering education and practice. However, in recent decades as mechanical drawing (drafting) education of civil engineers has disappeared from the undergraduate curriculum and the use of the digital computer as a drawing tool (CAD) appeared and grew, teaching of the trial-wedge method has tended to disappear from academic instruction as well as from the toolbox of design professionals in practice. Nevertheless, it remains a viable tool for solving this type of problem, especially when the geometry and surcharge loading of the retained soil are complex as noted above.

However, for the purposes of this monograph, which are to focus on illustrating general concepts, only a solution for the relatively simplistic geometry shown in Figure 4.7 will be pursued in detail. This development and concomitant computer-based solution methodology, including some worked examples, are presented in detail in Appendix C.

4.4.2.3.3 Solution for Seismic Loading

The trial-wedge method can easily be extended to include seismic loading and was, in fact, the methodology used to develop the original Mononobe-Okabe (M-O) Method that is discussed in Appendix A. Figure 4.9 illustrates the force system acting on each trial wedge (bordered and shaded in red in this case) under seismic loading and is essentially Figure 4.7 for the gravity-load case modified to include the seismic components.

Comparing Figure 4.9 to Figure 4.7, it is clear that the primary addition to the geotechnical components is the horizontal inertial force generated by the horizontal component of seismic acceleration acting on the mass of the material within the failure wedge. As is well known, only the outward (to the left in Figure 4.9) direction of horizontal acceleration needs to be considered in any practical application.

A secondary seismic effect is that the weight of the trial wedge is affected by the vertical component of seismic acceleration which, as noted earlier, can either be ignored or considered as desired. In any event, note that in the seismic-load case the traditional graphical solution of the trial wedge involves a force quadrilateral instead of a triangle and involves four vector lines-of-action. Note that in this case two vector magnitudes (the vertical and horizontal components of failure-wedge weight) are known a priori.

As with the gravity-load case, the seismic-load case can, in principle, be solved by graphical construction to find the trial wedge that produces the largest value of P^*_i that is assumed to be the active earth force under seismic loading, P^*_{ae}. Again, as with the gravity-load case, for the purposes of this monograph only the simple geometry shown in Figure 4.9 was investigated using a numerical solution for this monograph. This work is also presented in Appendix C.

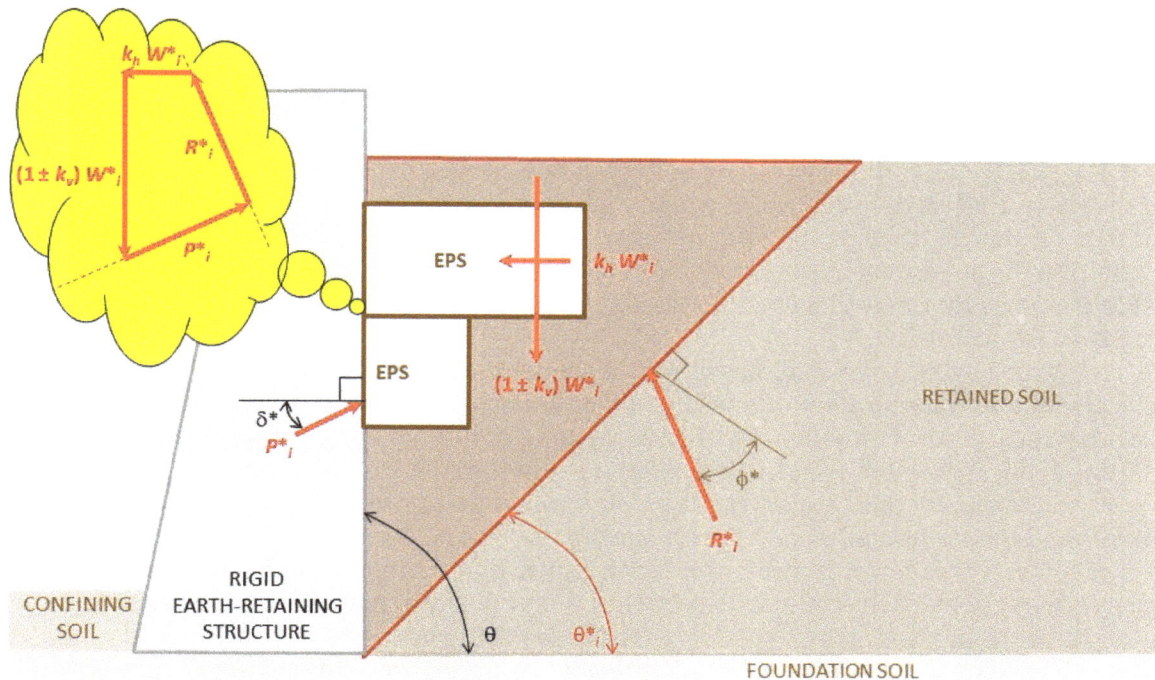

Figure 4.9. Lightweight-Fill Function - Partial-Depth Placement Alternative Seismic Loading - Analytical Model Based on Trial-Wedge Model.

4.4.2.3.4 Solution Assessment

Unlike the analytical methodology for the basic full-depth placement alternative that reflects more than 30 years of R&D and actual usage in Japan, the analytical methodology for partial-depth placement alternative with yielding RERS outlined above and developed in detail in Appendix C is believed to be the first of its kind. It is certainly the first time that this particular analytical model and associated analytical methodology developed by the writer has been published.

Although all key components of the proposed partial-depth methodology are based on a logical application of soil mechanics theory and do not rely on empiricisms, good practice dictates that the proposed methodology be properly vetted and verified before it is used in project-specific applications with confidence. A comprehensive R&D program similar to that suggested earlier in this chapter for the full-depth placement alternative should be conducted for the partial-depth alternative as well. Such a program should also include the necessary research to clearly define the transition from partial- to full-depth applications and how the ULS behavioral mechanisms transition in the process.

Furthermore, because the partial-depth placement analytical methodology as it exists at present only produces the lateral earth <u>force</u> on a RERS, additional work is required to develop a way to forecast the distribution of lateral earth <u>pressures</u> along the inside face of a RERS.

4.4.2.4 Analytical Method for Non-Yielding Conditions

Developing an analytical model and associated solution for the partial-depth placement alternative with non-yielding RERSs is much more challenging than it was for yielding RERSs. The primary reason is that the failure-wedge model developed in Appendix C that was used in the preceding sections for both gravity and seismic loading of yielding RERSs cannot, in theory at least, be used for non-yielding RERSs. This is because by definition the at-rest earth-pressure state that exists within the retained soil of non-yielding RERSs implies that non-ULS conditions exist within the retained soil so no ULS mechanism can even be postulated.

That having been said, based on the current state of knowledge for the full-depth placement alternative where it is assumed that under gravity loads at least the same results develop regardless of whether the RERS is yielding or non-yielding, it is possible that under partial-depth placement conditions for non-yielding RERSs that an analytical methodology could be developed that applies the full-depth methodology to the upper portion with EPS blocks and the traditional at-rest lateral earth pressures below that.

However, there are two additional issues that further complicate developing an analytical model for non-yielding RERSs. One is that in any actual construction there will be vertical shear stresses along the interface between the inside face of the RERS and retained soil. These stresses, which have historically been ignored in routine practice, will, nevertheless, exist and affect the observed lateral pressure distribution on a RERS. This has already been mentioned as part of the discussion for the basic full-depth placement. How the presence of partial-depth placement of EPS blocks will affect these shear stresses remains to be seen.

The other additional issue is how to deal with retained-soil surfaces that are planar but non-horizontal. To the best of the writer's knowledge, there is no exact solution for this problem for the basic all-soil/no-geofoam case so not having a well-defined baseline starting point (as there is for the yielding case) creates a knowledge void. However, some years ago the writer developed a relatively simple analytical model to approximate at-rest, sloped-ground conditions under gravity loading (Horvath 1990a) that may be of use with geofoam applications.

With regard to seismic loading, even the baseline all-soil/no-geofoam case for non-yielding RERSs has historically been a complex problem to deal with. Furthermore, this is an area that is currently undergoing considerable rethinking with regard to continued use of Wood's solution as discussed in Appendix A. Consequently, it is unclear at the present time what a logical starting point and basis of reference might be for developing an analytical methodology for the partial-depth placement alternative with non-yielding RERSs.

4.5 APPLICATION TO GEOFOAM WALLS

As mentioned earlier in this monograph, what can be viewed as the lightweight-fill function carried to its logical conclusion with ERSs is what is referred to colloquially as a geofoam wall. This is illustrated conceptually in cross-section in Figure 4.10.

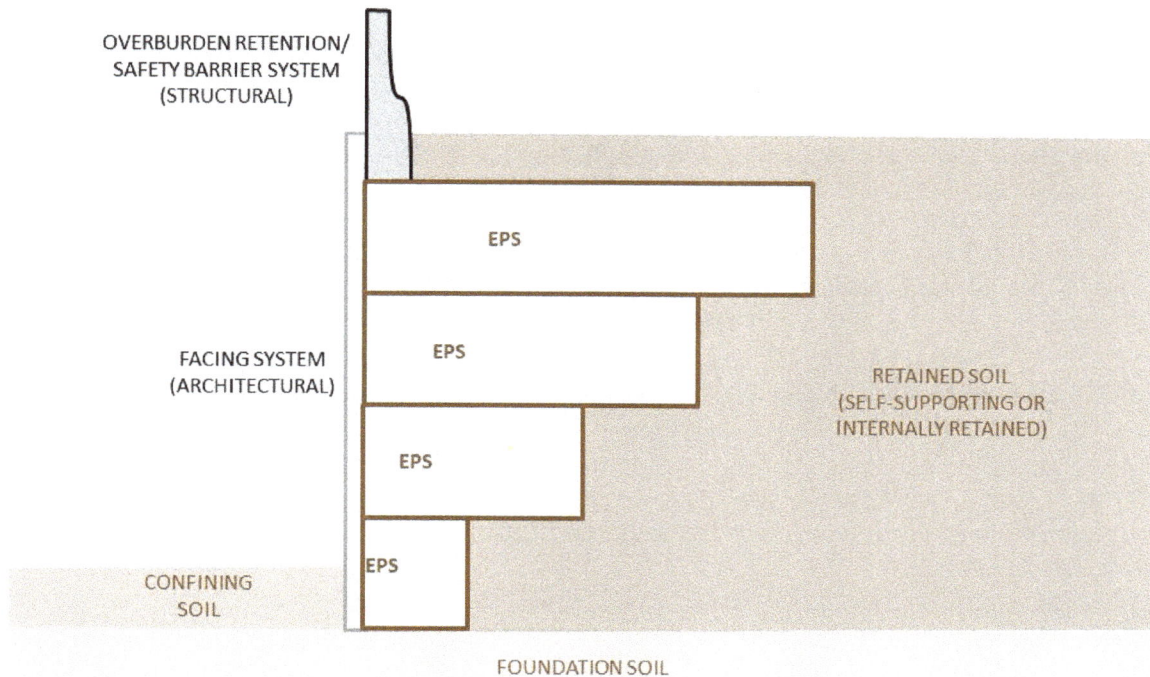

OVERBURDEN RETENTION/
SAFETY BARRIER SYSTEM
(STRUCTURAL)

EPS

FACING SYSTEM
(ARCHITECTURAL)

EPS

RETAINED SOIL
(SELF-SUPPORTING OR
INTERNALLY RETAINED)

EPS

EPS

CONFINING
SOIL

FOUNDATION SOIL

Figure 4.10. Geofoam-Wall Concept.

The key elements of a geofoam wall are:

- First and foremost, the retained soil adjacent to the assemblage of EPS blocks must be rendered self-retaining either by virtue of being at a sufficiently flat slope to be self-supporting through its inherent shear strength or internally retained using geosynthetics (MSE in a *reinforced soil slope*, RSS, configuration), soil nailing, etc.

- The soil placed on top of the assemblage of EPS blocks must also be retained. This is typically done with either a precast-PCC barrier (necessary in any case when a roadway is located above the geofoam wall), a MSEW, or other arrangement.

- The assemblage of EPS blocks is inherently both self-stable and provides a self-stable exposed vertical face to the overall geo-structure. Thus, whatever facing system is placed on the exposed face of the EPS blocks is, in principle, purely architectural in nature although in some cases, as a result of either choice or ignorance, an overly robust structural facing system has been used and assumed by the design professionals involved to provide (unnecessary) lateral support of some kind.

Despite the term 'geofoam wall' and the obvious implication that it is an ERS concept, the geo-structure illustrated in Figure 4.10 actually evolved from the original, circa-1970 lightweight-fill application of EPS-block geofoam as a material for embankments and related earthworks. The earliest embankments utilizing EPS blocks were slope-sided, a holdover from traditional soil embankments. When it was realized relatively quickly in the use of the geotechnology that EPS blocks are a true solid and thus inherently self-stable when stacked vertically, the design of earthworks incorporating EPS blocks rapidly morphed into vertical-faced earthworks being the basic design of choice for many reasons related to design and construction economy (less right-of-way required, less EPS required to purchase, fewer EPS blocks to place, etc.).

Nevertheless, even though the concept of geofoam walls evolved from traditional geofoam earthworks and not ERSs, they are mentioned here as they can also be viewed as a logical extension of the lightweight-fill function of ERSs. However, because the development history of geofoam walls is tied to earthwork applications, the analysis and design of geofoam walls has been adequately covered in publications related to the use of EPS-block geofoam for embankments (Stark et al. 2004a, 2004b) and is not covered further in this monograph.

4.6 APPLICATION TO FLEXIBLE EARTH-RETAINING STRUCTURES

While the concept of using the lightweight-fill function to reduce lateral pressures on ERSs is independent of the specific type of ERS, up to now the material presented in this chapter has focused on RERSs as they dominate actual as well as potential applications of this geotechnology.

There are a number of reasons why use of use of the lightweight-fill function is much less common and with much less potential for FERSs:

- Most FERS systems are installed partially, if not completely, from the ground surface, with the retained soil never exposed so that placement of lightweight material within the retained soil would require additional construction steps that may not be feasible due to site-access issues.

- Many FERS applications are temporary so that the cost benefit of using relatively expensive geofoam materials is much more limited and much less likely.

- Many permanent FERS applications, such as bulkheads, have most of the ERS below the permanent groundwater level so that buoyancy of geofoam materials is a problem.

The only consistent use of geofoam with FERSs of which that the writer is aware is with anchored bulkheads in Japan as shown conceptually in Figure 4.11 as well as in a photo on the back cover of this monograph. The writer is not aware of any simplified method for analyzing the behavior of such a bulkhead within the framework of traditional analysis and design methods that are based on the classical Free-Earth Support Method which would normally be used for anchored bulkheads. Therefore, project-specific numerical analyses using the FEM would appear to be necessary.

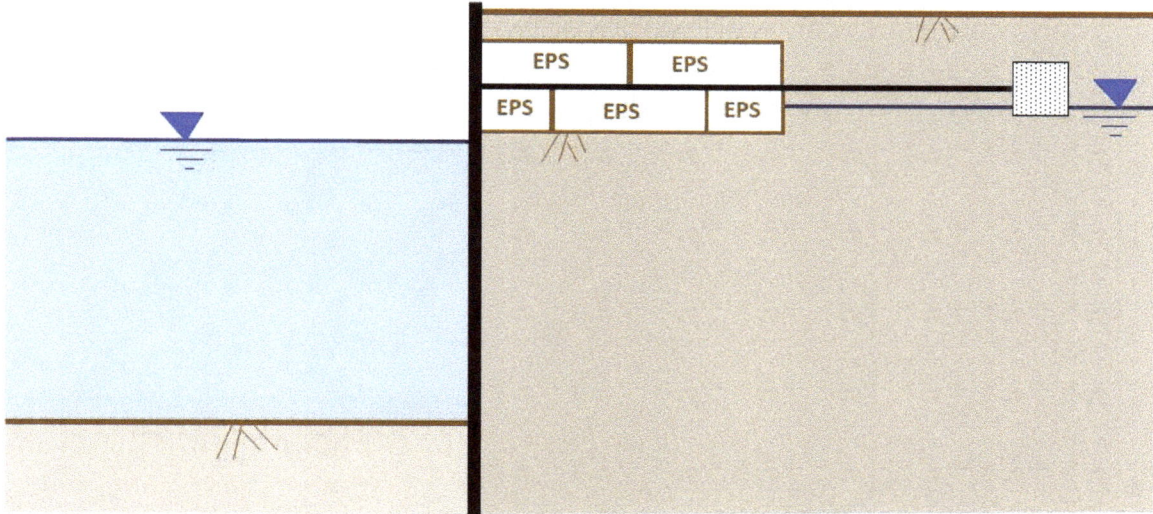

Figure 4.11. Lightweight-Fill Function - Flexible Earth-Retaining Structures Conceptual Application with Anchored Bulkheads.

This page intentionally left blank.

Chapter 5

Analysis and Design Based on the Compressible-Inclusion Function

5.1 INTRODUCTION

Available information indicates that the compressible-inclusion function has only been researched and used with RERSs to date. Therefore, this chapter will focus on RERS applications although a brief discussion of potential applications with FERSs is presented at the end of the chapter.

In the current states of knowledge and practice, the overall design process for the compressible-inclusion function applied to non-yielding and yielding RERSs is conceptually the same whether the Reduced Earth Pressure Wall (REP-Wall) or Zero Earth Pressure Wall (ZEP-Wall) concept is used (self-yielding RERSs are dealt with separately in Chapter 6). Therefore, this chapter is structured to first present the overall design process that is used regardless of specific application. Then the specifics of each application are addressed in separate sections.

As indicated in Chapter 3 and done in Chapter 4 for the lightweight-fill function, the contents of this chapter focus on the basic full-depth placement alternative as shown in Figures 3.7 and 3.9 for the REP- and ZEP-Wall concepts respectively. As with the lightweight-fill function, not only would full-depth placement be the choice in most applications, it is the alternative that has received the most research and development of analysis and design concepts to date.

However, the partial-depth placement alternative as shown in Figures 3.8 and 3.10 for the REP- and ZEP-Wall concepts respectively will receive limited presentation and discussion so that the current state of knowledge is defined for use in future research and development of this alternative. In addition, it turns out that the earliest known case history application of the compressible-inclusion function with a geofoam product used partial-depth placement so there is historical relevance and significance to discuss this as well.

5.2 ANALYSIS AND DESIGN BASICS

The essential, core design element of both REP- and ZEP-Wall applications is the chimney-orientation placement of a panel-shaped geosynthetic product between the inside face of the RERS and retained soil as shown in Figure 5.1 which is a composite of figures shown previously in Chapter 3. The primary, if not sole, geosynthetic function of this geosynthetic product is to act as the compressible inclusion. Assuming that the appropriate geofoam material has been specified as the key component of the geosynthetic product (a multifunctional geocomposite in most applications), this panel is typically relatively thin, of the order of 300 mm/12 in or less.

Note that the placement of geosynthetic tensile reinforcement within the retained soil is always optional and not a necessary component of the compressible-inclusion function. The sole purpose of using reinforcements is to increase the lateral stiffness of the retained soil so that there is greater lateral force reduction for a given magnitude of geofoam compression and concomitant lateral displacement of the retained soil.

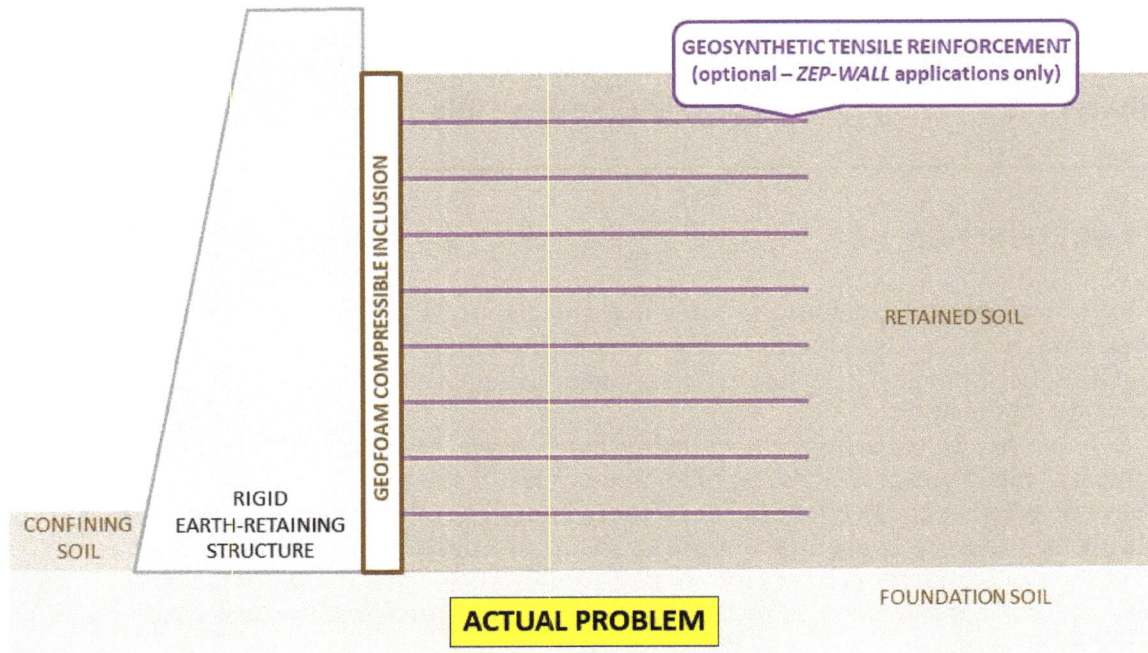

Figure 5.1. Generic Compressible-Inclusion Application – Basic Physical Components.

The single most important point to highlight before getting into the mechanics and historical evolution of past and present analysis and design methodologies for this function is that they have always been force-/stress-based, with consideration of displacements a necessary but secondary component. Research performed by the writer in preparation of this monograph suggests that a displacement-/strain-based approach and basis for analytical methodologies might be more appropriate and thus worthy of future research attention. This is addressed later in this chapter.

In any event, whether the retained soil is geosynthetically reinforced or not, the primary required outcome of the design process is very straightforward and simple in principle and involves 'sizing', i.e. determining the required thickness dimension, of the compressible inclusion for the desired level of lateral-force reduction. Unfortunately, experience to date indicates that "straightforward and simple" has not always translated into reality as it should, with significant conceptual errors made by both practitioners and researchers alike (Horvath 2010b).

The source of this disconnect appears to be that design professionals do not understand and appreciate several key precepts involving geofoam materials and products:

- the mechanical (stress-strain-time-temperature) properties of geofoam materials and products are, in general, very complex;

- from a material-property perspective, the compressible-inclusion function is fundamentally and profoundly different from the lightweight-fill function with which most design professionals are presumably much more familiar, at least at the present time. Specifically, the compressible-inclusion function is one of the few large-strain geofoam functions where the geofoam product is designed to be relatively compressible and can undergo compressive strains of the order of several or even tens of percent under service-load conditions. On the other hand, almost all other geofoam functions, including lightweight-fill, are small-strain functions where the design goal is completely the opposite, i.e. the geofoam product must be relatively stiff with operational compressive-strain levels typically less than 1%. Consequently, and very importantly, the well-known and widely published quasi-linear small-strain stiffness properties of EPS-block that are applicable to lightweight fills cannot be used for compressible-inclusion applications which are governed by highly non-linear large-strain properties (Horvath 2010b); and

- there is no, one geofoam material or product that is optimal for both large- and small-strain functional applications. Rather, geofoam products and the materials that comprise them need to be targeted to a specific geofoam function and functional application.

That there is no 'one-size-fits-all' geofoam material or product is something that was not appreciated in the early days (circa 1970s and 1980s) of geofoam development. For example, design professionals in then-technologically pioneering countries such as Norway and Japan thought that EPS-block of the same 'magic' density of 20 kg/m³ (1.25 pcf) could be used efficiently in almost every geofoam application from lightweight fill in earthworks to compressible inclusions over buried box culverts. Unfortunately, this simplistic thinking propagated around the world and persists to some extent to the present. As a result, when geofoam designs do not perform as intended there is a tendency to blame the concept when in reality the design professional is to blame for not correctly understanding the behavior of geofoam materials and products and, as a result, either using an inappropriate product in the faulted application or not interpreting the product's material behavior correctly.

In summary, the single most important aspect of using the compressible-inclusion function in practice is not only to use an application-appropriate geofoam material and product but, more importantly, to use its relevant mechanical properties in any analysis and design. This is hopefully understood more clearly by understanding how an actual application as shown in Figure 5.1 is physically visualized and modeled.

This is shown in Figure 5.2 which depicts how the lateral force vs. displacement behavior of both the compressible inclusion and retained soil (which may or may not be geosynthetically reinforced) can each be visualized as axial springs oriented horizontally and acting in series. These two springs are supported on the RERS which can act as either a rigid base (if non-yielding) or a displacing base (if yielding).

As depicted in the upper portion of Figure 5.2, the geofoam 'spring' is assumed to be initially unstressed and at maximum extension at its time of placement while the retained-soil 'spring' (whose stiffness can be significantly influenced by the presence of geosynthetic tensile reinforcement) is assumed to initially be at maximum stress based on the at-rest stress state with compaction-induced stresses included if desired. As depicted in the lower portion of Figure 5.2, the dual-spring system reaches equilibrium at some magnitude of lateral displacement of the compressible inclusion-retained soil interface at which compressive stresses in the geofoam (which increase from zero) and retained soil (which decrease from the at-rest value) are equal. Note that this lateral displacement is due to compression of the geofoam product alone (in the case of a non-yielding RERS) or compression of the geofoam product plus some lateral displacement of the ERS (in the case of a yielding RERS). In any event, the point emphasized here is that using the correct stiffness of the geofoam 'spring' is paramount if the problem is to be analyzed or designed correctly. This relates to the preceding comments about design professionals often using the incorrect stiffness of the geofoam 'spring' in their analytical models.

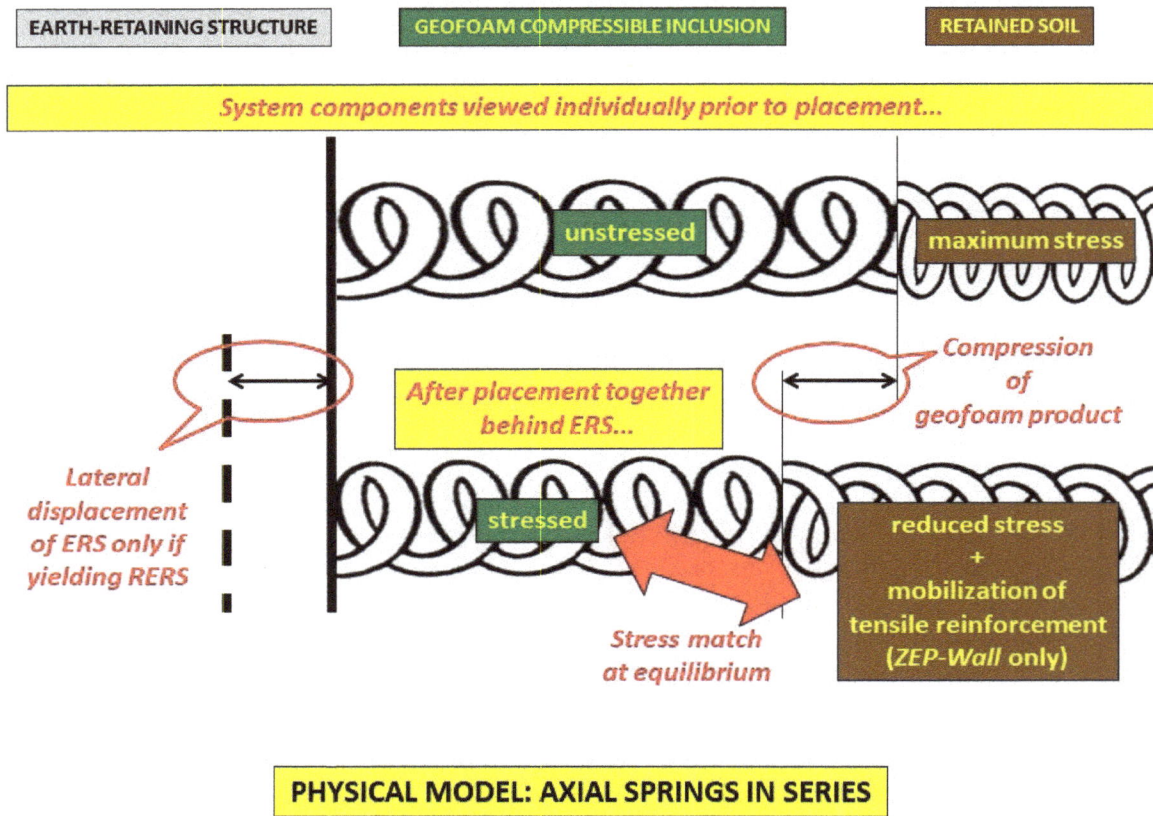

Figure 5.2. Generic Compressible-Inclusion Application - Physical Visualization and Model.

The physical visualization and model shown in Figure 5.2 can be translated into a mathematical model and concomitant solution as shown in Figures 5.3 and 5.4 for non-yielding and yielding RERSs respectively. Note that in these figures the force-displacement behavior of the problem components (the springs shown in Figure 5.2) is depicted in a generic curvilinear fashion (the stiffnesses of the springs are simply the slopes of the curves), although as will be seen subsequently linear or bi-linear approximations have often been used in routine practice for simplicity.

Note also that the ordinate (vertical axis) in Figures 5.3 and 5.4 can be scaled in different ways depending on the desired parameter used to illustrate a particular point. As will be seen subsequently, this parameter may be:

- lateral earth pressure,

- lateral earth pressure non-dimensionalized using atmospheric pressure, or

- a dimensionless parameter first defined by the writer in Horvath (2000) called the *relative lateral resultant force*, β, that is defined in Appendix B and used extensively in this monograph.

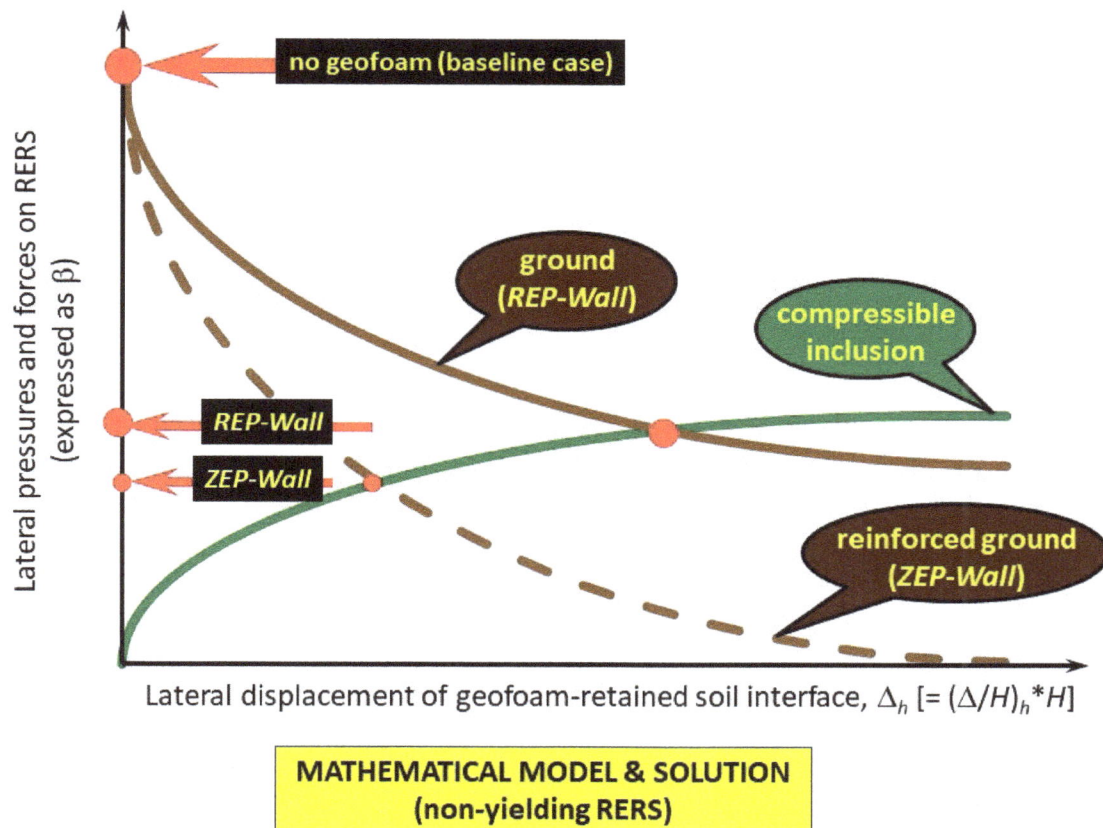

Figure 5.3. Generic Compressible-Inclusion Application - Mathematical Model and Solution/Non-Yielding RERS Case.

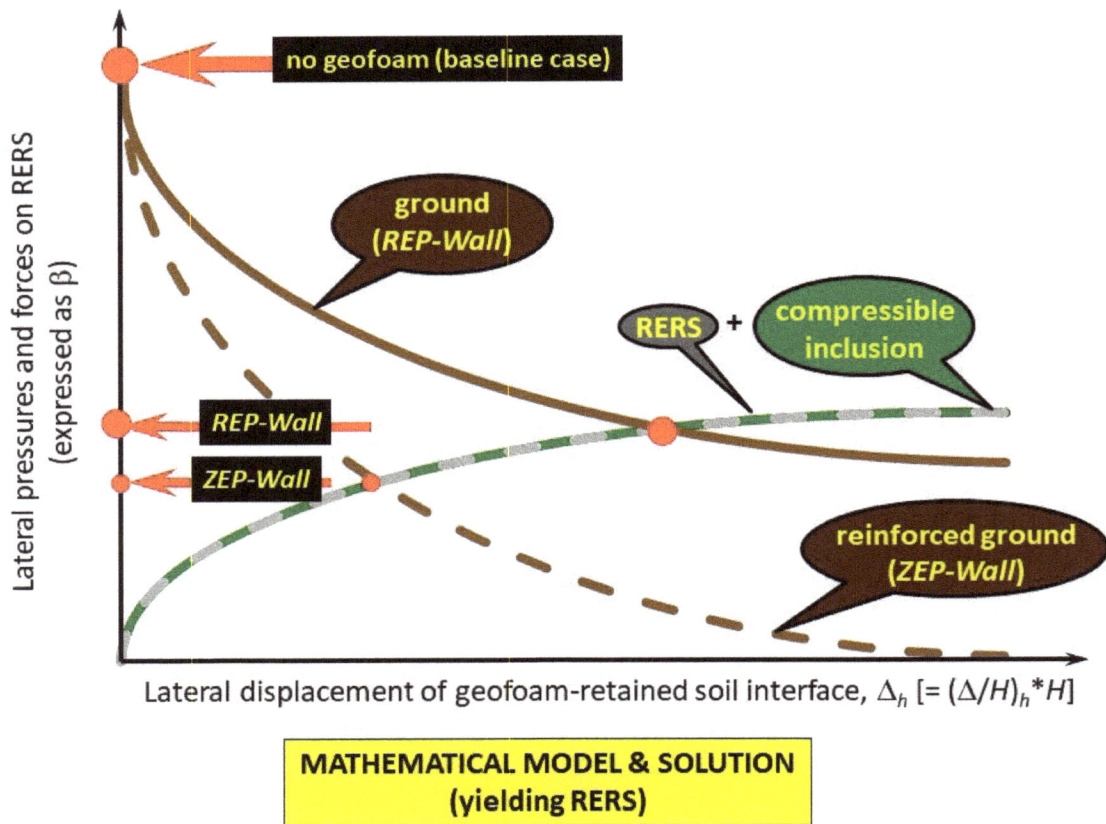

Figure 5.4. Generic Compressible-Inclusion Application -
Mathematical Model and Solution/Yielding RERS Case.

5.3 OVERVIEW OF THE DESIGN PROCESS

There are some significant differences in the overall design process when using the compressible-inclusion function compared to that discussed previously in Chapter 4 for the lightweight-fill function. To begin with, for the lightweight-fill function there is never a single 'correct' answer in terms of the required value of the angle θ^* for the stair-stepped geometry of EPS blocks and concomitant lateral pressure reduction. As shown conceptually in Figure 4.1, the optimal design should, in principle, always based on a project-specific cost vs. benefit assessment.

On the other hand, for the compressible-inclusion function there is always a unique 'correct' answer, at least for the REP-Wall concept. As shown in Figures 5.3. and 5.4, the resultant lateral earth force from an unreinforced retained-soil mass will always reach and maintain a minimum value that is typically defined using Coulomb's solution for the active earth pressure state based on constant-volume (critical-state) shear-strength. Thus, as will be seen subsequently in this chapter, design using the REP-Wall concept consists of choosing the appropriate geofoam product stiffness (for non-yielding RERSs) or combined ERS + geofoam stiffness (for yielding RERSs) that results in lateral displacement, Δ_h, equal to Δ_a where Δ_a is the lateral displacement required to mobilize the active earth pressure state. Note that Δ_a is typically expressed in the literature as the dimensionless ratio Δ_a/H (= $(\Delta/H)_a$) where H is the geotechnical height of the ERS.

As an aside, as will be seen and discussed at length, care must be taken when using tabulated $(\Delta/H)_a$ values that appear in text and reference books. This is because these values are based on very specific lateral-displacement modes that, as it turns out, do not correlate well with the actual patterns of lateral displacements observed along the interface between a geofoam compressible inclusion and retained soil whether unreinforced or reinforced. Consequently, judgment and subjectivity need to be exercised when using published $(\Delta/H)_a$ values with the compressible-inclusion function.

As a further aside and with reference to Figure 5.4, experience to date has been that use of the REP-Wall concept is usually not economically justified with yielding RERSs (which are, by definition, assumed to be inherently self-capable of mobilizing the active earth pressure state within the retained soil based on ERS displacement alone) unless one wants to ensure for structural-design purposes that the active state is achieved even under service loads[65]. Note, however, that because the ZEP-Wall concept can reduce lateral earth pressures below the active state, this concept is economically justified with yielding RERSs in addition to non-yielding RERSs.

Continuing on, note that although there is always a single-value 'correct' answer when using the REP-Wall concept there is still an economic (cost-benefit) aspect of design involving the compressible-inclusion function. This involves investigation as to whether or not further reductions in lateral earth pressures beyond the active state by using the ZEP-Wall concept are justified. Thus, an exercise similar to that shown qualitatively in Figure 4.1 is still worthwhile to carry out but in this case to evaluate the cost of the geofoam compressible inclusion plus tensile reinforcement vs. cost of the RERS (for new construction) or improve performance (for an existing ERS).

There are some design aspects that the compressible-inclusion function has in common with the lightweight-fill function. The most basic and fundamental one is assessment of a baseline case assuming no geosynthetics usage and using the traditional at-rest or active earth pressure state as appropriate depending on whether the RERS is non-

---

[65] See the discussion in Chapter 2 for an explanation of why this might be desirable.

yielding or yielding. This exercise will provide a technical performance and cost reference against which designs using geosynthetics can be compared.

In addition, where and when appropriate a separate design is also performed for seismic loads in addition to gravity loads. The results from the two designs are then compared to see which one governs the overall final design for the project. Note that although seismic loads are always larger in magnitude than gravity loads, they may not govern the final design because different safety margins (whether ASD safety factors or LRFD load and resistance factors) are often used for these different load cases. So, it is not known a priori whether gravity loads with a larger required margin of safety or seismic loads with a lower required margin of safety produces the more-critical final design.

In recent years the application of the REP-Wall concept specifically to reduce seismic loads on a non-yielding RERS has been called a *seismic buffer*[66]. This term did not originate with the writer and its origin is unclear to the writer. The only reason the term is mentioned here is to point out that there is nothing special or unique about geofoam used as a 'seismic buffer' as it is simply the basic REP-Wall concept under seismic loading.

5.4 ANALYTICAL METHODS FOR THE REDUCED EARTH PRESSURE WALL (REP-WALL) CONCEPT

5.4.1 Introduction

The key considerations when using the REP-Wall concept in practice are:

- Full- vs. partial-depth placement. As with the lightweight-fill function, full-depth placement is the first choice in most applications, whether for new construction or existing RERS, so almost all research to date has focused on this. Consequently, this chapter will focus on this alternative although partial-depth placement will be addressed to some degree.

- Gravity vs. seismic loading. The broad, overall need to consider both gravity and seismic loading where and when appropriate and to develop a design based on the more critical of the two is the same as that for the lightweight-fill function. However, an interesting development in recent years is that there have been projects (Inglis et al. 1996) where the primary, if not sole, reason for using a compressible inclusion was to reduce seismic loads. This very focused application interest has led to an uptick in research and the concomitant coining and subsequent rise in the use of the aforementioned term 'seismic buffer'. Unfortunately, in the writer's opinion much of the research into seismic buffers has been flawed to varying degrees (sometimes to the point of rendering the presented research of questionable value) by an improper understanding and assessment of the operational stiffness of the geofoam material and product used. This was noted earlier in this chapter and was discussed in detail in Horvath (2010b). The net result is that the current state of practice for analysis and design is more advanced for gravity loading as opposed to seismic loading despite the research resources expended on the latter.

- Load duration. Because the compressible-inclusion function is a large-strain geofoam function, this means that independent of the specific geofoam material and product used the operational strain levels will always be within the range where both plastic

---

[66] An additional, much less common synonymous term is *seismic isolation*.

(non-recoverable) strains under transient and cyclic loading (such as live and seismic loads) and creep strains under sustained loading (such as gravity loads) are significant[67]. Routine design for both transient/cyclic and sustained loading is generally based on the rapid-loading stiffness of the geofoam product (which is the de facto standard stiffness property both measured and reported in manufacturer's literature) because this generally reflects the most conservative upper-bound estimate of the geofoam product stiffness that will yield the most conservative design thickness for the compressible inclusion. However, it is considered good practice to check the tentative design using a temporally appropriate long-term stiffness that reflects consideration of material creep. As a minimum, this is to verify that the long-term strains of the compressible inclusion are within the operational limits of the geofoam material and product chosen. In addition, in more-advanced analyses using numerical methods such as the FEM the adequacy of overall system behavior in terms of soil stresses, ground-surface settlement, etc. can be checked for long-term conditions.

The remainder of this chapter is structured along the lines of these three considerations.

5.4.2 Full-Depth Placement Alternative

5.4.2.1 Gravity Loading

The development of relatively simple analytical methods for sizing compressible inclusions in full-depth REP-Wall applications dates back to at least the mid-1980s. From the start, the assumption was that the maximum reduction in lateral earth pressure that could be achieved with such a REP-Wall application was to some theoretical, well-defined, non-zero limit for both non-yielding and yielding RERSs as shown conceptually in Figures 5.3 and 5.4 respectively. Specifically, and as noted previously, it has always been assumed that this limiting lateral earth pressure is defined by the active state, regardless of whether the ERS is non-yielding or yielding. To date, there has been no compelling reason to assume otherwise although, as will be seen, the assumed distribution of lateral earth pressures associated with the active state has undergone significant change over time as a better understanding of the geomechanics involved has evolved as a result of research.

Consequently and also as noted previously, implicit in the development of REP-Wall analytical methods is that they, and thus the REP-Wall concept in general, are of use only for non-yielding RERSs where there is a clearly defined benefit in reducing lateral earth pressures from the at-rest state (which would be the logical basis for design in the absence of a compressible inclusion) to the active state. This reduction in resultant lateral earth force is typically of the order of 35% to 40% which is relatively significant. Because non-yielding RERSs do not, by definition, have any primary geotechnical ULS modes, the benefit of this reduction is entirely structural. This benefit can be realized as a reduced cost for new construction or increased structural safety margin for existing RERSs.

Although the analytical methodologies presented in this section focus on applications with non-yielding RERSs, there is, in principle, some potential benefit to using

---

[67] This is not an issue with the lightweight-fill function where the overarching design criterion in every application is to keep compressive stresses on the EPS blocks both within the quasi-elastic range so that plastic strains do not occur and also below the level at which creep strains become significant.

116

the REP-Wall concept with yielding RERSs as noted previously. Although yielding RERSs are, by definition, assumed to be inherently capable of mobilizing the active state on their own without a compressible inclusion, in many, perhaps most, actual cases they do not do so or do so completely under service-load conditions due to excess geotechnical capacity/resistance ('safety') that is intentionally built into the RERS-ground system. This margin-of-safety is very often greater than that assumed 'on paper' because of conservative assumptions regarding soil shear-strength parameters.

As discussed at length in Chapter 2, the only potential issue with incomplete mobilization of the active earth-pressure state for yielding RERSs is that the margin-of-safety of the structural components of the RERS is less than intended or believed as these components are typically designed based on active earth pressures. Thus, using a compressible inclusion with a yielding RERS would add to the 'spring' of inherent RERS displacement as shown conceptually in Figures 5.2 and 5.4 and ensure that the structural components of the RERS are always subjected to active earth pressures. The use of the REP-Wall concept with yielding RERSs is explored to a limited degree subsequently.

As noted above, the analytical methodology for REP-Wall application to non-yielding RERSs has gone through several iterations since the 1980s as understanding of the soil behavior within the retained soil has evolved as the result of ongoing research. Because additional, future modifications are likely to broaden the analytical methodology to include factors such as:

- yielding RERSs,

- the presence of surface-surcharge loads,

- initial compaction-induced lateral pressures, and possibly other considerations

it is useful to trace the evolution of this methodology to date in summary form.

For this purpose, reference will be made to Figure 5.5. In this figure, the lateral earth pressures corresponding to the normally consolidated at-rest state, p_o, are shown with the red triangle and the horizontal component of the lateral earth pressures corresponding to the active state, p_{ah}, with the green triangle. Both earth pressure states are depicted using the traditional hydrostatic distribution. The curve depicted with a yellow dashed line will be defined subsequently. As has been used consistently throughout this monograph, H is the geotechnical height of the ERS.

To begin with, all versions of the analytical methodology for REP-Walls developed to date share the same analytical model and several common assumptions and elements that are illustrated in Figure 5.6:

- The force-displacement behaviors for the key system components are shown qualitatively in this figure. The difference from what was shown more generally in Figure 5.3 is that the compressible inclusion is assumed to have linear behavior and the retained soil bi-linear behavior as opposed to the generalized curvilinear behavior for each shown in Figure 5.3. The assumption of linear behavior for the geofoam product is actually quite accurate, assuming that R-EPS (the most technically efficient material for almost all REP-Wall applications) or a geocomposite that uses R-EPS as its primary component is used. The bi-linear assumption for retained-soil behavior is also reasonable, assuming that the non-linear 'dip' that occurs as the shear strength of the soil transitions from peak to constant-volume (critical-state) conditions is neglected.

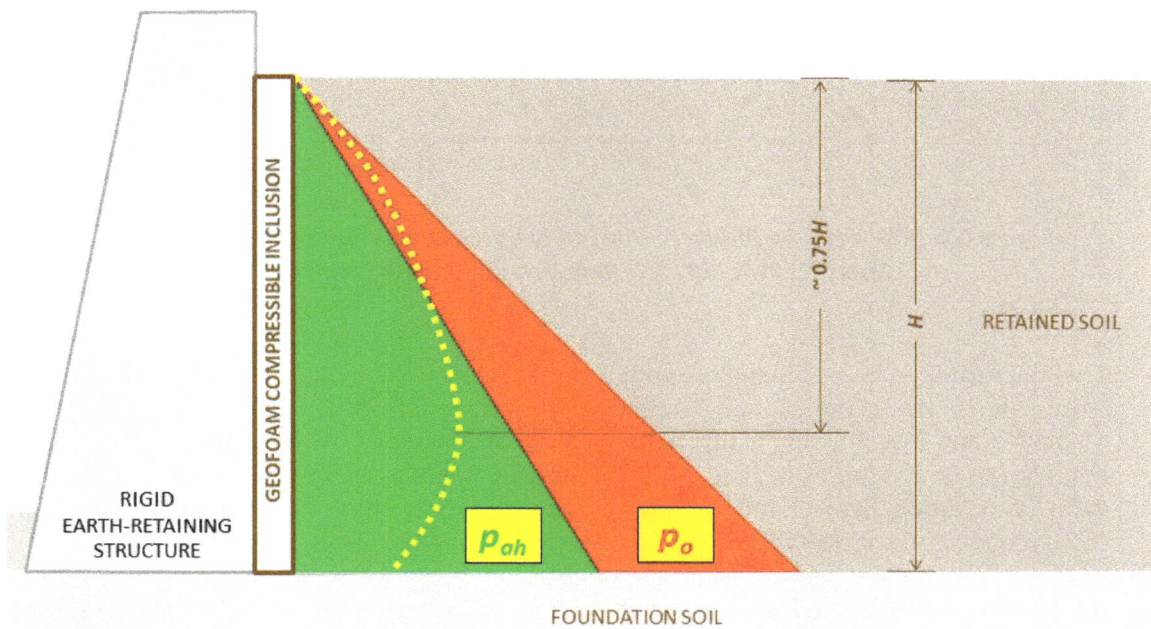

Figure 5.5. REP-Wall Application – Assumed Lateral Earth Pressures for Analytical Model Development.

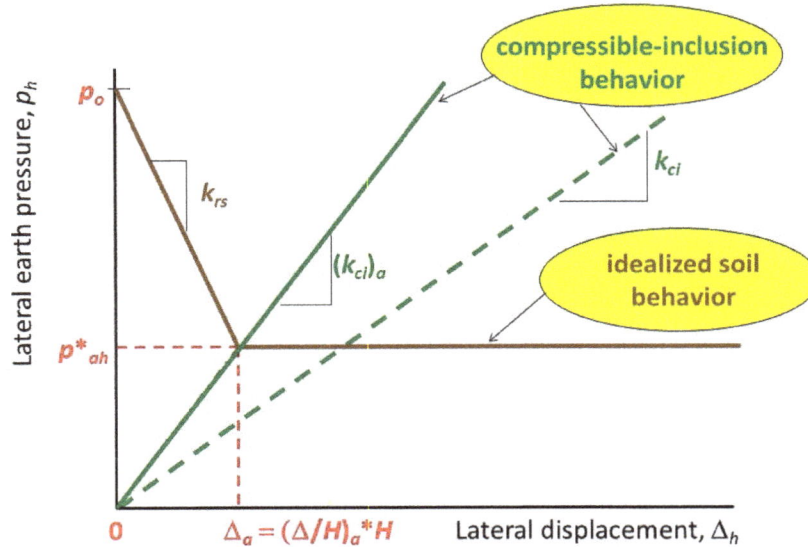

Figure 5.6. REP-Wall Application - Assumed Force-Displacement Behaviors for Analytical Model Development for Gravity Loading.

- The magnitude of lateral displacement, Δ_a, at which the theoretical active earth pressure state within the retained soil is mobilized, starting from an assumed zero lateral displacement and the normally consolidated at-rest state, is calculated for a given application and value of H using the dimensionless ratio $(\Delta/H)_a$. The magnitude of this ratio is assumed beforehand, generally based on published values such as found in Clough and Duncan (1991).

- Δ_a together with p_o and p^*_{ah} (the latter is a particular value of p_{ah} that is defined subsequently) can be used to calculate an equivalent axial-spring stiffness (see Figure 5.2), k_{rs}, for the retained soil. However, this is not done routinely as k_{rs} has not been used explicitly for any of the analytical methodologies developed to date.

- The generic spring stiffness of a compressible inclusion is shown graphically by the green dashed line in Figure 5.6 and conceptually in Figure 5.2 is defined as:

$$k_{ci} = E_{ci}/t_{ci} \qquad (5.1)$$

where E_{ci} is the operational value of Young's modulus of the geofoam product[68] and t_{ci} is the thickness of the geofoam product, i.e. the dimension in the direction of load application (horizontal in this case). Note that although k_{ci} has dimensions of force per

---

[68] Note that the modulus of the geofoam <u>product</u> is not necessarily the same as the modulus of the geofoam <u>material</u> that comprises the product. First of all, the product may be a geocomposite that consists of two or more different materials, each with their own modulus. In this case, the overall product modulus is the composite modulus of two springs in series. Alternatively, the product might consist of one material but that material might have holes through it (actually used in some seismic-buffer research) or be otherwise discontinuous in which case the operative modulus of the overall product is something less than that of the solid material between the holes or discontinuities.

unit length cubed they are better visualized as force per unit length (the dimensions of an axial spring) per unit area (i.e. length squared) of the compressible inclusion.

- The minimum required thickness of a compressible inclusion required to mobilize the active earth pressure state within the retained soil in a specific application is defined as $(t_{ci})_a$ and is determined by matching k_{ci} to a lateral earth pressure, p^*_{ah}, for an assumed magnitude of spring compression equal to Δ_a. The specific value of k_{ci} at which this match occurs is defined as:

$$(k_{ci})_a = E_{ci}/(t_{ci})_a \qquad (5.2)$$

and is illustrated by the green solid line in Figure 5.6.

- The way in which the various evolutionary generations of REP-Wall analytical methodologies differ is in the specific magnitude of p^*_{ah} used as the match point between the retained-soil 'spring' and compressible-inclusion 'spring'. However, the basic equation that is solved is the same in all cases. The overall equation to be solved is:

$$p^*_{ah} = (k_{ci})_a \cdot \Delta_a = [E_{ci}/(t_{ci})_a] \cdot \Delta_a \qquad (5.3)$$

which produces:

$$(t_{ci})_a = (E_{ci} \cdot \Delta_a)/p^*_{ah}. \qquad (5.4)$$

It is useful for the discussion that follows to express Equation 5.4 as:

$$\left(t_{ci}\right)_a = \frac{E_{ci} \cdot \Delta_a}{K_h \cdot \gamma_{eff} \cdot K_\sigma \cdot H} = \frac{E_{ci} \cdot \left(\Delta/H\right)_a}{K_\sigma \cdot K_h \cdot \gamma_{eff}}. \qquad (5.5)$$

Equation 5.5 essentially expands the denominator on the right-hand side of Equation 5.4 (i.e. the p^*_{ah} term) into its fundamental components: the vertical effective stress (= $\gamma_{eff} \cdot z$) at some depth z (= $K_\sigma \cdot H$, with K_σ being an assumed, dimensionless parameter within the limits $0 < K_\sigma \le 1$) within the retained soil, times a lateral earth pressure coefficient, K_h, that is also assumed.

Having defined the basic and common elements of the overall analytical methodology for design of a REP-Wall application with non-yielding RERSs under gravity loading with no surface surcharge that has been used since at least the mid-1980s to the present, the various evolutionary versions of this methodology over the last three decades are now addressed. As will be seen, all versions use the basic format of Equation 5.5 and vary only in the specific values of K_σ and K_h assumed:

1. Partos and Kazaniwsky (1987) deserve credit for defining the first version of the analytical methodology although the project on which they applied it was actually a partial-depth application that is discussed in some detail in a later section. Their seminal version of the methodology is thus referred to as Version 1.0 in various earlier published works by the writer. Partos and Kazaniwsky assumed $p^*_{ah} = p_o$ at a depth H, i.e. the bottom of the at-rest pressure triangle shown in Figure 5.5. Within the context of Equation 5.5, this means that they assumed $K_h = K_o$ and $K_\sigma = 1$. Note that using p_o and Δ_a

together is theoretically inconsistent as is obvious from Figure 5.6. The writer believes the reason this was done was the authors' intuitive belief that using the largest value of lateral earth pressure possible would produce the most conservative estimate of compressible-inclusion thickness. However, Equation 5.4 clearly shows that the opposite is true, i.e. the greater the magnitude of p^*_{ah} the smaller (and less conservative) the calculated value of $(t_{ci})_a$ will be. Nevertheless, the case history to which Partos and Kazaniwsky applied their methodology was apparently successful in that it met their design objectives of lateral earth pressure reduction (as confirmed by field instrumentation). However, in the writer's opinion this was largely (and fortuitously) due to three important facts. One, the authors used what appears to be relatively conservative soil properties in their lateral earth pressure calculations so the calculated pressures were greater than the actual to begin with. Two, they applied their engineering experience and judgement, most likely due to the then-novel nature of the application, and arbitrarily increased the calculated thickness of the compressible inclusion by a not-insignificant amount. Specifically, they used a 10-inch (254-mm) thick panel of geofoam when their calculations indicated that 6 inches (152 mm) would suffice. Three, creep of the compressible inclusion worked in their favor as the long-term stiffness of the specific geofoam product used was, based on independent research performed years later by the writer, approximately one-half that of the rapid-loading stiffness that Partos and Kazaniwsky used in their design calculations. Thus, in the long-term, the compressible inclusion was behaviorally about twice as thick as they thought it was and more than three times thicker than their calculations indicated was required. When all these factors are considered, the net end result in terms of actual, long-term performance was satisfactory in terms of t_{ci}. This is, as already noted, quite fortuitous as subsequent research by the writer indicated that the Version 1.0 analytical method used by Partos and Kazaniwsky can underestimate the required value of $(t_{ci})_a$ by a factor of more than two (specifically, 2.7 for the problem analyzed in Horvath (2000) as summarized in Table 3.5 of that report).

2. For the sake of presenting as complete a historical record on the subject as is known to the writer, mention is made at this point of fundamental research into the use of compressible inclusions that was conducted in the U.K. in the latter decades of the 20th century that resulted in numerous publications in the late 1980s and 1990s, i.e. around the same time and thereafter of the publication of the Partos and Kazaniwsky conference paper. While this U.K. research, along with the work of Partos and Kazaniwsky, was intellectually inspirational to the writer, the U.K. researchers did not carry through on their work in terms of exploring commercial materials and products that could be used as a compressible inclusion in practical, full-scale applications, nor did they develop any analytical methodology for quantifying the benefit of using a compressible inclusion or calculating $(t_{ci})_a$. Thus this U.K. research was of a useful, proof-of-concept nature but with little-to-no follow-on or end-use value in the writer's opinion. However, it does appear to have been the inspiration for follow-on research in Canada that was published by Karpurapu and Bathurst (1992). Their work, which is outlined in Appendix B of this monograph, produced a methodology that can be used to estimate $(t_{ci})_a$ although it does not follow the explicit form of Equation 5.4 so is considered an independent, dead-end analytical methodology for the purposes of this monograph. The writer's research to date indicates that the results in Karpurapu and Bathurst are more or less the same as Version 1.1 of the analytical methodology defined by Equation 5.4 that is discussed in the following item.

3. The writer's research into geofoam compressible inclusions began in the latter part of 1987 and, as previously noted, was inspired collectively by the aforementioned published work of Partos and Kazaniwsky as well as that in the U.K. that was ongoing contemporaneously. The writer's initial, years-long efforts were focused on finding commercially available materials that could be used for compressible inclusions as a theoretical concept that could not be implemented in practice (the path followed by researchers in the U.K.) made little sense to the writer. However, the writer also devoted time to a critical assessment and subsequent improvement of the analytical methodology used by Partos and Kazaniwsky. Based on initial research, the writer concluded that $p^*_{ah} = p_{ah}$ at a depth H, i.e. the bottom of the green triangle shown in Figure 5.5, was more rational to use compared to that used by Partos and Kazaniwsky as it properly linked p^*_{ah} and Δ_a. This results in a calculated value of $(t_{ci})_a$ that is greater as can readily be seen from Equation 5.4. Within the context of Equation 5.5, the change made by the writer means $K_h = K_a \cdot \cos\delta$ (δ in this case is the interface friction angle of the retained soil against the geofoam product which is typically taken to be the ϕ angle of the retained soil) in lieu of $K_h = K_o$ as used in Version 1.0 and $K_\sigma = 1$ (unchanged from Version 1.0). However, due to publication delays of several years that were beyond the writer's control and were long even by scholarly journal standards in the pre-Internet era, the writer's revised analytical methodology was not published until 1997 (Horvath 1997b, republished in Horvath 1998b) and is referred to as Version 1.1 in earlier published works by the writer. As noted in Item 2, the results obtained using Version 1.1 are more or less identical to those obtained using the results presented in Karpurapu and Bathurst (1992) which is not surprising as these authors assumed the classical hydrostatic (triangular) distribution of active earth pressures in their work.

4. Continued research by the writer, primarily using FE analyses, indicated that the distribution of lateral earth pressure that develops when a full-depth compressible inclusion is used is always curved in shape as shown by the yellow dotted curve in Figure 5.5. In hindsight, this is not surprising as this type of curved pressure distribution is now recognized as being broadly consistent with theoretical results from horizontal-arching theory under the mode of pure translation. When horizontal arching develops, the lateral earth pressure typically peaks at a depth of approximately $0.75 \cdot H$ as shown in Figure 5.5. As a result, the writer's Version 1.1 method described in Item 3 above was subsequently amended to using $p^*_{ah} = p_{ah}$ at a depth $0.75 \cdot H$, i.e. the depth of the peak of the horizontal component of the active earth pressure curve estimated using horizontal-arching theory. Within the context of Equation 5.5, this means $K_h = K_a \cdot \cos\delta$ (unchanged from Version 1.1) and $K_\sigma = 0.75$ (reduced from 1.0 in both Versions 1.1 and 1.0). The rationale for this revision was presented and discussed in Horvath (2000) where it was referred to as Version 1.2, a nomenclature also used in subsequent published works by the writer.

5. Also presented in Appendix A of Horvath (2000) is an alternative to the developmental methodology summarized in Equations 5.1 through 5.5. This alternative is based on using the resultant earth <u>force</u> as opposed to a peak value of lateral earth <u>pressure</u> when matching the compressible-inclusion and retained-soil stiffnesses shown in Figure 5.6. More specifically, the relative lateral resultant force, β, that was defined previously is used. It is of relevance to note that this alternative derivation for the basic case under discussion here (full-depth REP-Wall, gravity loading, no surface surcharge) yielded the exact same results as Version 1.2 of the original derivational methodology that was

discussed above in Item 4. Nevertheless, this alternative force-based derivation was assigned a unique nomenclature, Version 2.0, as it was based on an entirely different solution approach to the problem. As noted in Horvath (2000), the primary reason for developing Version 2.0 was to have an analytical methodology for the REP-Wall application that could easily be extended to the ZEP-Wall application so that that there was a common analytical basis for both versions of compressible-inclusion applications. As will be seen later in this chapter, in the ZEP-Wall application the distribution of lateral earth pressures is not geometrically simple and straightforward as shown in Figure 5.5 for the REP-Wall case so it is far easier to use resultant forces, not distributed lateral pressures, as the logical basis for a simplified analytical model for the ZEP-Wall case. An additional benefit of using resultant force as opposed to lateral pressures is that it allows a straightforward extension to problem variations such as considering surface surcharges as well as seismic loading both without and with surface surcharges.

In summary up to this point, the current state of knowledge results in the following equation for sizing the minimum thickness, $(t_{ci})_a$, of the compressible inclusion required to mobilize the active earth pressure state within the retained soil, independent of whether a traditional pressure (stress)-based approach (Item 4 above) or more recent force-based approach (Item 5 above) is used in derivation:

$$(t_{ci})_a = \frac{E_{ci} \cdot (\Delta / H)_a}{0.75 \cdot K_a \cdot \cos\delta \cdot \gamma_{eff}} \qquad (5.6)$$

where it is emphasized that E_{ci} must be the secant Young's modulus of the geofoam <u>product</u> over the stress range relevant to the application and K_a must be the Coulomb coefficient of active earth pressure from Equation A.4 in Appendix A. Note that Equation 5.6 applies only to non-yielding RERSs under gravity loading and without a surface surcharge.

Although not a necessary element for the design process, it has proven useful in both research and practice to define and use two dimensionless parameters that relate to the stiffness (or its inverse, compressibility) of the compressible inclusion. The first parameter is defined using the notation K_{ci} and is called the *absolute compressible inclusion stiffness*[69]. It is defined mathematically as:

$$K_{ci} = \frac{E_{ci}}{t_{ci} \cdot \gamma_w} = \frac{k_{ci}}{\gamma_w} \qquad (5.7)$$

where γ_w is the unit weight of water[70] and all other variables have been defined previously. Any consistent set of units can be used for the variables in Equation 5.7.

Conceptually, K_{ci} is simply the equivalent axial-spring stiffness (dimensions of force per unit length) per unit area of the compressible inclusion as depicted in Figure 5.2 and defined previously in Equation 5.1 but non-dimensionalized for generality. The benefits of using this new parameter are several-fold:

---

[69] In the writer's earlier publications, the notation k_{ci} was used for this parameter. It is changed here to avoid conflict with the parameter defined in Equation 5.1.

[70] There is no physical significance to using this parameter here. It is used simply because it is a universally known material property with the dimensions of force per unit length cubed that are necessary for the non-dimensionalization process.

- Its inherent non-dimensional nature makes it attractive to use globally because it is not dependent on a particular system of units as is k_{ci}.

- It is both simpler to state in manufacturer's literature and for design professionals to use in analysis and design compared to dealing with two separate, dimensioned properties, i.e. product modulus, E_{ci}, and product thickness, t_{ci}.

- It simplifies and facilitates comparison between and among commercial products by design professionals because any commercial product can be defined using a single, dimensionless parameter.

- It is a unified problem parameter that allows research performed by different researchers using different compressible-inclusion materials and products to be easily and rationally assessed and compared within a common framework.

- It is unambiguous to measure, define, and use when a product consists of multiple materials and/or materials with intentional holes or other discontinuities. The former situation is quite common with commercial geocomposite products that are the products-of-choice in ERS applications. The latter situation, which to date has been more common in research, has created interpretative errors when the modulus of the intact material has been conflated with the modulus of the geometrically altered product (Horvath 2010b).

- It requires no more or more-complex laboratory testing for manufacturing quality control (MQC) and manufacturing quality assurance (MQA) purposes than would be required to measure product modulus.

- It is relatively easy and straightforward to measure in practice in a two-step process as it does not require independent measurement of product modulus and product thickness. Rather, a single representative specimen of the actual geofoam product is simply tested in uniaxial compression as would be necessary in any event to determine the product modulus. The first step in the calculation process is to determine k_{ci} which is simply the slope of the plotted curve of compressive normal stress vs. compressive normal displacement. The second step is to then calculate K_{ci} using Equation 5.6.

To illustrate the use of K_{ci}, Equation 5.6, which is the simplified solution for the basic REP-Wall case, can be replaced by combining this equation with Equation 5.7 to yield the largest allowable value of K_{ci} to mobilize the active earth pressure state within the retained soil, $(K_{ci})_a$, in a given application is:

$$(K_{ci})_a = \frac{E_{ci}}{(t_{ci})_a \cdot \gamma_w} = \frac{0.75 \cdot K_a \cdot \cos\delta \cdot \gamma_{eff}}{(\Delta/H)_a \cdot \gamma_w}. \tag{5.8}$$

Note again that if manufacturers' literature characterized compressible-inclusion products by stating their value of K_{ci} it would be a relatively simple matter in practice for a design professional to use Equation 5.8 to calculate the required $(K_{ci})_a$ for a given project application then easily 'shop' commercial geofoam products to find one with a $K_{ci} \le (K_{ci})_a$.

124

Moving on now to the other newly defined parameter, it uses the notation λ_{ci} and is called the *relative compressible inclusion stiffness*[71]. It is defined mathematically as:

$$\lambda_{ci} = \frac{E_{ci} \cdot H}{t_{ci} \cdot p_{atm}} = \frac{k_{ci} \cdot H}{p_{atm}} = \frac{K_{ci} \cdot H \cdot \gamma_w}{p_{atm}} \qquad (5.9)$$

where p_{atm} is atmospheric pressure (used here solely as a non-dimensionalization parameter) and all other variables have been defined previously.

Any consistent set of units can be used for the variables that define λ_{ci}. It follows that the largest allowable value of λ_{ci} to mobilize the active earth pressure state within the retained soil, $(\lambda_{ci})_a$, in a given application is:

$$(\lambda_{ci})_a = \frac{E_{ci} \cdot H}{(t_{ci})_a \cdot p_{atm}} = \frac{(K_{ci})_a \cdot H \cdot \gamma_w}{p_{atm}}. \qquad (5.10)$$

Conceptually, λ_{ci} is the axial-spring (visualized in Figure 5.2) stiffness per unit area of the compressible inclusion relative to the geotechnical height, H, of the ERS. The benefit and desirability of having this parameter in addition to the previously defined K_{ci} is that, as noted previously and shown conceptually in Figure 5.6, research has shown that the magnitude of lateral displacement, Δ_a, necessary to mobilize the active earth pressure state is linearly proportional to the geotechnical height, H, of an ERS and thus typically tabulated in the literature as non-dimensional $(\Delta/H)_a$ ratios for various types and consistencies of soil as noted previously. Therefore, it is logical to normalize the absolute stiffness of a specific compressible inclusion product (as defined by K_{ci}) to the geotechnical height of a specific ERS to define the relative stiffness of that product (as defined by λ_{ci}) in that particular application. Note that doing so clearly recognizes the fact that the same compressible-inclusion product (with a given value of K_{ci}) will have varying benefit depending on the specific ERS (as defined by its geotechnical height, H) with which it is used.

Exploring these thoughts further, note that the ratio $(H \cdot \gamma_w)/p_{atm}$ in Equation 5.10 is dimensionless and, for a given H, has the same value that is independent of the system of units used. Consequently, this ratio can be thought of as a *dimensionless height-correction factor, H_{ci}*, to relate the performance of a given compressible-inclusion product as defined by its unique value of K_{ci} to its performance in a given ERS application as defined by the parameter λ_{ci}.

A further illustration of the utility of the λ_{ci} parameter is to revisit Figure 5.6 by non-dimensionalizing the axes as shown in Figure 5.7. When this is done it can be shown that the equivalent spring stiffness of the compressible inclusion is replaced by λ_{ci} (and the equivalent lateral spring stiffness of the retained soil becomes λ_{rs}).

---

[71] In earlier publications by the writer, this parameter was called the *normalized compressible inclusion stiffness*. Only its name is changed here to avoid confusion with the previously defined <u>absolute</u> compressible inclusion stiffness, K_{ci}, and the fact that both K_{ci} and λ_{ci} are normalized (i.e. non-dimensionalized) for generality, albeit using different constants because of the different dimensions involved.

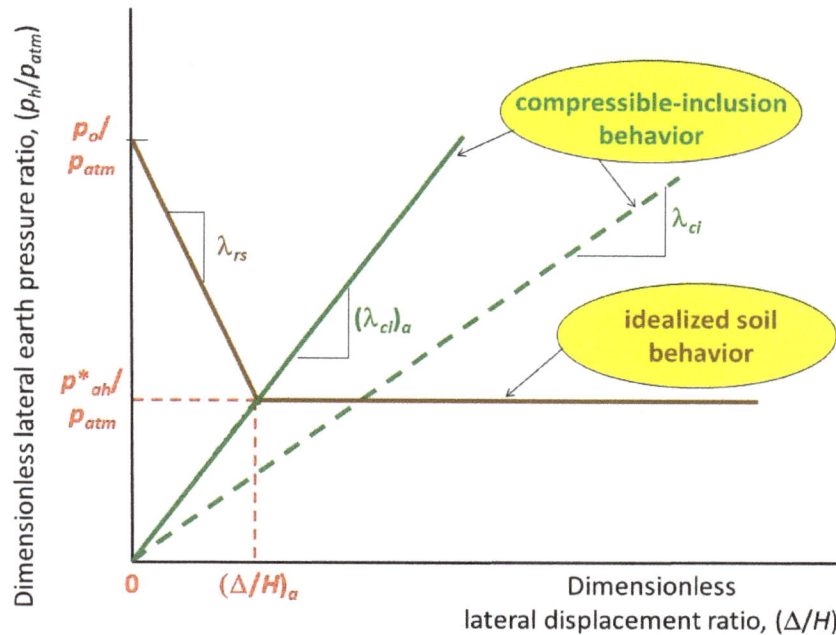

Figure 5.7. REP-Wall Application - Assumed Non-Dimensionalized Force-Displacement Behaviors for Analytical Model Development for Gravity Loading.

In summary, it is useful in both research and practice to always use both parameters, K_{ci} and λ_{ci}. The former allows for simple, absolute, comparison shopping between different compressible-inclusion products whereas the latter allows for rigorous analytical assessment of the efficacy of a specific compressible-inclusion product in a specific application. This is necessary because the efficacy of a compressible inclusion is always dependent on the geotechnical height, H, of the ERS. Note that calculating the dimensionless height-correction factor, H_{ci}, for a given value of H makes correlating K_{ci} and λ_{ci} for a project-specific application easier.

For reference purposes, the limiting values of both K_{ci} and λ_{ci} are:

- <u>zero</u> for the 'perfectly compressible' case of an infinitely soft compressible inclusion that allows unrestricted lateral displacement under zero lateral pressure and

- <u>infinity</u> for the 'perfectly rigid' case of an infinitely stiff (i.e. no) compressible inclusion that allows no lateral displacement under any magnitude of lateral earth pressure.

Quantitatively, then, the smaller the values of both K_{ci} and λ_{ci} the more compressible the compressible inclusion is. Note that this is opposite the trend effect of compressible-inclusion thickness, t_{ci}.

Once the parameters K_{ci}, H_{ci}, and λ_{ci} have seen greater use in manufacturers' literature as well as practice and research, empirical relationships will likely emerge as to what range of parameter values are required not only for the basic full-depth REP-Wall gravity-load case considered here but also other loading conditions as well as the ZEP-Wall application. One way of doing this might be to construct plots along the lines of Figure 5.6

126

but using non-dimensionalized variables for the axes and either K_{ci} or λ_{ci} in lieu of k_{ci}, with a corresponding non-dimensionalized retained-soil stiffness, K_{rs} or λ_{rs}.

One final comment before leaving the subject of gravity loading is that the focus of this section has been on design, i.e. determination of $(t_{ci})_a$. Note that the concepts presented here can be used to perform analyses as well, i.e. estimate what the lateral earth pressure or lateral resultant force reduction would be for a given t_{ci}. One could simply construct a plot similar to Figure 5.6 using a calculated k_{ci}, K_{ci}, or λ_{ci} based on the known t_{ci} using Equation 5.1 and see where it intersected the idealized lateral earth pressure curve for the retained soil.

5.4.2.2 Compaction-Induced Lateral Earth Pressures

A subset of REP-Wall applications involving gravity loading is the potential for reducing and even eliminating compaction-induced lateral pressures on non-yielding RERSs where such pressures are generally assumed to be inherently non-dissipating. This is the opposite of what happens with yielding RERSs that can displace as necessary to self-dissipate compaction effects.

The fact that normal construction compaction of fill and backfill material placed behind both yielding and non-yielding RERSs can result in total post-construction lateral earth pressures that are well in excess of the basic active or at-rest earth pressure states along the upper portion of the inside face of an ERS has been known for decades (Broms 1971, Ingold 1979). This phenomenon was the subject of extensive research in the 1980s (Seed and Duncan 1983; Duncan and Seed 1986; Duncan et al. 1991, 1993) and again more recently (Chen and Fang 2008). The net result of all this research has led to a more-detailed understanding of the very complex geomechanics behavior involved as well as the analytical means to accurately estimate both the magnitude and distribution of these pressures.

When taken together with several case histories involving non-yielding RERSs that were significantly damaged by unanticipated compaction-induced lateral earth pressures (Ebeling et al. 1996), this collective research demonstrates conclusively that compaction effects should be considered in routine design practice, at least for non-yielding RERSs. The logic for this is based on the fact that research indicates that only a relatively small magnitude of RERS lateral displacement is required to remove compaction effects (Symons and Clayton 1972) combined with the fact that, by definition, yielding RERSs always have the inherent ability to translate and/or rotate as necessary to relieve lateral pressures of any source that are in excess of the lower-bound lateral pressure defined by the active earth pressure state.

Because the results of research into compaction-induced lateral earth pressures were emerging in the late-1980s to early 1990s timeframe contemporaneously with the writer's early research into geofoam compressible inclusions, the writer recognized early on that compressible inclusions could be used to reduce or remove completely the effects of these pressures on non-yielding RERSs (Horvath 1996b, 1996c, 1997b, 1998a, 1998b).

As a direct outcome of suggestions made by the writer, this concept was explored via a research program involving 1-g physical testing at a full-scale, instrumented-retaining-wall test facility at the Virginia Institute of Technology and State University (Virginia Tech) that was funded by the Commonwealth of Virginia. The details of the test facility are described in Virginia Tech (undated) and the results of the specific research related to compaction stresses and geofoam compressible inclusions were documented in Reeves and Filz (2000).

R-EPS manufactured by an EPS molder based in Virginia (hence the reason for funding by the Commonwealth) under the tradename *TerraFlex* was the geofoam material and product used for the compressible inclusion in the testing performed for the Virginia Tech study. As noted previously and discussed in greater detail in Chapter 8, this is the geofoam material of choice (whether used alone or as the primary component of a geocomposite product) for most compressible-inclusion applications involving ERSs.

The results of this proof-of-concept testing demonstrated the effectiveness of a compressible inclusion composed of R-EPS in reducing compaction-induced lateral earth pressures, even when the retained soil is compacted to a high degree of relative compaction using a hand-operated, vertical-ram ('jumping jack') compactor which produces much larger compaction forces (Chen and Fang 2008) compared to a hand-operated, vibrating-plate ('nervous turtle') compactor.

One particularly interesting outcome of the Virginia Tech research was that the measured results confirmed what had long been known based on laboratory testing of relatively small test specimens of commonly used geofoam materials under large-strain conditions, i.e. time-after-construction is a significant problem variable that should always be considered explicitly in order to properly assess any lateral earth pressure reduction involving a geofoam compressible inclusion, not just for the effects of soil compaction. This is due to the inherent significant creep behavior of R-EPS under all stress levels[72]. The primary effect of creep in this case is that the stiffness (modulus) of the compressible inclusion decreases over time in a non-linear fashion.

To elaborate on this point, most compressible-inclusion applications are designed based on the material stiffness (i.e. Young's modulus) measured in strain rate-controlled uniaxial-compression tests performed at a relatively rapid rate (10% strain per minute) as has always been the de facto RCPS industry standard. This is a holdover from the pre-geofoam uses of all types of RCPS that began back in the 1950s and 1960s when the stress-strain-time behavior of both EPS and XPS under load-bearing was neither fully understood nor appreciated, especially because most RCPS applications in construction then and now, such as for flat-roof or in-wall thermal insulation, are not load-bearing in nature.

In general, the more rapid the strain rate the stiffer the response (with concomitant larger modulus) of EPS and its derivative materials such as R-EPS and GPS-PB as well. So, while using a Young's modulus based on relatively rapid loading may be acceptable for routine design practice as it results in a conservative design (i.e. greater required thickness of the compressible-inclusion product than is likely necessary under sustained loading such as gravity loads), when evaluating measured results or analyzing the actual performance of an ERS it is necessary to use a Young's modulus for the compressible inclusion that is temporally consistent.

Murphy (1997) presented an example of the latter for the first known case history use of a geofoam compressible inclusion in an ERS application in the U.S. that used GPS-PB. This case history is discussed in detail later in this chapter. Although this was only a partial-

---

[72] As discussed in detail in Chapter 8, GPS-PB, which has been manufactured in the U.S. since the 1970s, is nominally a stand-alone geosynthetic sheet-drain material that is marketed under the tradenames *GeoTech Drainage Board* and, with a factory-laminated non-woven geotextile on one face, *GeoTech Drainage Panel*. Since the 1990s, GPS-PB is also used as part of a commercially produced (in the U.S.) geocomposite with R-EPS that is marketed under the tradename *GeoTech GeoInclusion*. This product is used when the geosynthetic functions of thermal insulation, drainage, and compressible inclusion are all desired. By sheer coincidence, the compressibility of GPS-PB under both short- and long-term loading is comparable to R-EPS. Creep of 'normal' block-molded EPS such as used in small-strain applications such as lightweight fills can also be significant but only when the elastic-limit stress is exceeded.

depth REP-Wall application, the issues explored concerning the time-dependent stiffness of the compressible inclusion are still relevant.

In conclusion, the research conducted at Virginia Tech documented by Reeves and Filz (2000) demonstrated that a compressible inclusion composed of R-EPS can reduce the residual near-surface effects of compaction behind non-yielding RERSs. Given that GPS-PB has mechanical behavior similar to R-EPS this conclusion can logically be extended to GPS-PB as well as to the *GeoTech GeoInclusion* geocomposite that consists of a synergistic combination of R-EPS and GPS-PB.

Note, however, that this conclusion as to the efficacy of a compressible inclusion in reducing and even eliminating compaction-induced lateral pressures should <u>not</u> be extended to <u>all</u> geofoam materials...normal block-molded EPS in particular...on a blanket basis. This is because normal EPS-block of any density is relatively much stiffer than either R-EPS or GPS-PB under the relatively low-stress conditions encountered near the top of an ERS where compaction effects are greatest. Therefore, the efficacy of normal EPS-block in reducing compaction stresses is still to be determined by research and will, as a minimum, depend on the product density used as well as many other factors (Horvath 2011).

In any event, the magnitude of lateral earth pressure reduction for compaction effects depends on both the thickness of the compressible inclusion and the time after construction. To date, no known analytical methodology for quantifying the relative reduction in compaction effects has been developed. This is due, at least in part, to the fact that such a methodology would have to take into account the specific compaction equipment ('plant' as it is sometimes called) used and how close that equipment was operated to the inside face of the ERS. This conclusion is based on the fact that these project-specific factors were identified in research by Duncan et al. (1991) as being important for estimating compaction-induced lateral earth pressures behind non-yielding RERSs for the baseline case of no compressible inclusion.

5.4.2.3 Seismic Loading

5.4.2.3.1 Overview

The analytical methodology for full-depth REP-Wall applications under seismic loading, the so-called seismic-buffer application, is still evolving. As noted previously, there has been a considerable increase in research related to this application that began in the 2000s. However, most of the published research has been of a proof-of-concept nature that simply presented measured results to demonstrate the efficacy of the seismic-buffer concept in reducing lateral earth pressures or forces relative to some non-geofoam baseline case as opposed to trying to use the measured results to create a behavioral model and from this an analytical methodology. Therefore, most of this published research by others will be discussed in a subsequent section that deals with solution assessment for seismic loading.

The discussion of analytical methodologies for seismic loading presented here is divided into three sections. The first two discuss two very different solution methodologies developed by the writer. The third discusses solution methodologies proposed by other researchers.

The one element that all three presentations have in common is that the analytical methods discussed were each developed assuming a non-yielding RERS. As with gravity loading, the conventional wisdom has always been that the REP-Wall application is of minimal technical value, and thus of little-to-no cost benefit, with yielding RERSs. This is because the compressible-inclusion benefit, whether under gravity or seismic loading, is simply additive to the intrinsic ability of a yielding RERS to displace laterally as shown in Figures 5.2 and 5.4. As noted previously for gravity loading, the only argument that can be made for using a compressible inclusion with a yielding RERS is to ensure that the active earth pressure state will exist on the structural components of the ERS, even under service-load conditions when the ERS has adequate safety margins against the geotechnical ULS.

5.4.2.3.2 Mononobe-Okabe Based Method

The simplest and most obvious analytical methodology for REP-Wall applications under seismic loading is to apply the classic Mononobe-Okabe (M-O) Method[73] that was developed for traditional yielding RERSs to the framework developed earlier in this chapter for full-depth REP-Wall applications under gravity loading. In light of subsequent discussions in this monograph, it is useful to restate that the M-O Method considers only the acceleration magnitude of a seismic event and ignores the frequency, f, (or its inverse, period, T) content of the ground motions. This is emphasized as the current state of knowledge suggests that the frequency content of the seismic motion can be an important variable in terms of the lateral seismic earth loads on a RERS, both non-yielding and yielding.

To begin with, it is useful to restate the generic relationship developed for gravity loading (Equation 5.5) but this time with revised notation to allow for problem variables under seismic loading to differ from the corresponding variables under gravity loading (whether they actually differ or not is unimportant at this point):

$$\left(t_{ci}\right)_{ae} = \frac{E_{ci} \cdot \left(\Delta / H\right)_{ae}}{K_{\sigma e} \cdot K_{he} \cdot \gamma_{eff}^*} . \tag{5.11}$$

Note that the effective soil unit weight for gravity loading, γ_{eff}, is modified here for vertical-acceleration effects if desired. Note also that E_{ci} is not given a different notation from that used for gravity loading as the value used for gravity loading is not necessarily unique to that type of loading. As noted previously, when designing for gravity loads, it is usually desirable to design conservatively, i.e. size the compressible inclusion for short-term loading but check the adequacy of the design for long-term loading to make sure that long-term compressive strains of the compressible-inclusion product are within its performance limits. Therefore, the implication is that a design professional should select a value or values for E_{ci} deemed appropriate for the loading case being considered, whether gravity or seismic.

The writer's initial effort to define the variables to use in Equation 5.11 was presented in Horvath (2008b) and is as follows:

$$\left(t_{ci}\right)_{ae} = \frac{E_{ci} \cdot \left(\Delta / H\right)_{ae}}{0.75 \cdot K_{ae} \cdot \cos\delta \cdot \gamma_{eff}^*} \tag{5.12}$$

---

[73] As discussed in detail in Appendix A, the term 'Mononobe-Okabe Method' has been and is currently used for methodologies that deviate from the original version in various ways. As a result, what is presented in some published works as the M-O Method is actually not the original, 'true' method but some modified version of it. In this monograph, the term 'Mononobe-Okabe Method' or 'M-O Method' when used alone will only refer to the original version. Later derivative/modified versions will be clearly identified as such.

130

where K_{ae} is the coefficient of active earth pressure under seismic loading for the M-O Method (Equation A.13 in Appendix A). Within the context of Equation 5.11, this means:

- $K_h = K_{ae} \cdot \cos\delta$,

- $K_{\sigma e} = K_\sigma = 0.75$, and

- $(\Delta/H)_{ae} = (\Delta/H)_a$.

Several comments concerning Equation 5.12 are in order. First and foremost, in terms of the potential impact on overall results is that in light of recent research that is discussed in some detail in Appendix A, it may be desirable to use K_{ae} in Equation 5.12 that is calculated using the Seed-Whitman (S-W) Method instead of the M-O Method. The S-W Method appears to yield results that are much more in line with observed reality, especially at higher ground-acceleration levels. However, the problem with doing so is that the S-W Method was crafted assuming a very simplistic geometry for the yielding RERS (vertical inside face of the ERS, horizontal retained-soil ground surface, no surface surcharge). Consequently, dealing with problem geometries that deviate from these assumptions is problematic.

Next is that the assumption that $K_{\sigma e} = K_\sigma = 0.75$ is open to debate. The reason is that as discussed previously for gravity loading, the selection of $K_\sigma = 0.75$ was based either directly or indirectly (depending on whether Equation 5.6 was developed using Version 1.2 based on pressures (stresses) or Version 2.0 based on resultant forces) on the observation that lateral earth pressures have the depth-wise variation defined generically by the yellow dotted curve in Figure 5.5.

Strictly speaking, this assumption concerning $K_{\sigma e}$ is only correct for gravity loading. At the present time, there is conflicting evidence (discussed in Appendix A) as to what lateral pressure distribution is most appropriate for seismic loading although the evidence to date suggests that the ability of the RERS to displace laterally, whether due to rigid-body yielding of the entire ERS or flexibility and concomitant deformation of a nominally non-yielding RERS, is an important factor. It also appears that the geotechnical height, H, of the ERS may play a role as with higher RERSs the traditional assumption that the entire mass of retained soil displaces laterally as a single rigid body loses validity.

At the present time, it appears reasonable to assume that a non-yielding RERS with a compressible inclusion behaves as a pseudo-yielding RERS under seismic loading as it does under gravity loading. The original M-O Method as well as recent research suggests that the lateral earth pressure distribution under seismic loading is triangular in distribution which would make $K_{\sigma e} = 1$. On the other hand, the S-W Method as well as composite methods based on elements of the M-O and S-W methods mixed together use different pressure distributions for the gravity and seismic-increment components. This produces a $K_{\sigma e} < 1$, with a value that would be dependent on the specific application and the relative magnitude of the gravity and seismic load components.

In addition to these traditional rigid-body solutions, there is the body of work produced by Veletsos, Younan, and their associates that approached the seismic-loading problem of a RERS with variable yielding potential using the theory of elasticity. Their work as it is relevant to compressible inclusions is discussed the following section. Of relevance to the discussion here is that their work showed that the lateral earth pressure under seismic loading has a curved distribution, with the peak magnitude of the curve:

Lateral Pressure Reduction on Earth-Retaining Structures Using Geofoam
John S. Horvath, Ph.D., P.E., Life Member.ASCE

- moving downward and the overall curve approaching the shape of a triangle ($K_{\sigma e} \to 1$) as the yielding of a RERS transitions to the limiting yielding condition and

- moving upward ($K_{\sigma e} < 1$) as the yielding of a RERS transitions to the limiting non-yielding condition.

Thus, the conclusion drawn from their work is that the value of $K_{\sigma e}$ is dependent on the relative magnitude of yielding provided by the compressible inclusion but is likely closer to a value of 1 under seismic loading as opposed to the 0.75 under gravity loading. However, the smaller the value of $K_{\sigma e}$ the more conservative the final result in terms of the calculated value of $(t_{ci})_{ae}$ from Equation 5.11. Consequently, until such time as research provides a more-definitive answer, it would seem to be reasonable to conservatively use $K_{\sigma e}$ = 0.75 as reflected in Equation 5.12 which is a particular case of Equation 5.11.

The third and final comment concerns the assumption that $(\Delta/H)_{ae} = (\Delta/H)_a$. The writer is not aware of any research to suggest that this is not reasonable.

Although not shown here, in principle it is straightforward to construct plots similar to Figures 5.6 and 5.7 for seismic loading. Alternatively, both the gravity and seismic loading cases could be superimposed on the same plot. This would facilitate comparison of the two cases as well as readily allow for analysis to see how a compressible inclusion designed for one load case would perform in the other.

5.4.2.3.3 Extended Veletsos-Younan Method

The work of Veletsos, Younan, et al. that relates to their elasticity-based model of how a RERS responds to seismic loading is discussed in some detail in Appendix A. At various times over the course of about a decade, these researchers considered several variations of this model in terms of the assumed rigid-body displacement or deformation mode of the RERS as well as boundary conditions.

Also presented in detail in Appendix A, the writer used one particular combination of displacement and boundary conditions (a RERS that is free to rotate as a rigid body about its base) as the basis for a model of a non-yielding RERS with a compressible inclusion that compresses in a rotational manner. The writer termed the analytical methodology developed using this model the Extended Veletsos-Younan Method for analyzing the seismic response of a non-yielding RERS with a compressible inclusion. As discussed in Appendix A, the solution parameters d_θ and R_θ in the original Veletsos-Younan Method can be expressed in terms of the dimensionless parameters K_{ci} and λ_{ci} that reflect compressible-inclusion stiffness and were defined earlier in this chapter.

The primary benefit of the writer's Extended Veletsos-Younan Method is that, for any given problem, it allows a rigorous estimation of loads on the RERS over a continuous range of compressible-inclusion stiffnesses, including the baseline reference case of no compressible inclusion. This can be done only approximately using a methodology based on rigid-body mechanics such as the M-O Method.

Another significant benefit is that the Extended Veletsos-Younan Method considers the frequency of the input motion as part of the solution process. Recent work by Bathurst and his associates showed that the efficacy of a geofoam compressible inclusion used with non-yielding RERSs is somewhat dependent on the frequency of the input motion, something that a methodology based on rigid-body mechanics such as the M-O Method does not consider as pointed out earlier in this chapter.

However, there are some significant pragmatic limitations at present to using the Extended Veletsos-Younan Method on a routine basis in practice, with the most significant being that the necessary solutions are not readily available beyond the original publications. In addition, selection of the appropriate single-value modulus for the retained soil is problematic as it is well-known that modulus degrades (diminishes) from its initial, small-strain value non-linearly as a function of strain. Nevertheless, the Extended Veletsos-Younan Method is, in concept, an interesting solution alternative explored for non-yielding RERSs with a compressible inclusion subjected to seismic loading.

5.4.2.3.4 Other Methods

As noted previously, the early years of the 21st century saw a significant increase in research related to the seismic-buffer concept, with this term evolving contemporaneously to be the preferred colloquial term for the full-depth REP-Wall concept when used primarily to reduce seismic loads on non-yielding RERSs. With few exceptions, this research was performed under the overall direction of Prof. Richard J. Bathurst of the GeoEngineering Centre at Queen's University-Royal Military College of Canada (Queen's-RMC) in Kingston, Ontario. This is noteworthy because it means that, in reality, most of the published research to date on the subject of seismic buffers reflects a single viewpoint, understanding, and interpretation. There are both positives and negatives to this.

Also, as noted previously, most of this published research has been in the form of presenting and comparing results from both physical and numerical models, specifically:

- laboratory testing of a small-scale (1 m/3.3 ft high) model on a shake table (which is taken to be the 'ground truth') and

- analyses using *FLAC* FDM software (Zarnani and Bathurst 2005, 2008, 2009a, 2009b, 2011).

Consequently, the bulk of the discussion of this work is presented later in this chapter in a section that deals with solution assessment.

A noteworthy exception that is discussed here is one attempt (in two distinct versions) by Bathurst and colleagues to develop what purports to be a 'simplified' analytical methodology for potential use in routine practice. However, as will be seen, use of this methodology is actually quite complex in that it requires proprietary, purpose-written computer software with relatively sophisticated input for solution.

In a broad sense, the conceptual model used as the basis for this 'simplified' analytical method, which will be referred to hereinafter as the *Q-RMC Wedge Method* (Bathurst et al. refer to it as the *Displacement Model*) for reasons that will become obvious, is broadly similar to that assumed in the M-O Method for yielding RERSs. Specifically, the Q-RMC Wedge Method assumes that a rigid, triangular-shaped failure wedge of soil forms within the retained soil and that this wedge is able to displace laterally under horizontal seismic acceleration. However, in this case the wedge can displace laterally not because the ERS can displace (as in the original M-O Method) but because of the presence of a full-depth compressible inclusion. Note that this is conceptually the same as the analytical model and concomitant methodology developed by the writer that is an extension of the M-O Method and was discussed previously.

One deviation from the writer's Mononobe-Okabe-based method presented earlier is that some vertical displacement of the soil wedge is also assumed to occur in the Q-RMC

Wedge Method, not because of vertical seismic acceleration (which is ignored) but because the wedge is assumed to slide along an assumed planar intra-soil failure surface and thus has a vector component of vertical displacement as it displaces back and forth laterally.

However, what really distinguishes the Q-RMC Wedge Method from the writer's method based on the M-O Method is how the resultant forces that develop between the retained soil and RERS and along the intra-soil failure plane are calculated. The writer's M-O based method uses rigid-body statics as was done with the original M-O Method for yielding RERSs. On the other hand, the Q-RMC Wedge Method models these forces using linear axial springs which means that displacement magnitudes must be evaluated explicitly as part of the solution process. Thus, the overall problem setup for the Q-RMC Wedge Method is identical to that shown previously in Figure 5.3 where both the compressible inclusion and retained-soil mass (wedge in this case) are springs and the unique solution is defined by force <u>and</u> displacement equilibrium between the two springs.

What is referred to in this monograph as Version 1 of Q-RMC Wedge Method assumes that the soil wedge acts as a single rigid mass. The development of this version and presentation of calculated results are contained in Bathurst et al. (2007a), with an overview in Bathurst and Zarnani (2013). In general, the comparison between calculated results and those measured in the aforementioned shake-table tests were fair except at higher accelerations where the Q-RMC Wedge Method overestimated both lateral displacements as well as maximum forces on the ERS.

Version 2 of the Q-RMC Wedge Method followed soon thereafter. The significant difference from Version 1 is that the soil wedge is divided into multiple rigid layers with both shear (horizontal) and normal (vertical) forces between layers modeled as linear axial springs so that forces could vary between layers. However, a single lateral and vertical displacement of the overall soil wedge was still assumed.

Although the agreement between Version 2 calculated and measured (shake table) results appeared to be better than with Version 1, Version 2 has had minimal exposure in the English-language literature as the detailed treatment of the subject was published in the Chinese language with only an English abstract (Wang and Bathurst 2008a). What appears to be a somewhat shorter English-language version was published in a relatively obscure conference publication (Wang and Bathurst 2008b).

In any event, although the Q-RMC Wedge Method appeared to show some promise, it was apparently considered a dead-end pursuit although it was mentioned in the relatively recent state of practice summary of seismic-buffer research by Bathurst and Zarnani (2013). This may reflect the fact that the Q-RMC Wedge Method really did not fit the mold of a true 'simplified' methodology amenable for routine use by practitioners. As noted previously, its use requires application-specific computer software that is purpose-written and thus proprietary although the computer code can, in principle, be cloned based on what appears in the published papers in terms of necessary program logic. However, the required input is fairly sophisticated, not the least of which is the fact that an explicit time history of dynamic motion must be input, not just a peak ground acceleration or coefficient.

5.4.2.4 Solution Assessment

5.4.2.4.1 Overview

As with the lightweight-fill function that was addressed in Chapter 4, assessing the accuracy of analytical methodologies for the compressible-inclusion function is an essential validation step for acceptance and concomitant use of these methodologies in practice. As noted in Chapter 4, the writer deems instrumented, full-scale field installations and an assessment of the behavior of EPS-block under biaxial loading as essential components of a multi-component R&D program for evaluating lightweight-fill methodologies. Consequently, in the absence of such R&D outcomes at the present time, the solution assessment presented in Chapter 4 was largely qualitative and descriptive as opposed to quantitative. This is because funding and undertaking the necessary validation tasks is clearly something to be done by stakeholders such as the EPS industry and government agencies who have the necessary critical mass (incentive + resources + opportunities) to do this.

However, a limited quantitative solution assessment is presented in this chapter for compressible-inclusion applications, beginning here with the baseline full-depth placement case of REP-Wall applications. This is because the geofoam products typically used in such applications consist of a single, relatively thin panel that is subjected to uniaxial stresses in the horizontal direction only. Consequently, experience going back to the late 1980s has shown that the compressible-inclusion function can be modeled much more easily and reliably using numerical analyses for both gravity and dynamic (pseudo-seismic) loading conditions.

Note that this is not to say that results from well-instrumented, full-scale installations involving the compressible-inclusion function would not be useful to complement these numerical analyses. It is simply that field measurements are not judged to be as critical and essential for verification of compressible-inclusion analytical methodologies as they are for lightweight-fill analytical methodologies.

The solution assessment for the full-depth REP-Wall case that is presented here is separated into gravity and seismic loading as the assessment process to date has been distinctly different for each type of loading. Interestingly, although the use of the REP-Wall concept to reduce seismic loads (i.e. the seismic-buffer concept) was explored in detail approximately two decades after the use to reduce gravity loading had been well-established, the verification resources devoted to seismic buffers in terms of both physical testing and numerical modeling has been substantially greater to date than for gravity loading.

5.4.2.4.2 Gravity Loading

As discussed in Chapter 4 concerning solution assessment for the lightweight-fill function, the 'gold standard' for any such assessment is measured results from some type of meaningful physical testing that either directly (by virtue of physical dimensions) or indirectly (by virtue of artificial acceleration in a centrifuge) replicates a full-scale ERS. Unfortunately, as with the lightweight-fill function, there is a dearth of such quality testing targeted at full-depth REP-Wall applications under gravity loading. As discussed in the following section, most physical testing (and only involving relatively small models under 1-g conditions at that) of REP-Wall applications has occurred only relatively recently and focused on the response to dynamic loading in an effort to quantify the load-reduction benefit of seismic-buffer applications. The relatively few studies of value involving full-scale ERSs that focused on gravity loading involved partial-depth applications and are discussed later in this chapter.

Consequently, the only assessment of the full-depth REP-Wall application for gravity loading presented here is based on FE analyses performed specifically for this monograph. The FE mesh and material properties used are discussed in detail in Appendix E that also contains the results for both non-yielding and yielding RERSs for the baseline all-soil/no-geofoam case. These baseline cases were used to verify the basic performance and accuracy of the FE mesh and material properties as the calculated results could be compared to classical closed-form solutions.

To begin with, Figure 5.8 shows the specific version of the FE mesh used for the full-depth REP-Wall analyses presented and discussed in this section. With additional reference to Figure E.1 in Appendix E, note that the portion of the mesh between $X = 0$ and $X = 1$ metres and for the full height (Y-axis) of 6 metres consisted of bar elements oriented horizontally (depicted by the yellow dotted lines in Figure 5.8 and dark-red dotted lines in Figure E.1). These functioned as linear-elastic axial springs that were variously used to model the geofoam compressible inclusion, ERS yielding, or some combination of the two depending on the specific analysis performed.

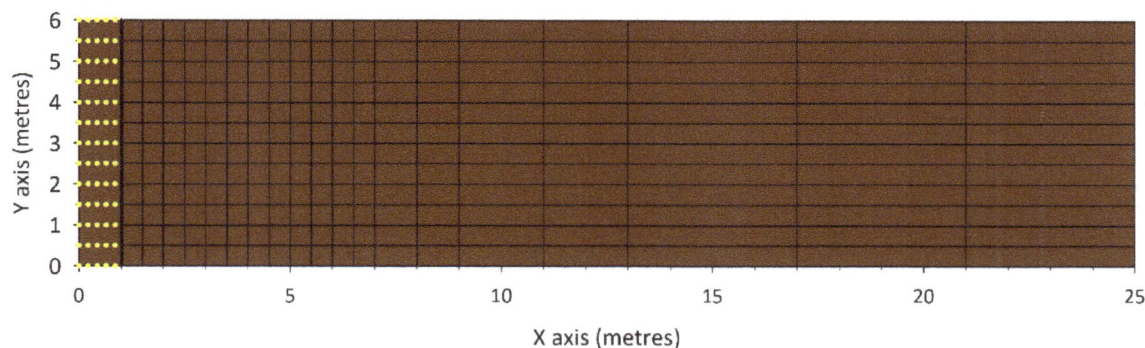

Figure 5.8. REP-Wall/Full-Depth Placement/Non-Yielding RERS - Finite-Element Mesh.

Two suites of FE analyses were performed for the assessments presented in this section. The first involved varying the thickness of the compressible inclusion for the basic case of a non-yielding RERS[74] assuming rapid-loading/short-term compressible-inclusion stiffness of a commercially available product as would normally be used in practice for gravity loading. The primary purpose of this suite of analyses was to explore the relationship between the resultant lateral earth force on the ERS and lateral displacement to see if the bi-linear model for this as shown in Figure 5.6 and other figures is a reasonable approximation of reality.

As discussed previously, this assumed bi-linear behavior is an essential element of simplified analytical methods for REP-Wall analysis that date back to the 1980s as it implies that there is always a unique, single-valued 'correct' answer for every application. This unique result is obtained from the minimum lateral earth force and concomitant unique minimum thickness of a compressible inclusion that can be used to produce this force. A corollary to this is that a compressible-inclusion thickness greater than this minimum value would not be 'bad' per se but would not produce further benefit in terms of load reduction. This overall thinking is reflected in the writer's current version of the simplified analytical method as expressed in Equation 5.6.

The issue of the effect of varying the compressible-inclusion thickness was actually explored to some degree in Horvath (2000) that used the same basic FE mesh and material properties (at least in terms of the ERS portion) as used in the analyses performed for this

---

[74] The RERS was not modeled explicitly in the mesh shown in Figure 5.8 but was simulated by fixing all nodes along the $X = 0$ boundary against X and Y displacement.

monograph. The conclusion in Horvath (2000) was that the solution of Equation 5.6 for the problem solved both then and for this monograph, $(t_{ci})_a$ = 175 mm (7 in), was reasonably verified by FE analyses that considered t_{ci} = 100/200/400 mm (4/8/16 in) in addition to the baseline all-soil/no-geofoam case.

Nevertheless, the decision was made to re-do these analyses for use in this monograph. To begin with, the FE analyses in Horvath (2000) used a very simplistic and approximate methodology to mimic residual soil-compaction stresses in the simulated construction that was part of the FE analyses. As noted in Appendix E, residual lateral stresses from compaction were not simulated for any of the analyses performed for and presented in this monograph. This was done so as not to unnecessarily complicate and obfuscate the basic soil behavior that is of greater interest in the present work. In addition, it was desired to analyze a greater number and range of compressible-inclusion thicknesses for this monograph so that various relationships involving the key problem variables and calculated outcomes could be more fully explored than in Horvath (2000). The outcomes from this new suite of analyses are now presented and discussed.

To begin with, as discussed in Appendix E, the FE software used for the analyses performed for this as well as earlier studies by the writer, *SSTIPNH*, allows for direct determination of lateral pressures acting on an ERS (and, from this, indirect calculation of lateral resultant force) in at least two ways:

- using results from the 1-D interface elements placed along the inside face of the ERS where it is in contact the retained soil and

- from the first column of 2-D 'solid' elements within the retained soil adjacent to the ERS-retained soil interface.

As discussed in Appendix E, there are pros and cons to each alternative. Consequently, the writer has found that whenever investigating a new problem it is useful to compare results from both types of elements before selecting one to use for any broader range of assessments. This is illustrated in Appendix E for the baseline cases of non-yielding and yielding RERSs without a compressible inclusion.

In the specific case of the compressible inclusions analyzed for this monograph, there is actually a third alternative that can be used. This involves the calculated forces in the bar (= axial spring) elements used to model the 1-D compressibility of the geofoam that are part of the FE output[75]. These are the elements depicted using the dotted yellow lines in Figure 5.8. In this case, the resultant lateral force can be determined directly as it is simply the sum of all the bar forces. Note that in this case it is the lateral pressures acting on the compressible inclusion that are calculated indirectly (they are the bar forces divided by the area supported by each bar).

Figure 5.9 shows the depth-wise distribution of normalized lateral earth pressures based on the three different FE element types described above for the case of t_{ci} = $(t_{ci})_a$ = 175 mm (7 in) noted previously. Also shown in this figure are the theoretical at-rest and horizontal component of active earth pressures, the latter based on Coulomb's classical solution and concomitant assumed hydrostatic pressure distribution.

---

[75] In earlier studies by the writer such as Horvath (2000), the compressible inclusion was modeled using 2-D solid elements. The reason for modeling the compressible inclusion differently in the analyses performed for this monograph was to provide for greater flexibility in the use of the FE mesh, specifically, to allow the same bar elements to be used to model ERS yielding for problems not involving a compressible inclusion.

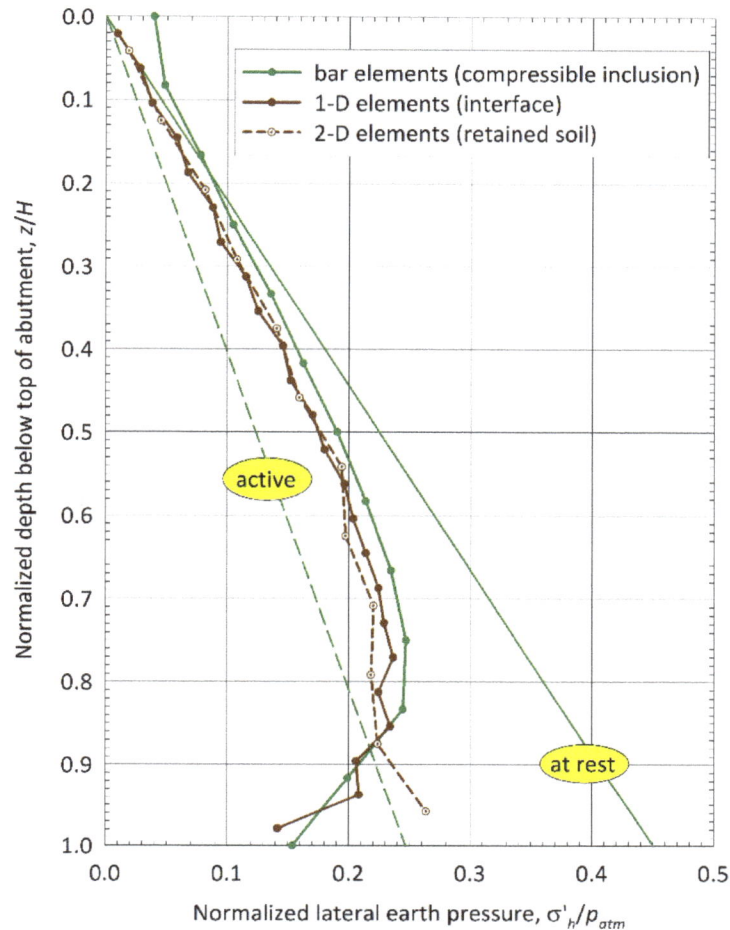

**Figure 5.9. REP-Wall/Full-Depth Placement/Non-Yielding RERS
Calculated Lateral Earth Pressures with 175 mm (7 in) Compressible Inclusion.**

One conclusion drawn from this figure is that the results from the three element types are in overall good agreement (note that because the compressible-inclusion material is a true solid it can sustain non-zero stresses at the ground surface, unlike the retained soil). Based on these results, the 1-D interface elements were chosen to best represent lateral earth pressures from the retained soil acting on the compressible inclusion while the bar elements inherently best represent the lateral pressures transmitted through the compressible inclusion and acting on the inside face of the ERS.

Figure 5.10 shows the variation in lateral earth pressures acting on the compressible inclusion as a function of compressible-inclusion thickness. The baseline all-soil/no-geofoam reference case presented in Appendix E is also included. A discussion of why this baseline case has pressures somewhat above the theoretical at-rest value along the lower half of the ERS can be found in Appendix E. As noted in the preceding paragraph, all results (except for the theoretical at-rest and active hydrostatic distributions) were based on results from the 1-D interface elements.

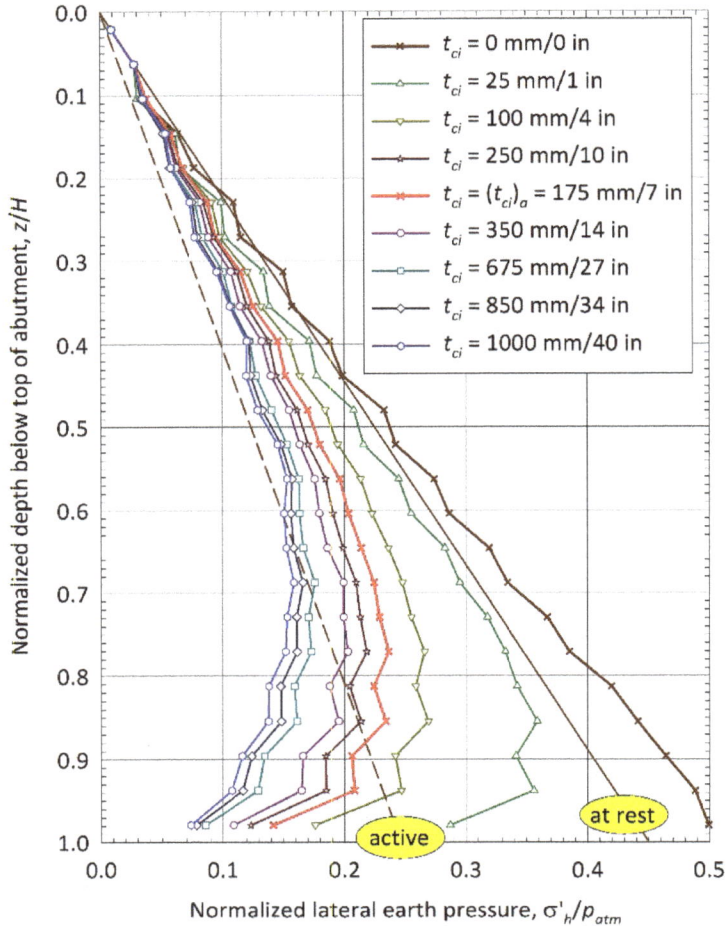

**Figure 5.10. REP-Wall/Full-Depth Placement/Non-Yielding RERS
Calculated Lateral Earth Pressures as a Function of Compressible-Inclusion Thickness.**

There are a number of items of interest in this figure:

- Additional reductions of lateral earth pressure occur for compressible-inclusion thicknesses beyond the $t_{ci} = (t_{ci})_a = 175$ mm (7 in) value produced by the current version of the writer's simplified analytical method (Equation 5.6).

- The relationship between lateral pressure reduction and compressible-inclusion thickness is non-linear, with significant benefit occurring initially with modest thickness followed by rapidly diminishing benefit with additional thickness.

- No matter what thickness of compressible inclusion is used, lateral earth pressures do not decrease below the theoretical active state for the upper 60% of the ERS. This is consistent with horizontal-arching theory as shown in Figure E.5 in Appendix E.

To explore some of these issues further, Figure 5.11 shows the variation of the dimensionless relative lateral resultant force, β, with compressible-inclusion thickness, t_{ci}, normalized to H and expressed as a percent. Although results from the calculated bar-element forces are preferred for the reasons noted previously, for the sake of completeness values based on the 1-D and 2-D elements are shown as well along with the theoretical at-rest and Coulomb-active values from Appendix E.

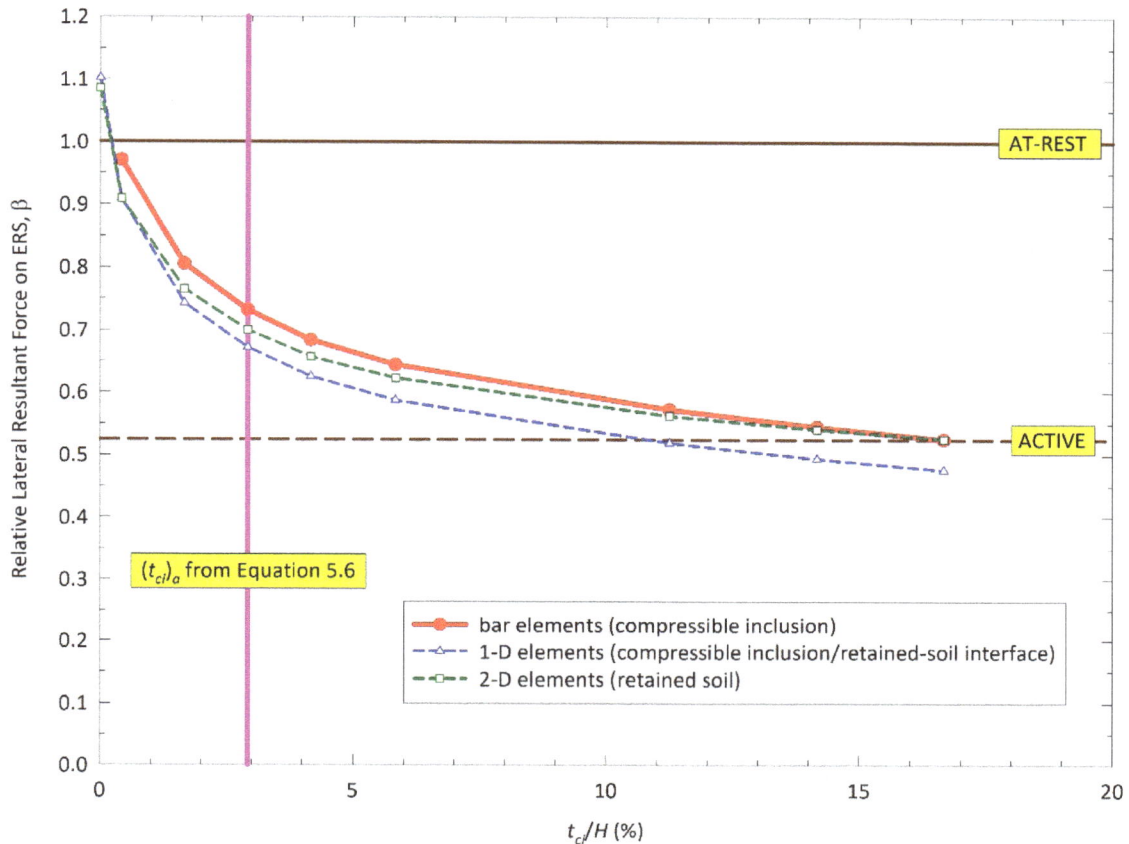

Figure 5.11. REP-Wall/Full-Depth Placement/Non-Yielding RERS Calculated Relationship between β and t_{ci}/H.

The minimum compressible-inclusion thickness necessary to achieve the active earth pressure state, $(t_{ci})_a$, based on the current version of the writer's simplified analytical method (Equation 5.6) is also highlighted in Figure 5.11. The non-linearity of the relationship between β and (t_{ci}/H) that was qualitatively apparent in Figure 5.10 is more clearly indicated in Figure 5.11.

The shape of the curves in Figure 5.11 suggests that replotting using semi-log scaling might be useful and this is shown in Figure 5.12. The virtually linear relationship between β and (t_{ci}/H) is quite clear. While the specific numerical relationship between these variables is unique to the problem analyzed, the linearity of the relationship suggests that a potentially fruitful area of research could be to develop an empirical analytical model relating β and $\log_{10}(t_{ci}/H)$.

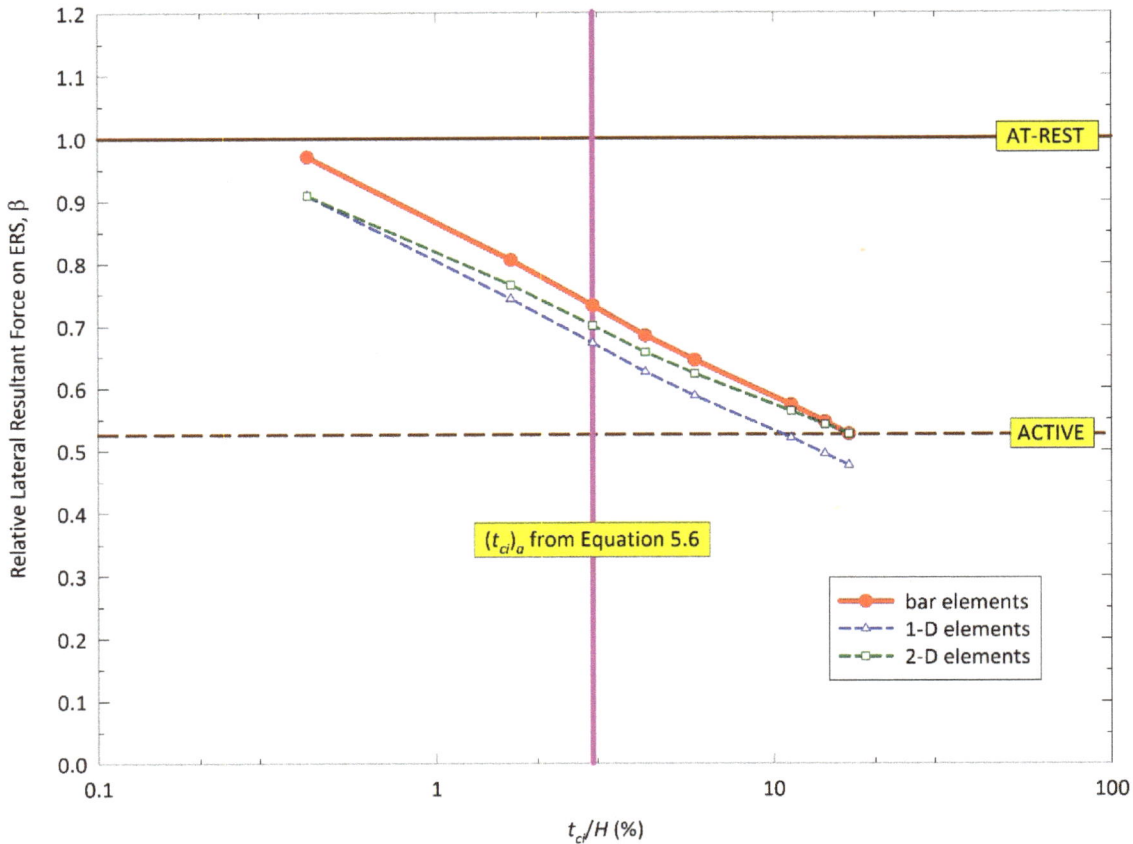

**Figure 5.12. REP-Wall/Full-Depth Placement/Non-Yielding RERS
Calculated Relationship between β and log₁₀ t_{ci}/H.**

However, more-promising lines of research for a relationship between β and the compressible inclusion involve the variables K_{ci} and λ_{ci} defined earlier in this chapter as each of these variables relates directly to the stiffness (compressibility) of the compressible inclusion which t_{ci} alone does not.

Considering first the absolute stiffness variable, K_{ci}, Figures 5.13 and 5.14 show the relationship between β and K_{ci} and β and log₁₀ K_{ci} respectively. Note that only the results from the bar elements used to simulate the compressible inclusion in the FE mesh are shown as the results from the 1-D and 2-D elements are similar and do not add anything to the discussion. Note also that the baseline all-soil/no-geofoam case would have $K_{ci} = \infty$ and the other limiting case of infinite compressibility would have $K_{ci} = 0$.

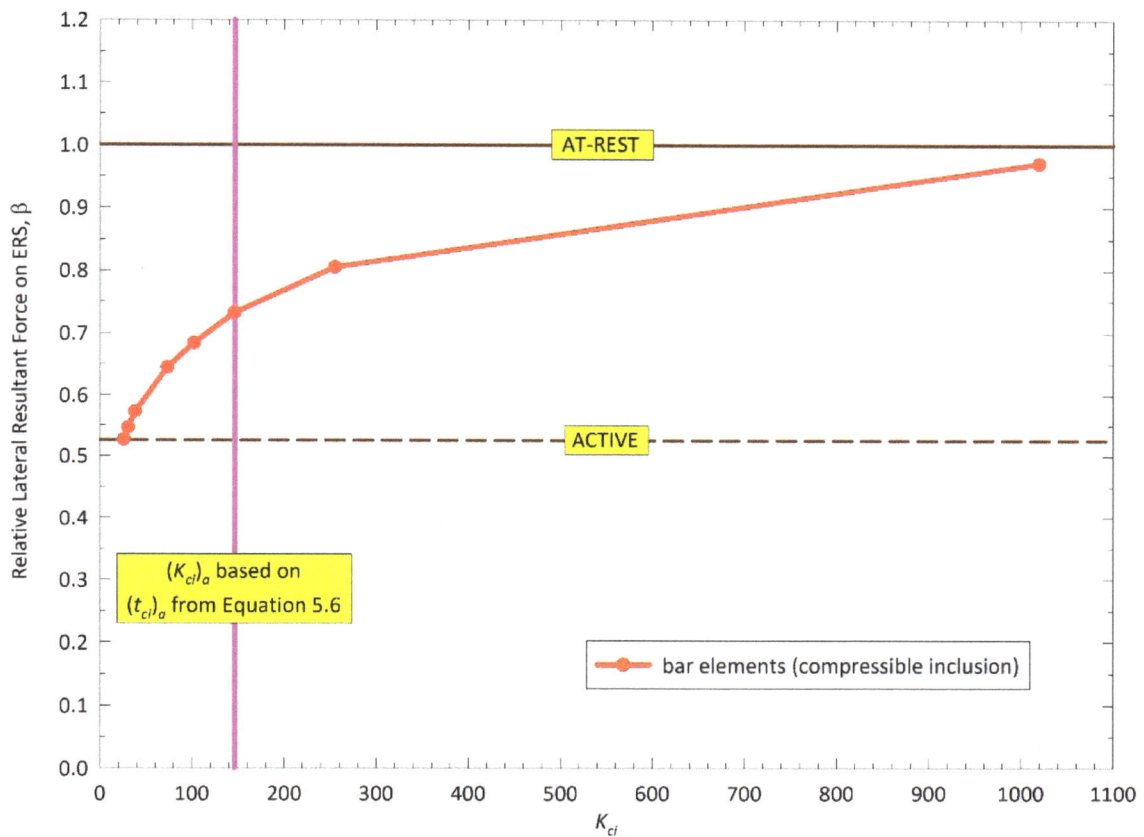

Figure 5.13. REP-Wall/Full-Depth Placement/Non-Yielding RERS Calculated Relationship between β and K_{ci}.

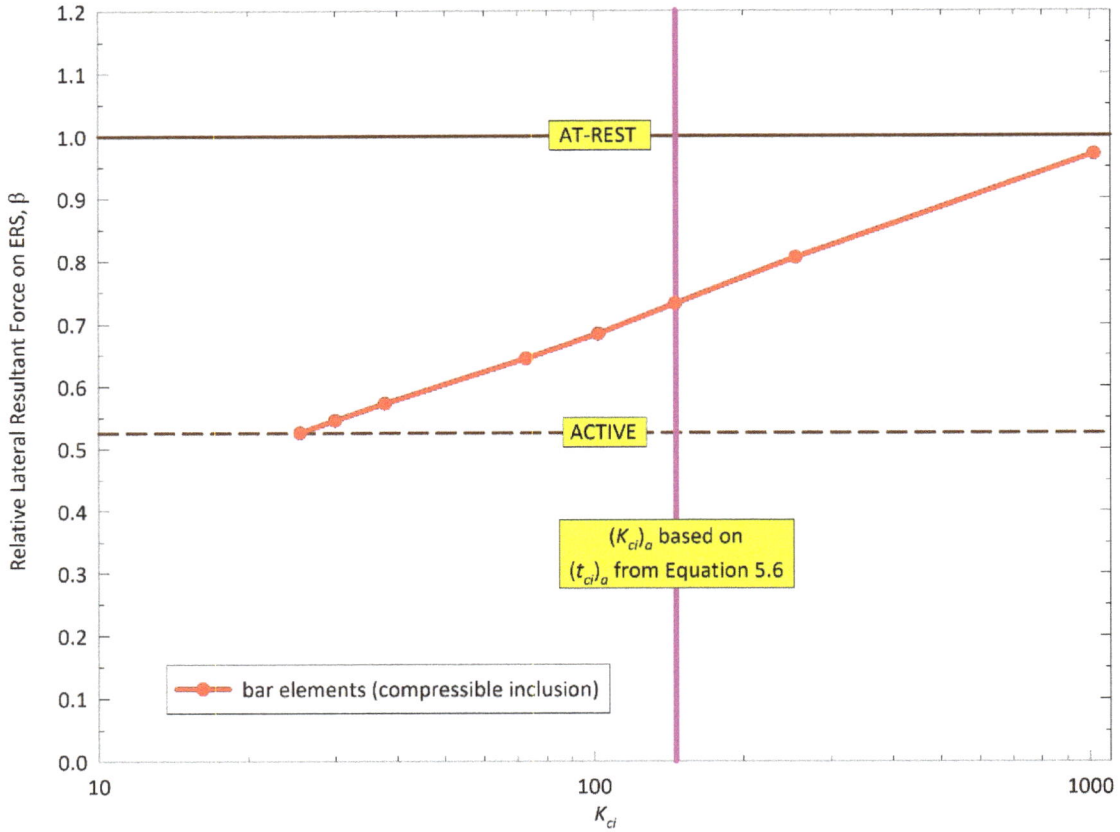

Figure 5.14. REP-Wall/Full-Depth Placement/Non-Yielding RERS Relationship between β and $\log_{10} K_{ci}$.

As an aside, Figures 5.13 and 5.14 illustrate the potential utility of the K_{ci} variable in routine practice. As noted earlier in this chapter, this variable reflects the <u>absolute</u> stiffness (compressibility) of a compressible inclusion that geofoam manufacturers and geosynthetics distributors could use in technical literature when marketing their compressible-inclusion products. A design professional could generate a curve similar to that shown in Figures 5.13 and 5.14 for a project-specific application; select a level of lateral earth force reduction they wanted and concomitant β-value that was needed to achieve this; and then simply 'shop' for a commercially available geofoam product that had a K_{ci} value equal to or less than that required to produce this β-value.

Figures 5.15 and 5.16 are similar to the preceding two figures but this time using λ_{ci} as the dimensionless relative compressible-inclusion stiffness variable against which β is plotted. Similar limiting cases of $\lambda_{ci} = \infty$ and $\lambda_{ci} = 0$ for infinite rigidity and infinite compressibility respectively apply here as well. The essentially linear relationship for the semi-log plots for both K_{ci} and λ_{ci} is noted. This observed behavior may prove useful in the future for developing simplified analytical models.

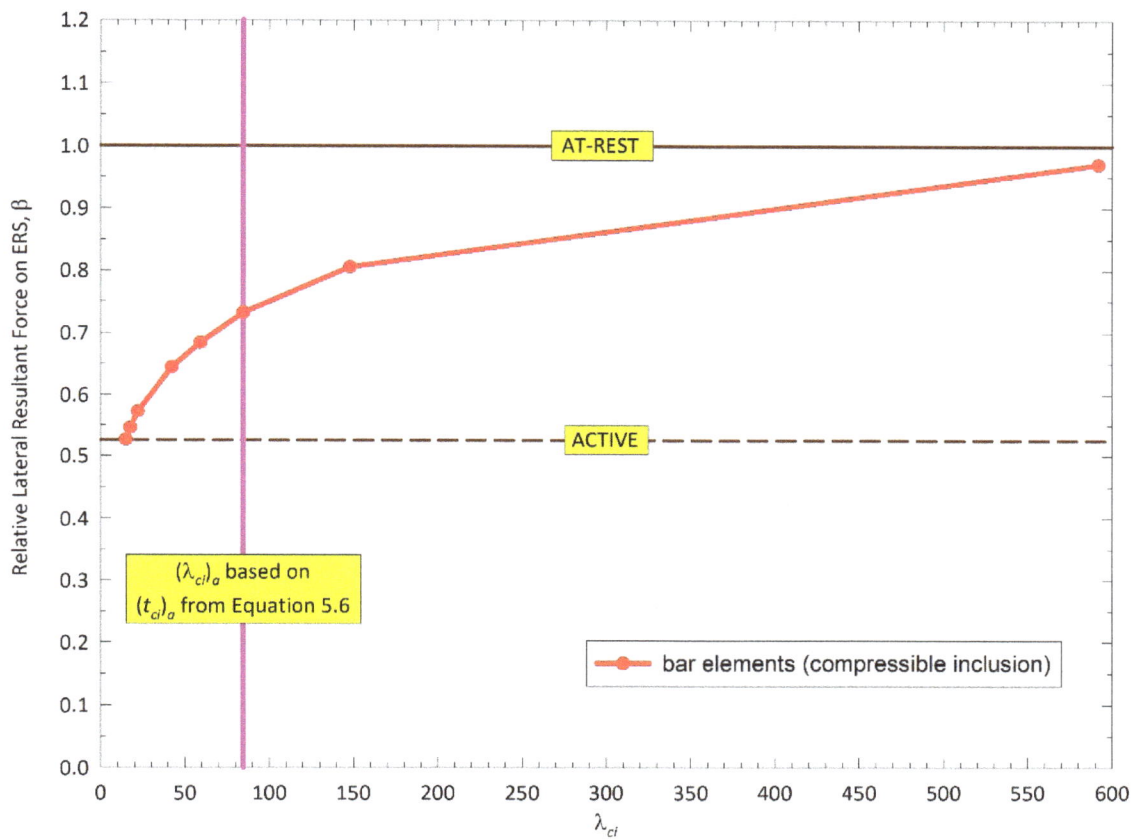

Figure 5.15. REP-Wall/Full-Depth Placement/Non-Yielding RERS
Relationship between β and λ$_{ci}$.

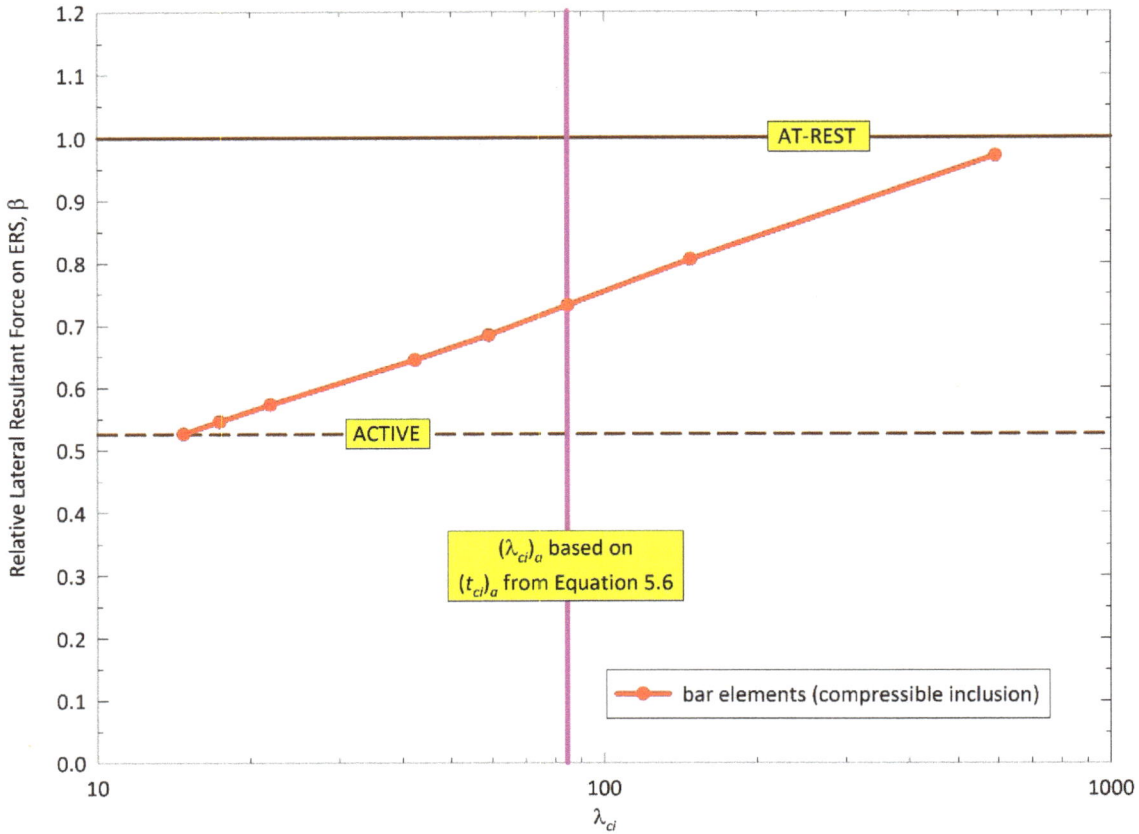

**Figure 5.16. REP-Wall/Full-Depth Placement/Non-Yielding RERS
Relationship between β and log₁₀ λ_{ci}.**

Before proceeding further in the discussion of these results, it is useful to recall that at the outset of the current discussion it was noted that the results presented in the preceding Figures 5.10 through 5.16 inclusive were for analyses based on using the rapid-loading/short-term stiffness of an actual compressible-inclusion material and product. This has always

been and remains the state of practice for REP-Wall applications as it is judged to be justifiably conservative to design for the worst-case scenario in terms of the speed at which a non-yielding RERS with a compressible inclusion will be backfilled or filled during construction. The rapid-loading/short-term modulus is also logical for use when assessing behavior under seismic loading.

However, also as noted earlier, it is considered part of the same good practice to check such a design using the long-term stiffness properties of the compressible-inclusion product to ensure that behavior remains within the operational stress-strain limits of the product. For this particular geofoam product, the long-term stiffness (which is typically defined using results from extended laboratory test durations of 10,000 hours or almost 14 months) is approximately one-half that measured in the standard rapid-loading tests. From a practical perspective, this means that in the long term the compressible inclusion would behave as if it were twice as thick as in the short term. So, for example, and with reference to Figure 5.10, this means that in the long term the compressible inclusion that was 175 mm (7 in) thick would behave as though it were 350 mm (14 in) thick.

With regard to the K_{ci} and λ_{ci} parameters, given the fundamental linear relationship of each with the operational value of Young's modulus of the overall compressible-inclusion product, each of these parameters would reduce in value by a factor of two, i.e. K_{ci} from 146 to 73 and λ_{ci} from 84 to 42, again assuming a compressible-inclusion that was 175 mm (7 in) thick initially. Note that in Figures 5.13 through 5.16 it is very easy to visualize these changes as they would simply follow the red curved lines in each case.

Continuing now with the discussion of results based on the rapid-loading/short-term stiffness of the compressible inclusion, the writer has found that continuous color-contoured plots of stress level[76], S, can be visually informative in situations where the stress field is complex and has large variation due to developing areas of soil failure (shear-strength mobilization). This was illustrated for the baseline all-soil case of a yielding RERS that was explored in Appendix E.

Therefore, it is of interest to develop similar plots for the basic full-depth REP-Wall case currently under consideration. In this particular case, the patterns of shear-strength mobilization within the retained soil can be seen as the thickness of the compressible inclusion increases. For reference purposes, note that stress level is inversely proportional to the ASD concept of factor-of-safety so a stress level equal to zero corresponds to a safety factor of infinity while a stress level equal to one corresponds to a safety factor equal to one.

Figures 5.17a through 5.17i show the results for all compressible-inclusion thicknesses, t_{ci}, studied and reflected in the preceding Figures 5.10 through 5.16 inclusive, beginning with the baseline all-soil/no-geofoam case previously considered in Appendix E. Also indicated for each thickness analyzed are the values of the two dimensionless stiffness parameters, K_{ci} and λ_{ci}.

---

[76] This dimensionless parameter is discussed in detail in Appendix E. In simple terms, it is the ratio of mobilized shear stress to available shear strength and varies from a value of 0 (isotropic stress conditions) to 1 (fully mobilized shear strength or soil 'failure').

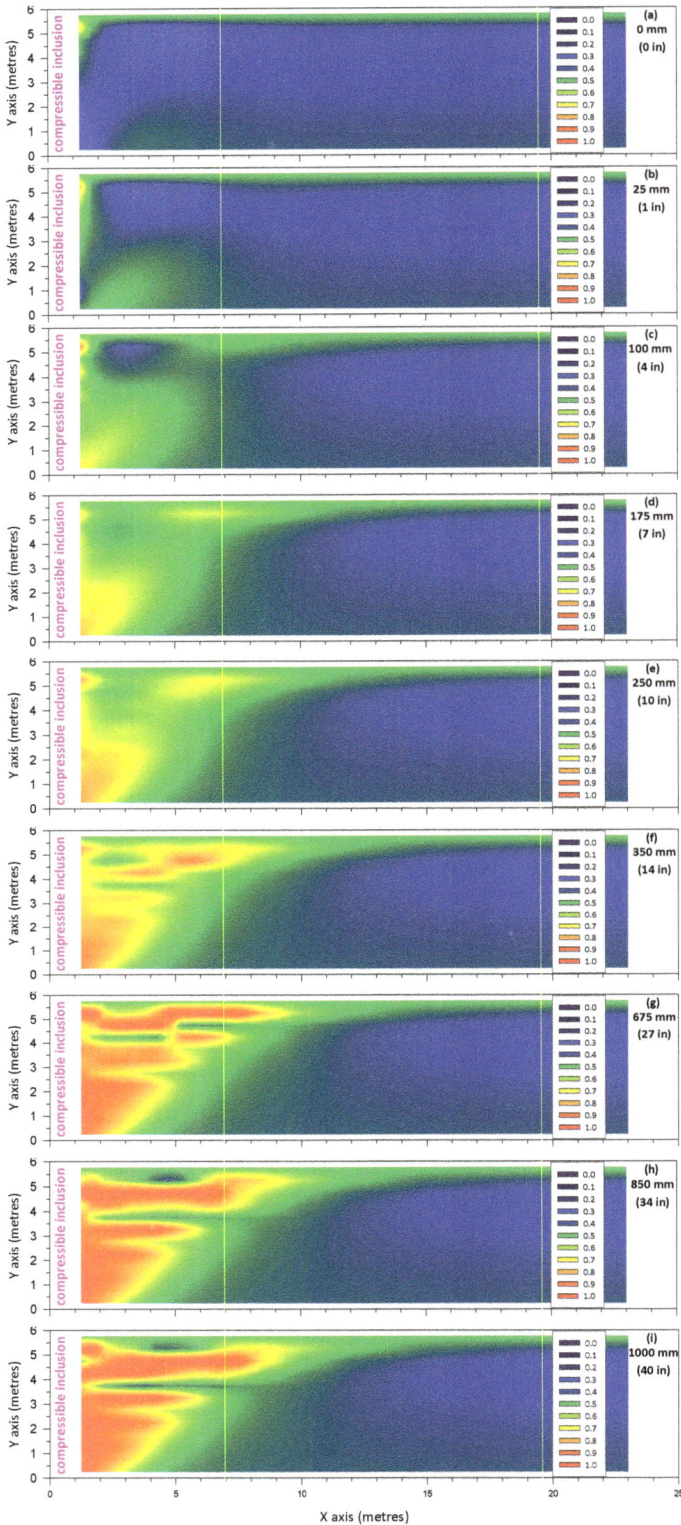

(a)
$t_{ci} = 0$ mm (0 in)
$K_{ci} = \infty$
$\lambda_{ci} = \infty$

(b)
$t_{ci} = 25$ mm (1 in)
$K_{ci} = 1019$
$\lambda_{ci} = 592$

(c)
$t_{ci} = 100$ mm (4 in)
$K_{ci} = 255$
$\lambda_{ci} = 148$

(d)
$t_{ci} = (t_{ci})_a = 175$ mm (7 in)
$K_{ci} = 146$
$\lambda_{ci} = 85$

(e)
$t_{ci} = 250$ mm (10 in)
$K_{ci} = 102$
$\lambda_{ci} = 59$

(f)
$t_{ci} = 350$ mm (14 in)
$K_{ci} = 73$
$\lambda_{ci} = 42$

(g)
$t_{ci} = 675$ mm (27 in)
$K_{ci} = 38$
$\lambda_{ci} = 22$

(h)
$t_{ci} = 850$ mm (34 in)
$K_{ci} = 30$
$\lambda_{ci} = 17$

(i)
$t_{ci} = 1000$ mm (40 in)
$K_{ci} = 25$
$\lambda_{ci} = 15$

Figure 5.17. REP-Wall/Full-Depth Placement/Non-Yielding RERS
Stress Levels Within Retained Soil as a Function of Compressible-Inclusion Thickness.

As noted previously, the minimum compressible-inclusion thickness required to mobilize the active earth pressure state, $(t_{ci})_a$, using the current version of the writer's simplified analytical method (Equation 5.6) appears to underestimate that thickness, assuming that the FE analyses are correct. This was an unanticipated outcome during the preparation of this monograph as a very similar problem analyzed for the writer's past study of integral-abutment bridges (Horvath 2000) indicated that the calculated $t_{ci} = (t_{ci})_a$ =175 mm (7 in) should have been adequate to fully mobilize the active state. While Figures 5.17a-i illustrating patterns of shear-strength mobilization give some insight into this, as it turns out the greater insight comes from examining lateral displacements and strains within the compressible inclusion as shown in the following figures.

To begin with, Figures 5.18a and 5.18b show, respectively, the lateral compressive strains (engineering) within the compressible inclusion, ε_{hci}, and non-dimensionalized lateral displacements, Δ_{ci}/H, along the compressible inclusion-retained soil interface, both expressed in percent and both as a function of non-dimensionalized depth.

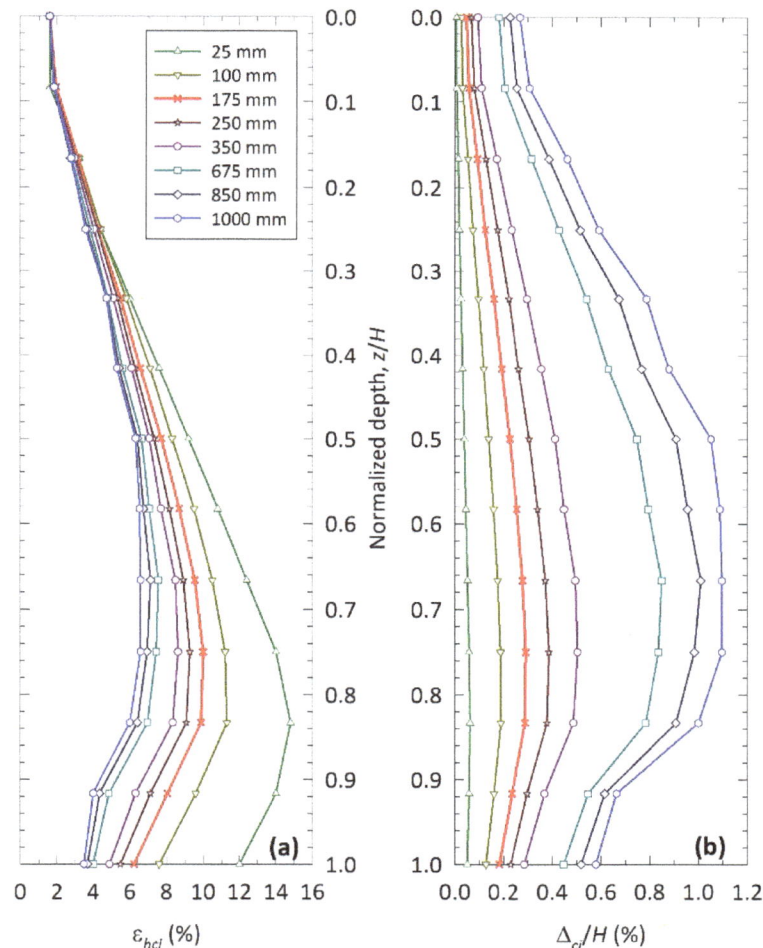

Figure 5.18. REP-Wall/Full-Depth Placement/Non-Yielding RERS Compressible-Inclusion Lateral Strains and Displacements as a Function of Compressible-Inclusion Thickness.

148

Before proceeding further in the discussion of these results, it is necessary to emphasize an important caveat in the conclusions that follow. The compressible inclusion in all FE analyses performed for this monograph was modeled as a linear-elastic material. Specifically, it was modeled based on a material (R-EPS) and product known to be commercially available in the U.S. The maximum calculated compressive strains in Figure 5.18a are well within the linear-elastic range for this material and product as demonstrated in laboratory testing performed on behalf of the manufacturer by an independent geotechnical engineering testing laboratory with international experience. Furthermore, the writer had an opportunity to review the test results produced by this laboratory.

The significance of emphasizing this point is that if 'normal' EPS were used as the compressible inclusion this material does not have the same linear behavior in unconfined axial compression as R-EPS, a point that was illustrated and emphasized in Horvath (2010b). Thus, both the compressive strain and lateral displacement profiles shown in Figure 5.18 would be different, especially at shallower depths where lateral earth pressures are smaller in magnitude and the low-stress stiffness of normal EPS would be most prominent. Furthermore, because compressive stiffness of normal EPS depends on its density, it is not possible to perform a generic suite of calculations that would be applicable in all cases.

In conclusion, it is thus important to appreciate that the results and conclusions that follow in the remainder of this section are applicable only for compressible-inclusion materials that exhibit linear-elastic behavior for compressive strains that begin at zero and extend beyond the limits of the strains shown in Figure 5.18. Normal EPS of any density does <u>not</u> fit this criterion.

Continuing on, as discussed in Appendix E, $(\Delta/H)_a$, the dimensionless ratio of minimum lateral displacement necessary to mobilize the active earth pressure state behind an ERS, is typically defined (e.g. Clough and Duncan 1991) assuming a RERS that displaces laterally in either the classical translation or rotation-about-bottom mode. This is an important detail as this ratio is an integral component of the simplified analytical method for full-depth REP-Wall applications that goes back to the 1980s and continues to the present in the form of the writer's current version of this analytical methodology (Equation 5.6).

The significance here is that it is obvious in Figure 5.18 that the calculated compression pattern of the compressible inclusion and, concomitant lateral-displacement pattern of the compressible inclusion-retained soil interface, are very different from a depth-wise compression/displacement pattern that is either depth-wise uniform in magnitude (as would occur with pure translation) or decreasing linearly with depth to zero from a maximum at the ground surface (as would occur with rotation-about-bottom). If anything, the calculated patterns for a compressible inclusion are closest to the classical rotation-about-top mode where the lateral displacements are zero at the top of the ERS and then increase linearly with depth to a maximum at the bottom of the ERS.

The implication of all this is that all prior versions of the simplified analytical method for full-depth REP-Wall applications appear to be inherently flawed. This is because the application of $(\Delta/H)_a$ ratios from Clough and Duncan (1991) did not account for this significant difference in the displacement pattern of the retained soil between that implied by use of the published $(\Delta/H)_a$ ratios and that actually occurring with a compressible inclusion of uniform initial thickness.

Attempted reconciliation of this disconnect between the implied and actual lateral-displacement patterns along the compressible inclusion-retained soil interface begins by noting that the ratio of the average Δ_{ci}/H to the peak Δ_{ci}/H all the cases shown in Figure 5.18b ranges from 0.6 to 0.7 with an overall average of 0.67 (= ⅔) for all cases. Stated another way, the average lateral displacement of the retained soil (and average compressive strain of the compressible inclusion as well) is approximately two-thirds that of the peak value. This is remarkably consistent considering the relatively large range of compressible-inclusion thicknesses analyzed (25 to 1000 mm/1 to 40 in).

For simplicity of use, this average value of lateral displacement can be interpreted as a quasi-uniform magnitude of lateral displacement for an equivalent mode of pure translation. Note, however, that this is an extreme approximation as the calculated distributions shown in Figure 5.18b are far from being depth-wise uniform.

Because the simplified analytical method, whether derived on the basis of lateral earth <u>pressure</u> as shown earlier in this chapter (Figure 5.6 and Equation 5.6) or lateral resultant earth <u>force</u> as shown in Appendix A of Horvath (2000), is based on a magnitude of displacement that is implied to exist over the entire geotechnical height of the ERS, it is clear that the lateral-displacement variable, Δ, in Equation 5.6 needs to be adjusted so that the <u>average</u>, not peak, lateral displacement/compression of the compressible inclusion is matched to the assumed lateral displacement of the retained-soil mass as reflected in the $(\Delta/H)_a$ ratio chosen by the design professional.

This goal can be achieved by multiplying the right-hand side of Equation 5.6 by a factor of 1.5 (= 1/2/3) which results in the following updated version of the writer's simplified analytical method:

$$(t_{ci})_a = \frac{2 \cdot E_{ci} \cdot (\Delta/H)_a}{K_a \cdot \cos\delta \cdot \gamma_{eff}}. \tag{5.17}$$

Note that this change would also affect the relationship for the absolute compressible inclusion stiffness, K_{ci}, required to mobilize the active earth-pressure state. Equation 5.8 would thus become

$$(K_{ci})_a = \frac{K_a \cdot \cos\delta \cdot \gamma_{eff}}{2 \cdot (\Delta/H)_a \cdot \gamma_w}. \tag{5.19}$$

However, the relationship for the relative compressible inclusion stiffness, λ_{ci}, required to mobilize the active earth-pressure state remains unchanged from that given previously but is restated here for ease of reference:

$$(\lambda_{ci})_a = \frac{(K_{ci})_a \cdot H \cdot \gamma_w}{p_{atm}}. \tag{5.10}$$

The results up to this point are summarized in Figure 5.19 that shows the calculated (using the FEM) relationship between the relative resultant lateral force, β, and Δ/H for both the peak and average values of Δ. Three cases are also shown for the simplified bi-linear model that is used with the writer's simplified analytical model:

- medium-dense sand per Clough and Duncan (1991) and Equation 5.6 that reflects peak Δ,

- medium-dense sand per Clough and Duncan (1991) and Equation 5.17 that reflects average Δ, and

- loose sand per Clough and Duncan (1991) and Equation 5.17 that reflects average Δ.

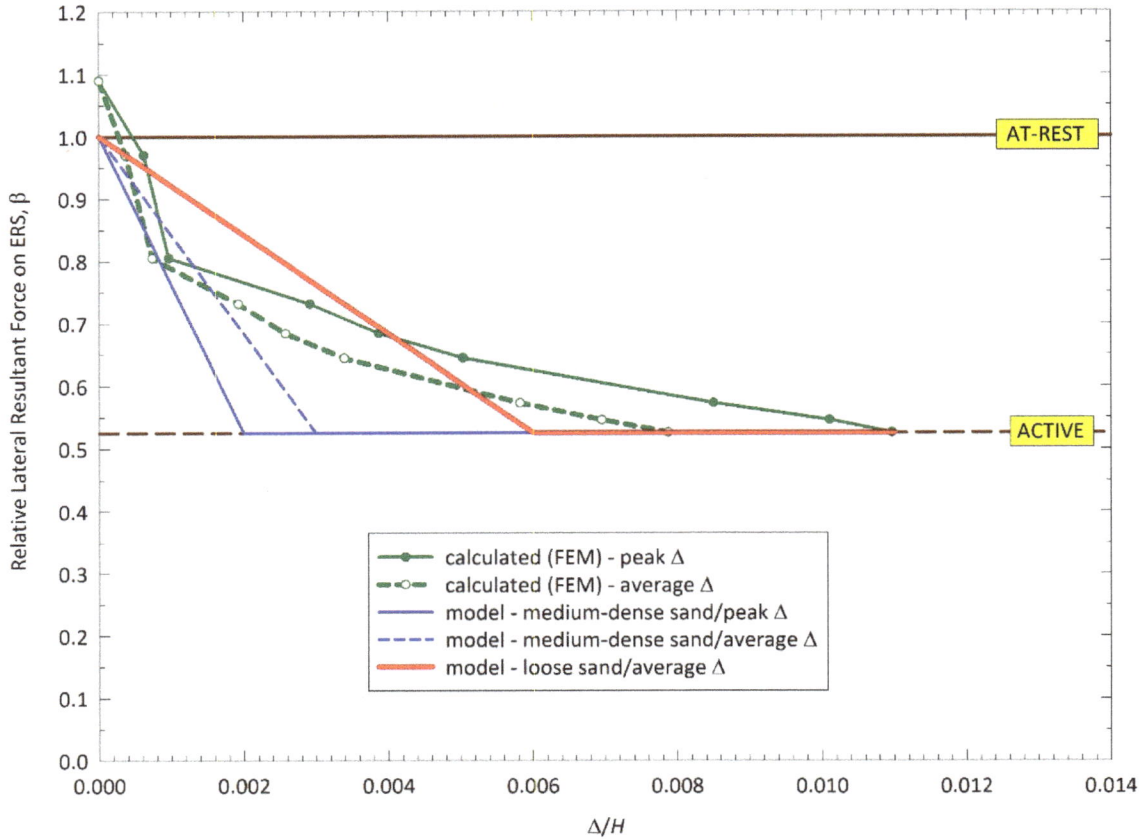

**Figure 5.19. REP-Wall/Full-Depth Placement/Non-Yielding RERS
Relationship between β and Δ/H for Finite-Element Model Used.**

Note that the first case (medium-dense sand/peak Δ) reflects the assumptions used in Equation 5.6 to arrive at the value of $(t_{ci})_a$ = 175 mm (7 in) that was used in the FE analyses presented earlier in this section. The third case (loose sand/average Δ) reflects revised assumptions based on Equation 5.17 plus results presented in Appendix E that suggest Clough and Duncan's "loose" value for $(\Delta/H)_a$ better reflects the behavior of the soil used in the FE analyses performed for this monograph. Note that for this third case, $(t_{ci})_a$ = 525 mm (21 in) which is three times the compressible-inclusion thickness estimated initially based on the aforementioned medium-dense sand/peak Δ assumptions.

The conclusion drawn from Figure 5.19 is that <u>average</u> lateral displacement of the compressible inclusion-retained soil interface, which is the same as the average compression of the compressible inclusion, is a better performance metric of the compressible inclusion-retained soil system than the <u>peak</u> value as has been implied in usage dating back to the 1980s to the present. The reason as stated previously is because the $(\Delta/H)_a$ ratios commonly found in the published literature assume the classical RERS displacement modes of either pure translation or rotation-about-bottom. As is apparent from Figure 5.18b, the lateral-displacement pattern of a compressible inclusion-retained soil interface mimics neither but the average is at least closer to pure translation than rotation-about-bottom[77].

Note in Figure 5.19 that the average calculated (FE) results shown with the green dashed curve agree best in terms of displacement necessary to mobilize the active earth pressure state with the model based on average lateral displacements for assumed loose soil conditions (red solid lines). There is still some discrepancy (overall, the simple bi-linear model is clearly a crude approximation compared to the calculated results) but this is likely due in part to the fact that the calculated results start from an earth-pressure state slightly greater (β = 1.09) than the theoretical at-rest state (β = 1.0) for reasons discussed in Appendix E.

To close out this lengthy discussion of solution assessment for the basic full-depth REP-Wall application, it is of interest to address several issues that can significantly impact conclusions that might be drawn from the preceding discussion. First and foremost is the aforementioned issue of temporal effects on the compressive stiffness (Young's modulus) of the compressible-inclusion product used. Depending on the specific geofoam material(s) that comprise the product, material creep can be significant even under relatively low stress levels. This is certainly true for the materials-of-choice in most applications (R-EPS used either alone or in combination with GPS-PB to provide a multi-functional product).

As noted several times previously, in routine practice the Young's modulus obtained for geofoam materials under the de facto standard rapid strain-rate-controlled loading (typically $\dot{\varepsilon}$ = 10% min[-1]) is used for design. For compressible-inclusion applications, this is generally very conservative as it overestimates the stiffness that will exist in the actual application, at least under gravity loading. The actual creep behavior of both R-EPS and GPD-PB are temporally non-linear, with compressive creep occurring at a decreasing rate that is broadly similar to the creep behavior of soils. After approximately one year, the Young's modulus for these geofoam materials is typically about one-half the value measured in the standard rapid-loading test.

Because of the linear relationship between required compressible-inclusion thickness, $(t_{ci})_a$, and material modulus, E_{ci}, in Equation 5.17 (as well as Equation 5.6 that this equation replaced), this means that after approximately one year the compressible inclusion behaves as if it were twice as thick as it was initially. The significant beneficial effect of time on the behavior of a compressible inclusion within a relatively short timeframe after construction raises the issue of whether or not this benefit should be considered more routinely in practice, especially since any change to basing design on average, as opposed to peak, lateral displacements as outlined above will, as a minimum, cause a 50% increase in calculated required thickness of the compressible-inclusion product. This is independent of whether it is assumed that the retained soil is dense, medium-dense, or loose in terms of its relative density.

The next issue considered is that of a yielding RERS. From an overall behavioral perspective in terms of its effect on the retained soil and the lateral earth pressures that the retained soil imparts on a RERS, yielding can be viewed as an equivalent compressible inclusion. This is because a certain magnitude of lateral displacement, Δ, due to yielding of the RERS will result in a certain reduction in lateral earth pressure acting on the RERS. This is the underlying logic behind the age-old practice of designing a yielding RERS for the active earth pressure state.

---

[77] There are no known (to the writer) published $(\Delta/H)_a$ correlations based on the classical yielding mode of rotation-about-top which, as noted earlier in this section, is arguably the best match with the calculated lateral-displacement patterns, at least in terms of the three classical RERS yielding modes that were presented and discussed in Chapter 2.

Although noted previously but not explored for this monograph, there may be situations in practice where use of a compressible inclusion with a yielding RERS is desirable to ensure complete mobilization of the active earth pressure state within the retained soil under service conditions. This would be particularly advantageous from a structural engineering perspective with regard to the ERS components for the reasons discussed in Chapter 2.

Estimating the compressible-inclusion thickness required for use with a yielding RERS requires first quantifying the yielding of a RERS under service-load conditions so that an equivalent compressible-inclusion thickness can be calculated for the yielding contribution. The is not a simple task which is why historically it has not been done in routine practice. Based on the present state of knowledge, a numerical analysis using the FEM or other analytical methodology would be required in order to make this assessment accurately. This requirement would seem to limit the practical potential for using the REP-Wall concept with yielding RERSs at the present time.

The final issue addressed is the use of a compressible inclusion of non-uniform depth-wise thickness. This issue has not been considered previously by the writer or anyone else to the best of the writer's knowledge, at least in any detail. However, the subject was raised as a question to the writer during a visit to Japan in 1996 (the Japanese engineers who raised the question were more interested in seismic loading as opposed to gravity loading) so perhaps it has been explored there.

In this regard, it is clear from the preceding discussion that the depth-wise compression profile of a compressible inclusion that has a uniform depth-wise thickness (and uniform stiffness as was noted earlier) is highly non-uniform (Figure 5.18a). This results in a depth-wise non-uniform variation in lateral displacement along the compressible inclusion-retained soil interface (Figure 5.18b). It is also clear from the preceding discussion that the average, not peak, value of this lateral compression is what determines the degree to which the active earth-pressure state is mobilized. This suggests that the most efficient compressible-inclusion configuration would be one in which the compressible-inclusion compression and concomitant lateral displacement of the retained soil approach depth-wise uniform distributions. Figure 5.18b indicates that this would require the compressible inclusion to be substantially thicker near its top than the bottom.

A non-linear depth-wise variation in compressible inclusion thickness would appear to be the most effective from a purely theoretical perspective. However, this would not be cost effective from either a manufacturing or constructability perspective. A pragmatic compromise might be to use a larger uniform thickness for the upper portion of the ERS with a smaller uniform thickness for the lower portion. This would appear to be a worthwhile subject for future research.

5.4.2.4.3 Seismic Loading

The discussion of solution assessment as it applies to full-depth REP-Wall application with seismic loading, a.k.a. the seismic-buffer concept, requires a different approach than that used in the preceding section for gravity loading. This is because the development of the seismic-buffer concept to date has followed a path unique to this application.

As has been discussed several times in this monograph, the traditional technological path taken, for the most part, with both the lightweight-fill and compressible-inclusion functions for reducing loads on ERSs has been first and foremost to hypothesize and develop relatively simple analytical models and concomitant solutions intended for use in routine practice. This was done for two reasons:

- to facilitate initial acceptance and use of these then-new geotechnologies in practice by design professionals and

- to provide an initial basis of comparison for academicians conducting fundamental research into these geotechnologies. This is because at some point the results obtained using these relatively simple methods need to be compared to measured results obtained in some type of physical testing, e.g. instrumented observation of actual structures; instrumented large-scale models under 1-g conditions; instrumented small-scale models under centrifuge (multi-g) conditions. As a fallback alternative in the absence of meaningful physical testing, numerical analyses such as the FEM can be used as at least a preliminary, first-order verification tool as has been the case in this monograph. Based on the outcomes of this 'reality check', the analytical methodology in question is either validated as being reasonably accurate for continued use; modified to provide better agreement with observations; or abandoned to be replaced with another, more-accurate method, with the results from physical testing and/or numerical analyses providing insight into the correct behavioral mechanism that needs to be incorporated into a new simplified analytical methodology.

Contrary to this approach, development of the seismic-buffer concept has, for the most part to date, focused primarily on what has been called "proof-of-concept" research, i.e. academic research geared toward demonstrating that a geofoam compressible inclusion can indeed reduce lateral earth loads on a non-yielding RERS under conditions of simulated seismic loading. This proof-of-concept was initially based on physical testing of a relatively small-scale (1 m/3.3 ft high) instrumented model subjected to shake-table testing under laboratory conditions at the Royal Military College of Canada (RMC[78]) in Kingston, Ontario.

More or less contemporaneous with these RMC shake-table tests, an initial suite of numerical analyses using the FDM and *FLAC* commercial software was performed with the goal of replicating the measured results from physical testing with calculated results from the numerical analyses. It appears that the intent was to calibrate the FDM model so that it could be used with confidence to perform a follow-on suite of parametric analyses of non-yielding RERSs of various heights and thus extend the very limited utility of the RMC shake-table testing.

In any event, essentially the path taken to date with the seismic-buffer concept has been, in a sense, to conduct the verification phase for simplified analytical methodologies before there was a method or methods to verify. This is certainly a viable approach to the overall problem as the results from laboratory testing and advanced numerical models can be used to provide guidance in the formulation of a physical model and from this a relatively simple, practice-oriented analytical methodology.

However, to date there appears to be relatively limited movement in that direction. The aforementioned Q-RMC Wedge Method does not appear to have evolved into something that is even remotely attractive for routine use in practice. The most practice-oriented outcome to date that has appeared in publication is in Zarnani and Bathurst (2009a), with a summary in Bathurst and Zarnani (2013), of what are called "preliminary design charts" relating *isolation efficiency* (defined in Appendix B) and the equivalent spring stiffness of the

---

[78] 'RMCC' is also used as an acronym for this institution. However, the writer's experience is that 'RMC' is more commonly used in geotechnical engineering literature.

compressible inclusion per unit area (essentially the parameter k_{cl} defined in Equation 5.1 although Bathurst and Zarnani labeled this parameter 'K'). Note that Bathurst and Zarnani's K parameter has dimensions using specific metric units so is not directly equivalent to the writer's proposed dimensionless K_{cl} parameter.

In all fairness, research to date indicates that seismic loading presents many more challenges compared to gravity loading when it comes to developing simplified analytical methods. The most significant issue is that the ability of a compressible inclusion in a given application to reduce seismic loads on a non-yielding RERS appears to be dependent on the frequency of the input motion relative to the natural frequency of the RERS-geofoam-soil system. This is consistent with the decades-long findings of many researchers for traditional yielding and non-yielding RERSs without a compressible that are subjected to seismic loading as discussed in Appendix A so should not come as a surprise. That having been said, that has not prevented the nearly 100-year-old Mononobe-Okabe Method or nearly 50-year-old Seed-Whitman Method from being widely used in practice to the present even though these methodologies do not consider the frequency content of the assumed seismic loading

However, the larger, overarching problem with using any of the results from the research reported by Bathurst et al. is that the writer believes the stiffness of the geofoam products used, whether in the original suite of RMC shake-table tests or as simulated in the most-recent suite of numerical analyses reported in Bathurst and Zarnani (2013) for simulated RERSs up to 9 metres (20 ft) high, has, with one exception, not been correctly assessed. Specifically, except for the one RMC shake-table test (out of six total with a compressible inclusion) that used R-EPS, all other tests and numerical simulations have used normal EPS-block in one form or another. Furthermore, in most cases the work of Bathurst et al. relied on empirical correlations for EPS-block stiffness under small-strain conditions as opposed to testing the actual material used under the large-strain conditions typical of compressible-inclusion applications being either actually tested or simulated numerically.

As explained in detail in Horvath (2010b), normal EPS-block has highly non-linear, inelastic behavior when loaded within the relatively large-strain range that is characteristic of all compressible-inclusion applications. While this does not preclude its being used as a compressible inclusion, there are three significant issues to consider:

- The single biggest mistake made by academicians and practitioners alike worldwide is to assume that density/unit weight of EPS-block is the only material property necessary to accurately estimate its primary stiffness property (Young's modulus). Unfortunately, in recent years this misconception has been fostered and promulgated by ASTM standards that use EPS density as the sole nomenclature basis for different grades or types of EPS-block for geofoam applications. As discussed in detail in Horvath (2011, 2012), there are numerous technical issues that can affect the mechanical (stress-strain-time-temperature) properties of EPS-block that are simply not reflected in the density/unit weight of the final product. As a result, EPS-block of a given density can exhibit a surprisingly large range of mechanical behavior depending on the exact way in which it was made.

- Separate from material-quality issues, the correct assessment of the operative Young's modulus of EPS-block is very complicated, especially when cyclic loading is involved, as the material is both non-linear and inelastic over a large-strain range. This makes both interpretation of the RMC shake-table test data and formulation of a constitutive model for use in numerical analyses very difficult.

- In general, normal EPS-block is inefficient and not cost-effective to use as a compressible inclusion, especially under relatively low-stress situations such as would have existed in the RMC shake-table tests where the simulated non-yielding RERS was only 1 metre (3.3. ft) high.

In consideration of these issues, the writer believes that the published works of Bathurst et al. may not have been properly assessed with regard to stiffness of the compressible inclusion. In particular, it calls into question the accuracy of the aforementioned "preliminary design charts" that have appeared in later publications by these authors.

In summary up to this point, it appears safe to say that the research to date into the use of the basic full-depth REP-Wall application of compressible inclusions, a.k.a. the seismic-buffer or seismic-isolation concept, to reduce seismic loads on non-yielding RERSs has at least been sufficient to qualitatively support the conclusion that relatively significant, and thus technically and economically useful, load reductions can be achieved compared to the baseline case of not using a compressible inclusion. The writer does, however, question the quantitative conclusions in the form of design charts published in Zarnani and Bathurst (2009a) and Bathurst and Zarnani (2013) because of legitimate questions and concerns over how the operational value of Young's modulus was assessed or evaluated in the process that produced these charts, combined with the fact that using EPS-block density alone as a necessary and sufficient metric of geofoam stiffness is seriously flawed.

It is also safe to say that, independent of the modulus issue, the relative magnitude of seismic-load reduction, whether expressed using the concept of 'isolation efficiency' as proposed by Bathurst et al. or by simply extending the more general β parameter as suggested by the writer in Appendix B, is a complex function of a specific application and the frequency content of the cyclic load applied to the system. At the present time, it appears that an advanced numerical-analysis technique such as the FDM or FEM is required to adequately consider all the factors that drive the final result.

That having been said, it is still of interest to explore the writer's simplified analytical method based on the M-O Method that was proposed earlier in this chapter using the RMC shake-table tests performed by Bathurst et al., even though the geotechnical height, H, of the simulated RERS in these tests was only 1 metre (3.3 ft). As noted previously, one of the six different compressible inclusions evaluated in this physical-testing program was the same R-EPS commercial product that the writer simulated in the gravity-loading FE analyses performed for this monograph. Some of these FE results were presented earlier in this chapter. Consequently, the stiffness of the compressible inclusion for that one RMC shake-table test is known reliably to the writer's satisfaction.

In addition, although the original M-O Method does not consider the frequency content of an earthquake, that has not prevented it from being the single most common analytical methodology, in one form or another, used in routine practice to the present as discussed in detail in Appendix A. Consequently, there is some justification for the writer's M-O Method-based simplified analytical method for the seismic-buffer application even though the writer's methodology does not consider frequency content.

To begin with, Equation 5.12, which is restated here for ease of reference, is used to calculate the required minimum thickness of the compressible inclusion under seismic loading, $(t_{ci})_{ae}$:

$$\left(t_{ci}\right)_{ae} = \frac{E_{ci} \cdot \left(\Delta / H\right)_{ae}}{0.75 \cdot K_{ae} \cdot \cos\delta \cdot \gamma_{eff}^{*}} . \tag{5.12}$$

Note that this is the original equation that was presented earlier in this chapter for use as part of the writer's simplified analytical method related to seismic loading. Unlike the writer's related equation for gravity loading, Equation 5.12 will not be further modified to account for the effect of the lateral-displacement pattern of the compressible inclusion-retained soil interface simply because there is uncertain information at this time as to what this pattern actually looks like under seismic loading. As discussed in some detail in Appendix A, the broad subject of seismic loading on both yielding and non-yielding RERSs is one that has been undergoing considerable reassessment in recent years after many decades of relatively little development and advancement.

The next step is defining the material properties of the soil and compressible inclusion that were used in the RMC shake-table tests so that the various parameters in Equation 5.12 can be evaluated. Information concerning the RMC shake-table tests can be found in Zarnani et al. (2005), Bathurst et al. (2007b), and Zarnani and Bathurst (2008), from which the following were obtained:

- γ^*_{eff} = 15.2 kN/m³ (96.9 lb/ft³)
- ϕ_{peak} = 51°
- ϕ_{cv} = 46°
- $\delta = \phi$ (assumed by writer)
- E_{ci} = 270 kN/m² (5.6 kips/ft² = 39 lb/in²).

Note that E_{ci} is the rapid-loading/short-term value for the R-EPS geofoam material and product that was used. While this value would typically be conservative for gravity loading, it was clearly appropriate for the dynamic loading conditions considered here.

The unusually high ϕ values require some elaboration and comment. Bathurst et al. used a sandy 'soil' that consisted of artificially crushed, graded, and blended olivine, a naturally occurring magnesium iron silicate with a specific gravity of solids, G_s, that is somewhat greater than that of natural quartz sands[79]. As a result, the retained soil used in all of the RMC shake-table tests they performed consisted of particles that were unusually and uniformly angular compared to that normally found in natural soils. As will be seen, this choice of fill material, combined with the fact that it was compacted to a dense consistency, was unfortunate as for the model used it appears that the full potential of the compressible-inclusion materials in terms of load reduction was not exploited.

The RMC shake-table test evaluated here was run with horizontal accelerations as high as $k_h \cong 0.75$ so the K_{ae} parameter in Equation 5.12 was evaluated using both the original M-O Method and the recommendations of Seed and Whitman (as detailed in Appendix A) as significant variations between the two methods were expected at higher accelerations. In addition, both the M-O and S-W Methods were evaluated using both ϕ_{peak} and ϕ_{cv} as it is unclear if sufficient lateral displacement of the retained soil in all cases was sufficient for strength conditions to pass over the peak and strain-soften to the constant-volume (critical-state) condition.

To begin with, Figure 5.20 shows various calculated values of K_h, the general coefficient of lateral earth pressure, as a function of assumed values of k_h, the horizontal acceleration coefficient, over the full spectrum of values used in the RMC shake-table test being evaluated here. Note that both K_{ae}, the lateral earth pressure coefficient for the

---

[79] Bathurst et al. reported that they intentionally avoided use of a siliceous soil for health-related issues involving silica dust in an indoor space.

seismic active state[80], as well as K_{onc}, the coefficient of lateral earth pressure at-rest under normally consolidated gravity-load (i.e. k_h = 0) conditions, are shown. The latter was calculated using the well-known equation:

$$K_{onc} = 1- \sin \phi. \tag{5.20}$$

Note that the values of K_{ae} shown for the M-O Method are based on $k_v = 0$. As discussed in Appendix A, for traditional RERSs it has long been established that k_v has relatively little effect on calculated values of K_{ae} so is often neglected in routine design methods. This is the reason that k_v is ignored completely in the S-W Method.

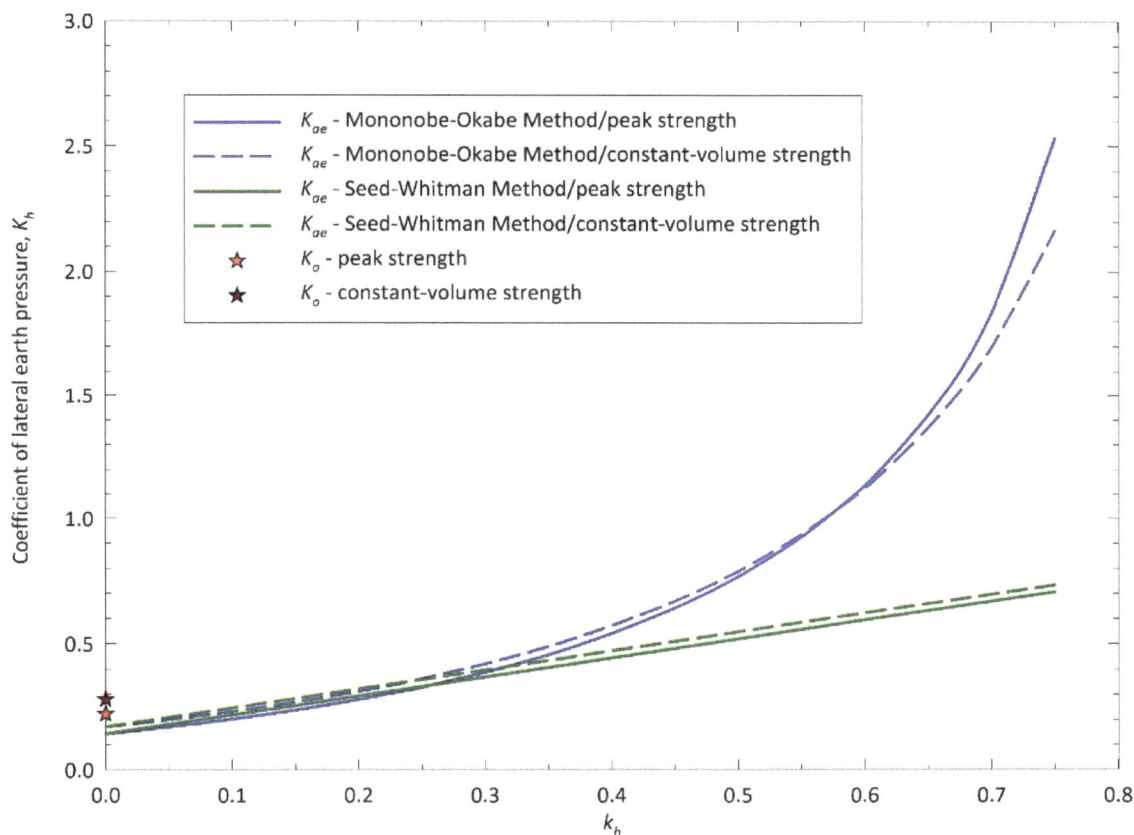

Figure 5.20. Relationship between K_h and k_h for RMC Shake-Table Test Using R-EPS Compressible Inclusion.

The results shown in this figure are entirely as expected based on the extensive discussion of seismic analyses in Appendix A. The M-O and S-W Methods are in general agreement for relatively low values of k_h up to approximately 0.3 but diverge increasingly for higher values beyond approximately 0.4. For either method, the results are relatively insensitive to whether the peak or constant-volume shear strength is used.

---

[80] Note that K_{ae} from both the M-O and S-W Methods is the same as Coulomb's K_a when $k_h = 0$.

Next shown in Figure 5.21 are the calculated values of $(t_{ci})_{ae}$, the minimum required thickness of the compressible inclusion in order to mobilize the active earth pressure state under seismic conditions, as a function of k_h. These calculated values are based on Equation 5.12 using the values of K_{ae} shown in Figure 5.20 and assuming $(\Delta/H)_{ae} = 0.001$ (= 0.1%) which is the published value (Clough and Duncan 1991) for dense sand. Note that for $H = 1$ metre (3.3 ft) as in this case this means that Δ_{ae}, the <u>average</u> value of lateral displacement required to mobilize the active earth pressure state under seismic conditions, is only 1 millimetre (0.04 in). While this value may seem relatively small in magnitude, it is relevant to note that this value is nonetheless realistic in this particular case as it is the same order of magnitude as average lateral displacements of the compressible inclusion-retained soil interface measured in the RMC shake-table tests as reported in Bathurst et al. (2007b).

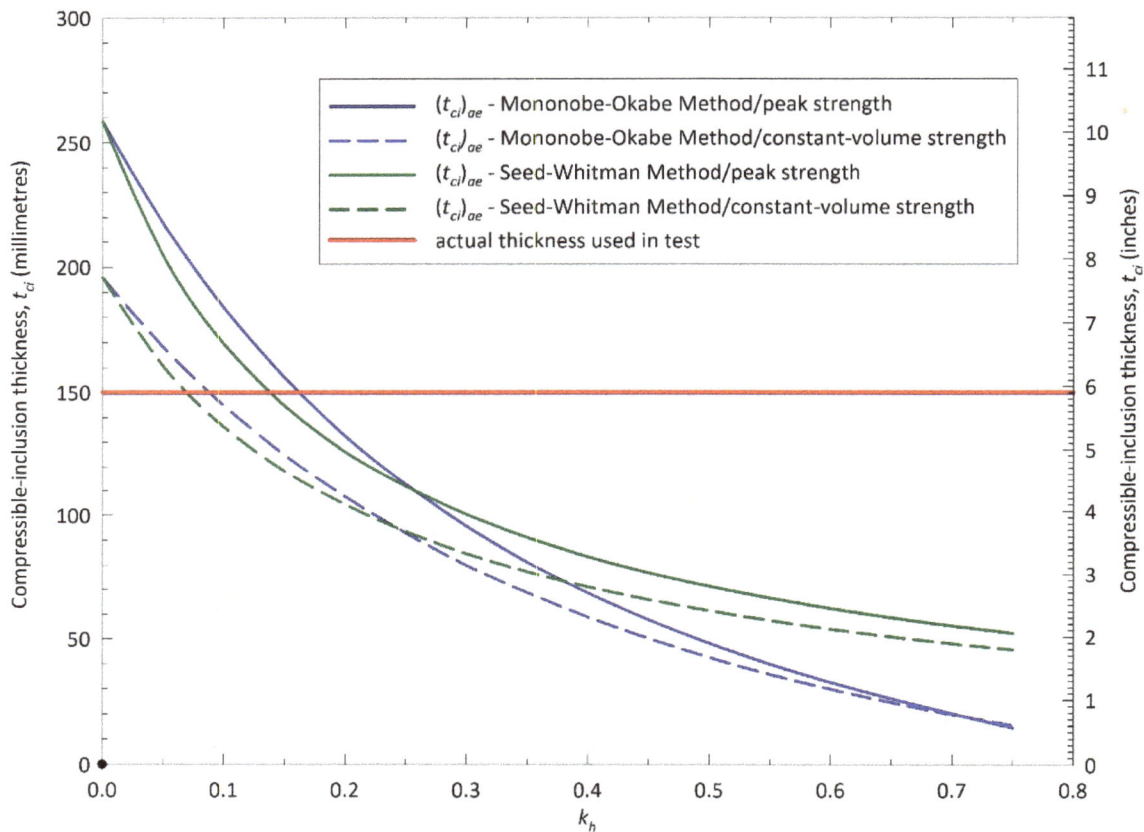

Figure 5.21. Relationship between t_{ci} and k_h for RMC Shake-Table Test Using R-EPS Compressible Inclusion.

Several aspects of the results shown in this figure are interesting:

- The calculated values of $(t_{ci})_{ae}$ are very sensitive to the assumed value of k_h, especially at lower levels of shaking.

- The calculated values of $(t_{ci})_{ae}$ are sensitive to whether the peak or constant-volume value of ϕ is assumed, again, especially at lower levels of shaking.

- The calculated values of $(t_{ci})_{ae}$ display increasing sensitivity to the particular analytical method (M-O vs. S-W) used to calculate K_{ae} as k_h increases.

- The actual thickness of the compressible inclusion used may not have been sufficient to allow full mobilization of the shear strength of the retained soil under relatively low levels of shaking. This appears to be due to a combination of the low height of the test box combined with the unusually high value of ϕ for the fill material used that resulted in relatively small values of lateral earth pressure on the compressible inclusion that barely caused it compress.

Lastly, the results shown in Figure 5.22 are the lateral resultant forces, P_h, acting on the simulated non-yielding RERS. Shown in this figure are:

- the calculated values, P_{ae}, using both the M-O and S-W Methods. For clarity, only the results assuming ϕ_{peak} are shown as there is very little difference with the results assuming ϕ_{cv};

- the calculated values, P_{oe}, using Wood's solution for a non-yielding RERS and assuming ϕ_{peak} conditions. This is to provide a baseline, all-soil/no-geofoam result for reference. As discussed in Appendix A, Wood's solution has, in recent years, been deprecated as an analytical methodology that is correct for non-yielding RERSs. However, the simple fact is that it was the widely accepted standard for such applications for decades and still remains in widespread use in routine practice. Consequently, there is some value to showing results from this analytical method; and

- the trend lines for the measured results in the RMC shake-table test for both the compressible inclusion under study as well as the baseline all-soil/no-geofoam test that was performed. Note that the actual measured results for both cases shown in Bathurst et al. (2007b) are somewhat scattered and non-linear. However, both cases show broad linear trends and it is the writer's interpretation of those trends that is presented in this figure.

The results shown in this figure are quantitatively inconclusive and thus unhelpful as a meaningful solution assessment of the full-depth REP-Wall concept under seismic loading (a.k.a. seismic buffer or seismic isolation). While the measured forces using a compressible inclusion are clearly smaller in magnitude compared to the baseline reference case of measured forces with no compressible inclusion, they are still considerably larger than forecast forces using the S-W Method, even at relatively high accelerations where analyses presented above indicate that the compressible inclusion should have been fully effective at mobilizing the active earth pressure state.

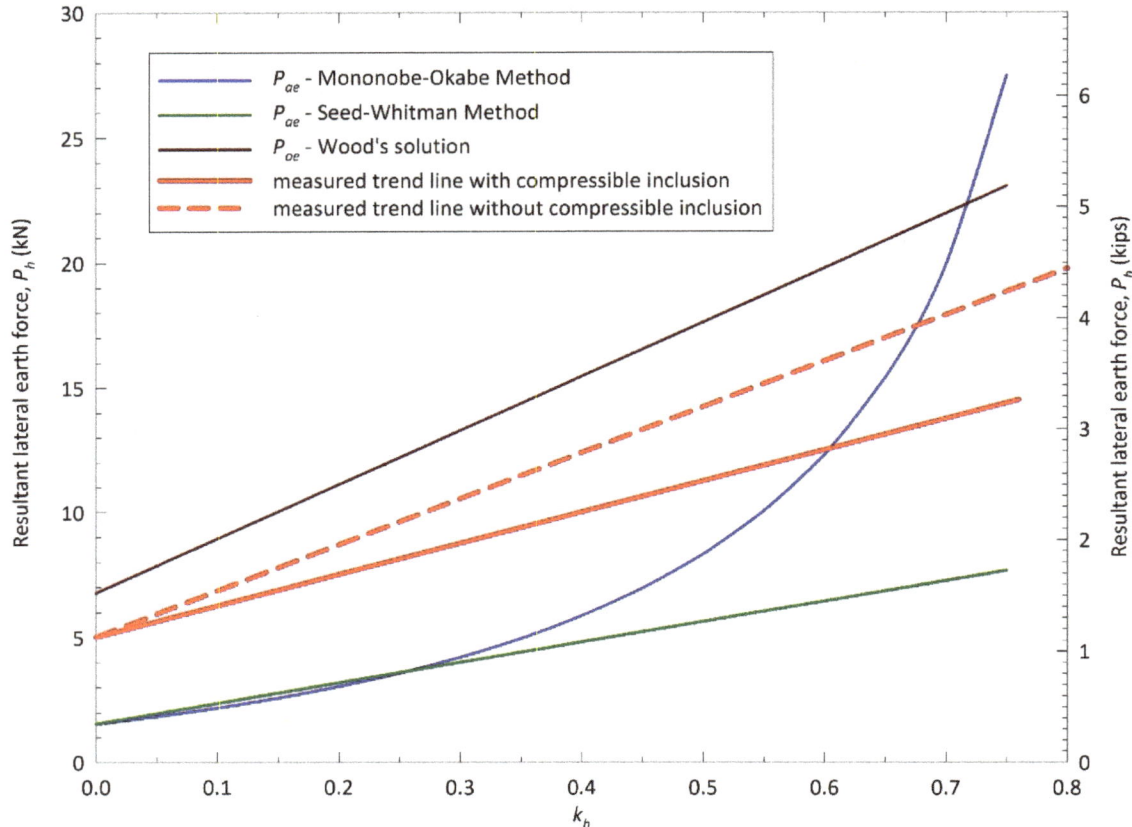

Figure 5.22. Relationship between P_h and k_h for RMC Shake-Table Test Using R-EPS Compressible Inclusion.

In the writer's opinion, this observed outcome may, at least in part, be an artifact of the small scale of the RMC shake-table tests. Specifically, the RMC model was constructed so that full friction existed between the bottom surface of the retained soil and the underlying bottom of the shake-box. This would have provided significant constraint against lateral displacement for the lower portion of the retained soil, a fact confirmed qualitatively at least by the writer's FE analyses performed for this monograph that used the same boundary condition for gravity loading. Because such a constraint is more or less independent of the geotechnical height, H, of a RERS, it would have had an overwhelming effect on the behavior of the relatively low-height (1 metre/3.3 ft) model used in the RMC shake-table tests. This was actually confirmed indirectly by the measured lateral displacements as reported in Bathurst et al. (2007b) that showed that the upper portion of the retained soil laterally displaced approximately twice as much as the lower portion throughout the entire range of accelerations.

The only unambiguous conclusion drawn from the results shown in Figure 5.22 is qualitative in nature. Specifically, the measured results qualitatively support both Wood's solution for non-yielding RERSs and the S-W Method for yielding RERSs (and, by extension, non-yielding RERSs with an appropriate compressible inclusion as in the case here) in that the resultant lateral force acting on RERSs increases more or less linearly with the magnitude of horizontal shaking. A corollary to this is that the long-recognized conclusion that the M-O Method forecasts unreasonable results in that they increase exponentially with increasing horizontal acceleration is also supported.

To conclude the discussion in this section, despite a substantial uptick in interest and concomitant research in the 21st century into seismic applications of the full-depth REP-Wall concept, the current state of knowledge remains unresolved and confusing to design professionals engaged in routine practice. While the RMC shake-table tests confirmed that there is indeed a benefit to using a compressible inclusion with non-yielding RERSs, for any number of reasons as outlined above the results of these tests are quantitatively of little or no value in the writer's opinion. While a simplified, practice-oriented analytical methodology based on the S-W Method is, in principle, available, it remains unproven (it should be clear that the M-O Method is unrealistic for k_h greater than ~0.3). This leaves as the only 'proven' analytical approach the use of a site- and application-specific analysis using advanced numerical methodologies such as the FDM and FEM.

5.4.3 Partial-Depth Placement Alternative

5.4.3.1 Introduction and Overview

As discussed in Chapter 3 and illustrated conceptually in Figure 3.8, the REP-Wall concept can, in principle, be used with a compressible inclusion that covers only a portion of the inside face of a RERS that is in contact with the retained soil. Such an application is referred to in this monograph as the *partial-depth placement alternative* of the REP-Wall concept.

To date, this alternative does not appear to have been studied in any formal or methodical manner. This is most likely due to the fact that there are many more potential variations in applications compared to the partial-depth placement alternative using the lightweight-fill function. In the latter case, no matter how much EPS-block is used the blocks must always be configured in such a way so as to minimize the vertical stresses acting on the assemblage of blocks from a combination of soil/pavement cover plus live loads as shown conceptually in Figure 3.6. On the other hand, with the partial-depth placement alternative using the compressible-inclusion function, the panel of geofoam can be placed anywhere along the inside face of the RERS.

Another possible reason why the partial-depth placement alternative of the REP-Wall concept has likely not been studied is that there is little incentive to do so as opposed to using the basic full-depth placement alternative. In new construction, there is little money to be saved by not using full-depth placement and in retrofitting an existing RERS, excavating for full-depth placement as opposed to partial-depth placement is unlikely to be problematic in most cases.

The net result is that at the present time there is no known simplified analytical methodology for the partial-depth placement alternative when applied to the REP-Wall concept, even for gravity loading. Consequently, the only reliable analytical methodology is to perform a project-specific assessment using an advanced numerical-analysis tool such as the FEM or FDM.

5.4.3.2 Case History Experience

Despite the lack of formal R&D to date into the partial-depth REP-Wall alternative, there is a case history application that was reasonably well instrumented at the time of its construction in the 1980s and subsequently researched in detail in the 1990s under the writer's direction using the FEM. By happenstance, this case history was also the first known use of a geofoam product specifically for its geosynthetic function as a compressible inclusion with the primary goal of reducing lateral earth pressures on a non-yielding RERS.

Details concerning this case history can be found in Partos and Kazaniwsky (1987). In summary, a proposed high-rise building in Center City (downtown) Philadelphia, Pennsylvania required a basement wall approximately 10 metres (33 ft) deep on one side but more than 2 metres (7 ft) less on the other. A conventional design based on at-rest lateral earth pressures was judged to create a substantial, permanent unbalanced lateral load on the substructure of this building. As a result, there was a desire during the design phase of the project to permanently reduce the lateral earth pressures on the high side of the building to better balance the overall lateral loading of the substructure.

Several alternatives to achieve this load reduction were considered and the then-novel use of a geofoam compressible inclusion to induce the active earth pressure state was chosen for use. For a variety of reasons, the compressible inclusion was only used over a 6.5-metre (21-ft) height, i.e. approximately two-thirds of the 10-metre (33-ft) geotechnical height, H, of the basement wall. The compressible inclusion was placed so that there was both an upper and lower portion of the basement wall without a compressible inclusion.

Given the broad ignorance among U.S. design professionals about geofoams in general that existed in the 1980s and the lack of application-specific geofoam compressible-inclusion products during that timeframe (in fact, the term 'compressible inclusion' did not even exist at that time), a commercially available geofoam drainage panel consisting of GPS-PB was used as the compressible-inclusion product. By pure happenstance, GPS-PB, at least as manufactured in the U.S. at that time, had uniaxial compression properties essentially identical to R-EPS which did not enter the commercial market in the U.S. until the 1990s, more than a decade after this case history project was undertaken.

Because this was such an innovative and novel solution at the time, Partos and Kazaniwsky had to develop an analytical methodology for forecasting the required thickness of the compressible inclusion. The result was described earlier in this chapter where it was referred to as Version 1.0 of the current simplified analytical methodology for full-depth REP-Wall applications under gravity loading.

Given the novelty of their design solution, Partos and Kazaniwsky prudently had several stress meters and extensometers installed to monitor the performance of the ERS-geofoam-retained soil system both during and for some time immediately after construction to confirm that their design goals were met. Results of these readings are presented in Partos and Kazaniwsky (1987).

The Partos and Kazaniwsky paper coincided temporally with the writer's entry into full-time academia and was one of the publications that motivated the writer to both conduct research into what was later defined (by the writer) as the compressible-inclusion function as well as to identify geofoam materials that could be used economically for this function. This led eventually to identifying R-EPS as the basic geofoam material-of-choice for this role as well as creation of a multi-functional geocomposite consisting of both R-EPS and GPS-PB when fluid drainage is also desired (as it typical with most ERSs).

Relevant to the present discussion is that during the 1990s, the writer supervised a graduate-student (Master-level) research project at Manhattan College by George P. Murphy to evaluate the Partos-Kazaniwsky case history in detail using the FEM[81]. The primary goal of this project was to evaluate the theoretical effect of the long-term behavior of a compressible inclusion as by then the significance of creep in all functional applications of geofoam, but especially large-strain ones such as compressible inclusion, had become well-

---

[81] Murphy used an earlier version of the writer's *SSTIPNH* program for his analyses.

known to and thus appreciated by the writer (Horvath 1998c). In particular, based on unfunded creep testing performed by the writer at Manhattan College, it was known that the GPS-PB geofoam material and product used in the Partos-Kazaniwsky case history was approximately twice as compressible under sustained (creep) uniaxial loading of 10,000 hours (slightly more than one year) compared to the de facto standard rapid-loading tests used to define the short-term stiffness of this and all other geofoam materials and products.

The writer had Murphy use the Partos-Kazaniwsky case history as the subject of Murphy's research. Although this was not a full-depth REP-Wall application as would have been desired, the advantages of using an actual case history with short-term field measurements that could also be compared to results from FE analyses using the standard short-term stiffness properties of the geofoam product outweighed any negatives.

The outcomes of Murphy's research were published in Murphy (1997) and are abstracted and discussed here. To begin with, Figure 5.23 shows the basic problem geometry of the Partos-Kazaniwsky case history site that formed the basis for Murphy's study. Note that the depth variable used in this and other figures in the present discussion is the site elevation shown in Partos and Kazaniwsky (1987) where Elevation 0 is believed to be approximately Mean Sea Level.

As discussed in Partos and Kazaniwsky (1987), their analyses indicated that a compressible-inclusion thickness of 6 inches (152 mm) would be adequate but they used their professional engineering judgment (no doubt wanting to be justifiably conservative in their design given the novelty of the application at the time) and increased the actual installed thickness to 10 inches (254 mm). Note that at that time the commercial product they used was being produced in panels that were 48 inches (1219 mm) square and 2 inches (51 mm) thick so the requisite thickness and height of the compressible inclusion was assembled on-site from such panels.

Figure 5.24 shows the following lateral earth pressures from theoretical stress states plus calculated results from Murphy's research:

- the theoretical normally consolidated at-rest state (as calculated by Murphy, not that shown in Figure 12 in Partos and Kazaniwsky (1987));

- the horizontal component of the theoretical active state based on Coulomb's solution (as calculated by Murphy but similar to that shown in Figure 12 in Partos and Kazaniwsky (1987));

- the calculated (FE) results assuming both the short-term and long-term stiffness of the compressible inclusion based on uniaxial laboratory testing; and

- the range of measured results obtained by Partos and Kazaniwsky over a period of four months after the end of construction in the 1980s. Note that these were made at the ERS-compressible inclusion interface.

To complement the results shown in Figure 5.24, Table 5.1 shows the lateral resultant forces per unit width of ERS (i.e. in the direction perpendicular to Figures 5.23 and 5.24 and expressed using the dimensionless β parameter) and their respective points of application referenced to the elevation datum used in both Figures 5.23 and 5.24.

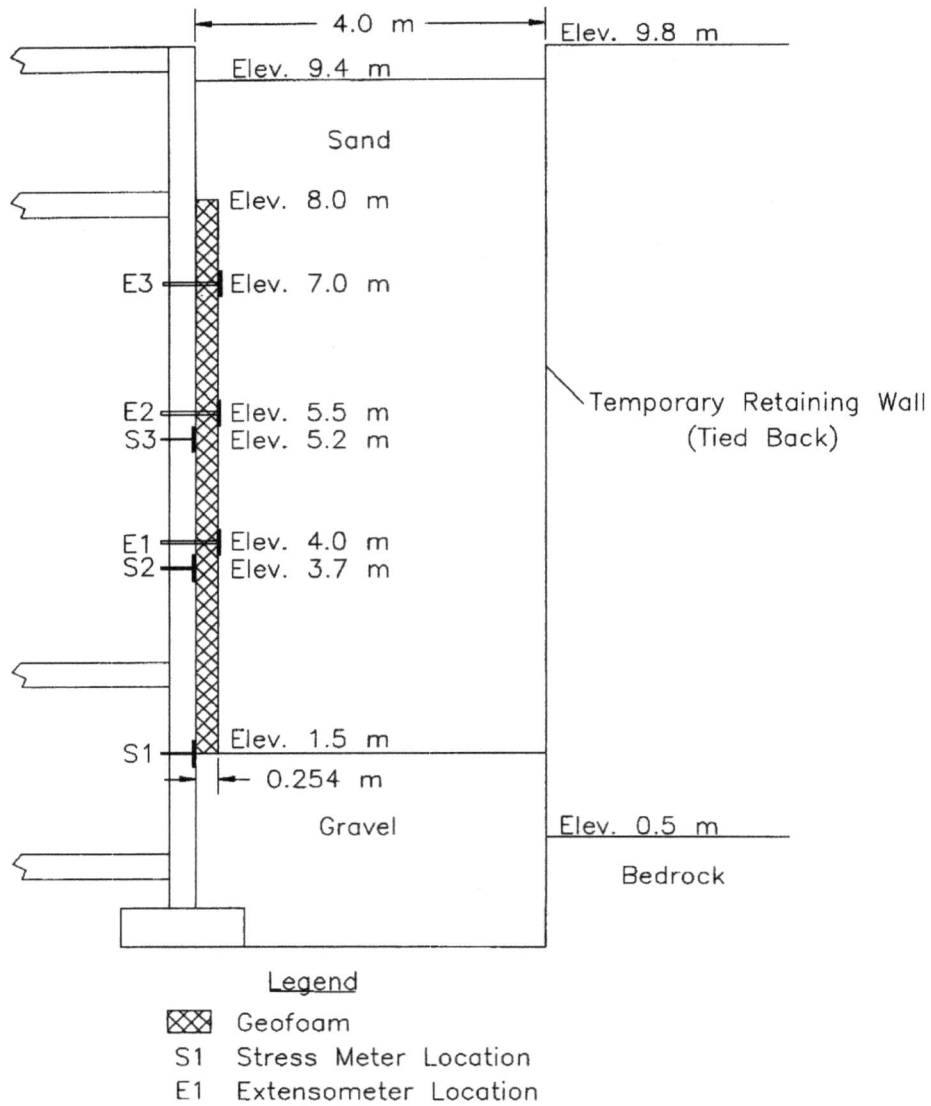

Figure 5.23. REP-Wall/Partial-Depth Placement/Non-Yielding RERS Partos-Kazaniwsky Case History - Overall Problem Geometry [from Murphy (1997)/Figure 1].

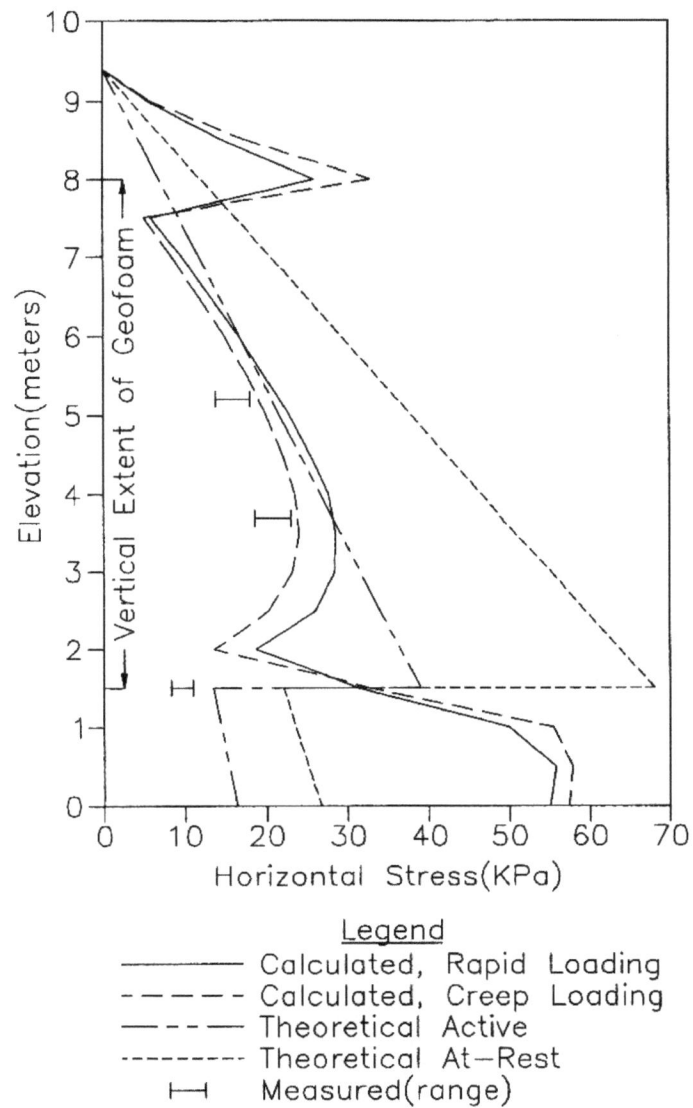

Figure 5.24. REP-Wall/Partial-Depth Placement/Non-Yielding RERS Partos-Kazaniwsky Case History - Lateral Earth Pressures [from Murphy (1997)/Figure 5].

**Table 5.1. REP-Wall/Partial-Depth Placement/Non-Yielding RERS
Partos-Kazaniwsky Case History - Lateral Resultant Forces
[modified from Murphy (1997)/Table 3].**

case	lateral resultant forces	
	β	elevation of point of application (metres)
at-rest	1	3.7
active	0.58	3.7
short-term geofoam stiffness	0.74	3.4
long-term geofoam stiffness	0.71	3.3

To begin with, Murphy's calculated lateral earth pressures shown in Figure 5.24 based on the standard short-term/rapid-loading stiffness of the compressible inclusion show overall good agreement with the measured results that were taken immediately after construction. There is some disagreement with the lowest measured result but this is likely due to the sharp boundary that existed in the actual construction between the gravel and sand backfill materials (as shown in Figure 5.23, the lowest stress meter was placed exactly at the interface between these backfill layers) as well as the sharp transition between no compressible inclusion and a compressible inclusion that occurred at the same point.

The classical 'horizontal trapdoor' distribution of arching effects is well represented by Murphy's calculated results. Thus, while the compressible inclusion appears to have been completely effective at inducing the active earth pressure state within its vertical limits this came at the cost of significant stress redistribution and concomitant increases in lateral pressures on either side of the compressible inclusion. As discussed in Chapter 2, when arching occurs within a soil mass in any orientation, earth forces do not 'disappear', they are simply redistributed. This is quite apparent in Murphy's calculated results shown in Figure 5.24. Note that these zones of localized very high earth pressures caused by the redistributed earth forces could have significant structural engineering outcomes in terms of resulting localized zones of very high flexural stresses that could cause cracking in the basement wall unless these moments were accounted for in design.

Nevertheless, it is of interest that the compressible inclusion in this case resulted in approximately two-thirds of the theoretical potential reduction in lateral resultant forces from the at-rest to the active earth pressure states. This is an interesting coincidence with the fact that the compressible inclusion covered approximately two-thirds the geotechnical height of the ERS. Whether such a one-to-one correlation between resultant-force reduction and usage of a compressible inclusion holds in general for all partial-depth REP-Wall applications is clearly something worth investigating in future research.

The final point of note with regard to the lateral earth pressures and resultant forces in the Partos-Kazaniwsky case history is that they were not greatly impacted by the different stiffnesses of the compressible inclusion, at least in terms of the calculated results (actual long-term stress measurements were attempted but proved to be unavailable). This is reflected in Figure 5.25 that shows the line-contoured stress levels, S, for the two analyzed cases (the contouring interval, ΔS, = 0.04 in both cases). The largest stress level within the retained soil in both cases occurred at and adjacent to the bottom of the compressible inclusion, at the sand-gravel interface, but only increased from 0.80 to 0.89 indicating the shear strength of the retained soil was not fully mobilized anywhere.

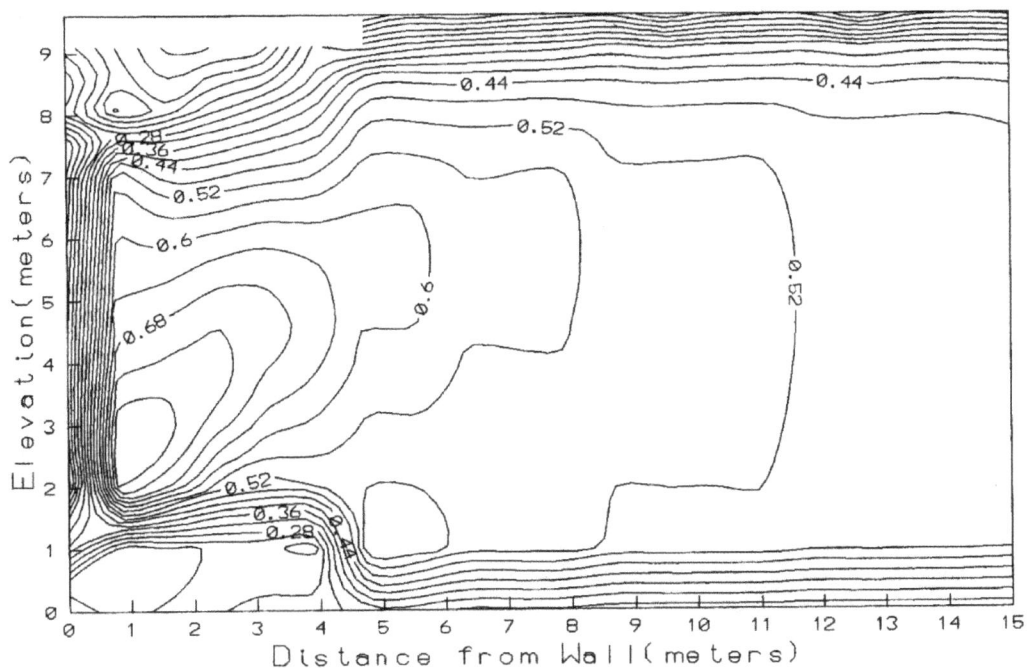

(a) Short-term (standard rapid-loading) compressible-inclusion stiffness.

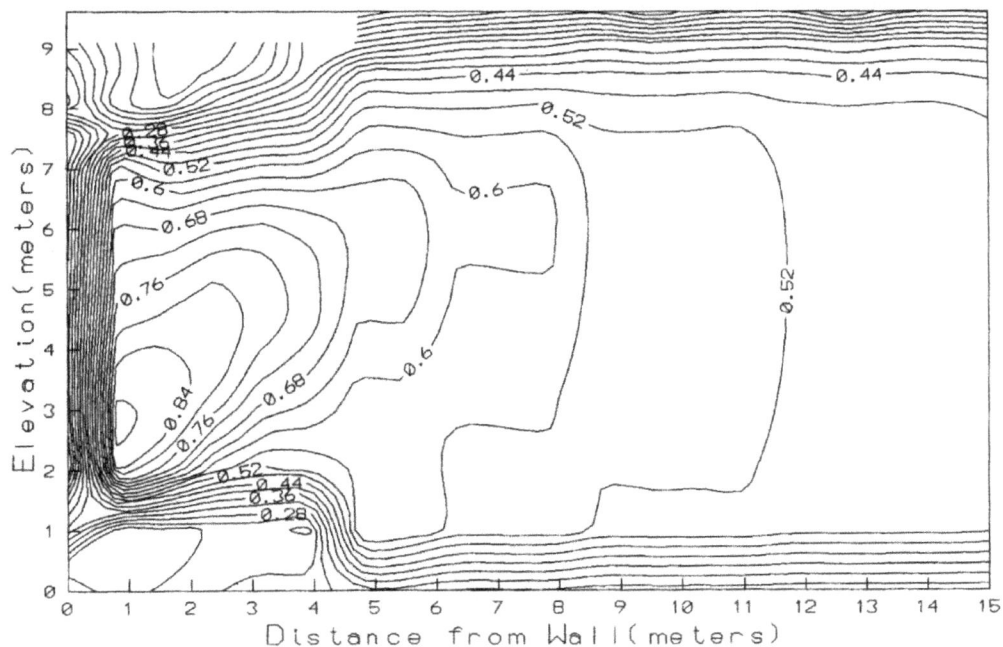

(b) Long-term compressible-inclusion stiffness.

Figure 5.25. REP-Wall/Partial-Depth Placement/Non-Yielding RERS
Partos-Kazaniwsky Case History - Calculated Stress Levels, *S*
[from Murphy (1997)/Figures 3 and 4].

168

The final issue considered is the lateral displacements of the compressible inclusion-retained soil interface. Figure 5.26 shows the calculated (FE) values for both the short- and long-term stiffnesses of the compressible inclusion as well as the measured values at the end of the post-construction monitoring period. Note that the ordinate (vertical axis) in this figure is scaled differently than the preceding figures as it only includes the vertical limits of the compressible inclusion, not the entire geotechnical height, *H*, of the ERS. This is because the lateral displacements are, by definition, zero above and below the compressible inclusion. Note also that Murphy (1997) was unclear as to the nature of the measured values which show overall good agreement with the short-term calculated values. This is despite the fact that Partos and Kazaniwsky (1987) noted that the displacement measurements likely under-represent the actual values as each extensometer was zeroed only sometime after backfill had progressed up to the level of that extensometer so that the device could be properly seated.

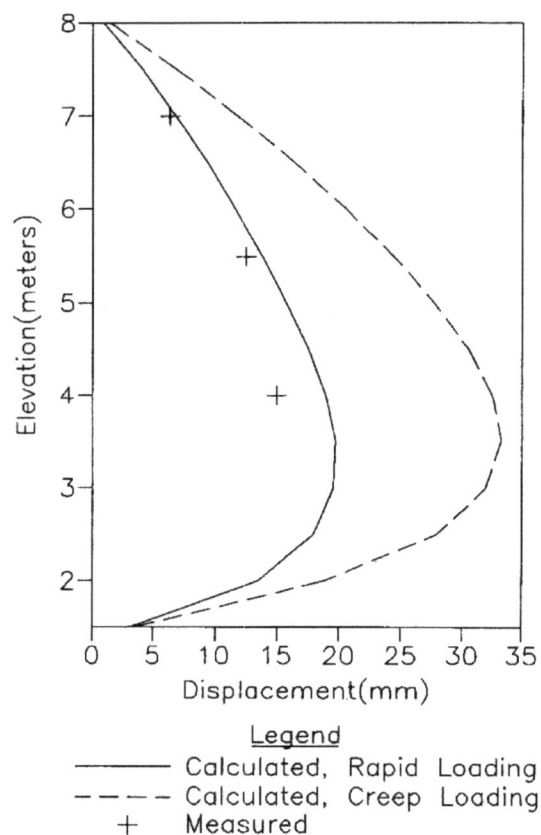

Figure 5.26. REP-Wall/Partial-Depth Placement/Non-Yielding RERS Partos-Kazaniwsky Case History - Lateral Displacements of Compressible Inclusion-Retained Soil Interface [from Murphy (1997)/Figure 6].

The results shown in Figure 5.26 are interesting for two reasons. First, the pattern of measurements is consistent with the pattern of the calculated values. When this qualitative outcome is combined with a similar conclusion drawn from Figure 5.24 with respect to lateral earth pressures, this supports a conclusion that the FE model used by Murphy correctly captured the soil-structure interaction (SSI) behavior involved.

Second, unlike lateral earth pressures (Figure 5.24) and the stress levels associated with them (Figure 5.25) that show a modest calculated change under long-term compressible-inclusion stiffness conditions, the calculated lateral displacements show a substantial increase between the short- and long-term results. However, this observation should not be assumed to be universal as this particular outcome is likely highly dependent on the particular elements of this case history. The contoured stress levels, S, in Figure 5.25a show that the shear strength of the retained soil was already substantially (up to 80%) mobilized under the short-term compressible-inclusion stiffness so the 50% reduction in compressible-inclusion stiffness resulting from long-term loading only increased this mobilization to 89% maximum as shown in Figure 5.25b.

However, over the same short- to long-term transition, substantial additional lateral displacements occurred, at least theoretically. This is conceptually similar to a statically indeterminate structure where a hinge develops in a structural member. The moments and concomitant flexural stresses in the hinge may not change much, if at all, with additional loading but substantial displacements of the hinge and concomitant deformation of the structural member can occur.

5.4.4 Synthesis of Relationship Between Full- and Partial-Depth Placement

Although there is still much to learn and much work to be done in terms of developing relatively simple analytical methodologies for use in routine practice, a number of useful conclusions and observations can still be drawn concerning the basic REP-Wall version of the compressible-inclusion function:

- Full-depth placement alternative/gravity loads: The primary goal of the basic full-depth placement alternative is and always has been to allow the development of the active earth pressure state adjacent to non-yielding RERSs under gravity loads. While this is generally achievable, when a compressible-inclusion product of uniform depth-wise thickness is used (as is typically the case) the depth-wise pattern of lateral displacement that develops along the compressible inclusion-retained soil interface differs substantially from any of the three traditional displacement modes: translation, rotation-about-bottom, and rotation-about-top. The displacement pattern that develops is curved with a peak closer to the bottom of the ERS than its top. The resulting distribution of lateral earth pressures that develops follows a more or less identical curve that is very close to that forecast using any one of a number of theories for horizontal arching. The most significant consequence of this distribution of lateral displacements is that care must be taken when basing a simplified analytical methodology on published values of $(\Delta/H)_a$, the dimensionless ratio of lateral displacement necessary to mobilize the active earth pressure state. This is because published values of this ratio are typically based on simple, conventional lateral-displacement modes such as translation and rotation-about-bottom. As a result, although a simplified, design-oriented methodology for the full-depth placement alternative under gravity loads has existed since at least the late 1980s, it is still a work-in-progress that is undergoing refinement as research continues to be performed and new information becomes available.

- <u>Full-depth placement alternative/seismic loads:</u> The extension of the full-depth placement alternative to cases involving seismic loading, the so-called 'seismic buffer' or 'seismic isolation' application, is complicated by the fact that frequency of the input motion appears to be a significant problem variable. In addition, there has not been any known published study to date of what the lateral-displacement pattern of the compressible inclusion-retained soil interface looks like under seismic loading. Limited data from the RMC shake-table tests suggests that the displacement pattern is highly non-uniform, in some case greater near the top compared to the bottom and in some cases the exact opposite although these observations may have been unduly influenced to the low height of the shake-box and the rough base incorporated into it. Furthermore, there is no known published study of $(\Delta/H)_{ae}$, the dimensionless ratio of lateral displacement necessary to mobilize the active earth pressure state under seismic loading. Taken together, this means that a simplified, design-oriented analytical methodology for seismic loading remains elusive at the present time and that designs should be based on site- and project-specific analyses performed using advanced numerical techniques such as the FEM and FDM.

- <u>Partial-depth placement alternative:</u> Research into the partial-depth placement alternative has, to date, been limited to a consideration of gravity loading only. Information from both an instrumented full-scale application and FE analyses indicate that behavior under this alternative is significantly different from that under the full-depth alternative. Specifically, using a compressible inclusion over only a portion of the inside face of a RERS triggers a completely different behavioral mechanism within the retained soil. Unlike the full-depth alternative that mobilizes development of an active earth pressure 'wedge', the partial-depth alternative is akin to building a trapdoor behind the ERS than can induce arching (horizontally in this case) but only over the vertical limits of the 'trapdoor'. Although lateral earth pressures over the region of the trapdoor are substantially reduced from the at-rest earth pressure state, they increase substantially above the at-rest values on either side of the trapdoor as would be expected. As a result, the net load-reduction value of a partial-depth placement application is not straightforward as the limited extent of the compressible inclusion involves a complex redistribution of lateral earth pressures in addition to a localized reduction in these pressures. Whether or not the localized areas of lateral pressure increase and concentration on either side of the trapdoor are problematic for the RERS structurally can only be determined on a project-specific, case-by-case basis. In any event, because of the infinite combination of variables related to a partial-depth placement in terms of what percentage of the inside face of the RERS is covered with a compressible inclusion and where along the inside face of the RERS the compressible inclusion is placed, it would appear that development of a simplified analytical methodology for the partial-depth placement alternative would be challenging. At the present time, it would appear necessary to perform an advanced numerical analysis using the FEM or FDM whenever a partial-depth application is being considered under gravity-load conditions. This would certainly be true as well whenever seismic loading was considered in addition.

5.5 ANALYTICAL METHODS FOR THE ZERO EARTH PRESSURE WALL (ZEP-WALL) CONCEPT

5.5.1 Introduction and Overview

After an initial burst of proof-of-concept research from the late 1980s to the early 1990s by the writer (Horvath 1990b) as well as by A. McGown, K. Z. Andrawes, et al. in the U.K., the ZEP-Wall concept has seen much less research, development, and use compared to the REP-Wall concept. In particular, the relatively extensive research in recent years into using the REP-Wall concept to reduce seismic loads on non-yielding RERSs has not been extended to including geosynthetically reinforced retained-soil masses. This is particularly surprising as the research group at RMC-Queen's led by Bathurst that has pioneered most of the seismic-buffer research performed worldwide to date has a decades-long history of being actively involved in R&D involving MSE, MSEWs, and SRWs, including developing analytical methodologies for both gravity and seismic loading of SRWs (see, for example, Bathurst and Alfaro 1996). So, the RMC-Queen's geotechnical research group is well-versed on the benefits of using geosynthetic tensile reinforcement within a retained-soil mass.

This overall lack of ZEP-Wall R&D is also somewhat surprising given its:

- contemporaneous origins with the REP-Wall concept;

- underlying basis in the MSE concept that is now well-proven and has been widely accepted for decades worldwide;

- having the attraction of efficiently and cost effectively (compared to the lightweight-fill function) reducing lateral pressures on both existing and new ERSs to essentially zero if desired; and

- being potentially of technical benefit even with yielding ERSs for which the REP-Wall concept offers no or minimal benefit as discussed previously.

Furthermore, as discussed in detail in Horvath (2000), the ZEP-Wall concept also has the potential to significantly improve the performance of both existing and new integral-abutment bridges, something that the REP-Wall concept cannot achieve alone in the absence of any other ground-modification strategy. This is discussed further in Chapter 6 dealing with self-yielding RERSs.

As will be seen in the following sections, what R&D has occurred to date for the ZEP-Wall application has focused on developing a relatively simple analytical model and concomitant methodology for the basic full-depth placement alternative under gravity loading. Therefore, the solution assessments performed for this monograph, all of which are numerical analyses based on the FEM, have focused on this combination of variables. However, there is some discussion of seismic loading as well as some preliminary investigation of the partial-depth placement alternative.

5.5.2 Full-Depth Placement Alternative

5.5.2.1 Gravity Loading

5.5.2.1.1 Non-Yielding Rigid Earth-Retaining Structures

The only known published works dealing with development of a relatively simple analytical methodology for use in routine practice for the basic ZEP-Wall case of full-depth placement and gravity loading are those by the writer. All efforts to date in this regard have been based conceptually on the same physical spring model used for the REP-Wall case, i.e. matching assumed spring stiffnesses of the compressible inclusion and retained-soil mass (internally reinforced in the ZEP-Wall case), and as shown in Figures 5.2 and 5.3. The key differences with the ZEP-Wall case are that the spring representing the retained-soil mass has its stiffness (spring constant) influenced by the presence of geosynthetic tensile reinforcement, and that in the limiting case of sufficient yielding the lateral pressures and resultant force on the ERS will decrease to zero as opposed to some limiting, non-zero value that is usually assumed to be the active earth pressure state in the REP-Wall case.

The writer's simplified analytical methodology for the ZEP-Wall has gone through evolution in thinking over the years although the basic conceptual framework of matching the behavior of the ground spring to that of the compressible-inclusion spring as shown qualitatively in Figure 5.3 has remained the same. The primary issue that has varied over time and is still unresolved to the writer's satisfaction is how to quantify the yielding vs. lateral-resistance behavior of the reinforced retained-soil mass as shown qualitatively in Figure 5.3. The reason is that there are no known simple Δ/H relationships in the published literature as in the REP-Wall case for any of the traditional displacement modes, i.e. translation and rotation.

The broad elements of the thought process used by the writer to develop a simplified ZEP-Wall analytical method begin by noting that on the one hand the limiting case of no yielding can be defined by classical at-rest conditions as it was for the REP-Wall case. This is because it is reasonable to assume that geosynthetic tensile reinforcement has no influence on lateral earth pressures in the absence of lateral displacement or deformation.

For the other limiting case of the minimum magnitude of yielding (lateral displacement) to result in zero lateral resistance, to date the writer has used knowledge gained from MSE (actually MSEW) research in the form of observed lateral displacement of MSEW facings as a useful place to start. Observation of the performance of actual MSEWs indicates that the magnitude of facing lateral displacement is affected not only by the height of the MSEW but also by the stiffness of the reinforcements, with metallic ('inextensible') vs. polymeric ('extensible') reinforcements producing significantly different results. In addition, the height-wise variation in facing lateral displacement is, in general, non-uniform and highly variable, and depends on the flexural stiffness of the facing treatment which, in practice, can vary from essentially rigid to almost perfectly flexible.

Unfortunately, lateral displacements of MSEW facing systems are rarely a design issue in routine practice so there has not been a lot of formal research of the subject in terms of being able to calculate either the magnitude(s) or height-wise pattern of these displacements accurately. As a result, there are just some published order-of-magnitude empirical relationships of peak facing displacements, Δ_{max}, as a function of height, H, of the MSEW structure, with different Δ_{max}/H ratios for metallic vs. polymeric reinforcements.

The writer's first attempt at developing a simplified analytical methodology was published in Horvath (1997b). It was based on lateral earth pressures, not resultant forces, and assumed that the distribution of lateral earth pressures was always trapezoidal in shape as a result of some unspecified combination of shear-strength mobilization within the reinforced soil mass plus soil-compaction effects near the ground surface as discussed earlier in this chapter. The addition of surface-surcharge effects was optional.

This assumed shape of this overall, combined lateral earth pressure 'envelope' was largely empirical in nature and based on measured results obtained in large-scale model tests performed in the U.K. using an instrumented retaining-wall test facility as reported in the published literature by K. Z. Andrawes and colleagues in the early 1990s. It was further assumed by the writer that the magnitudes of this pressure envelope changed as a function of yielding (lateral displacement), decreasing from a maximum at the at-rest/no-yield limiting case until all lateral pressures became zero at the other limiting case of some minimum required magnitude of yield (lateral displacement) based on the aforementioned MSEW research. The variation in the magnitudes of this pressure envelope was assumed to be linear between these two limiting cases as a function of Δ/H.

It is worth noting that this analytical methodology was presented in Horvath (1997b) only as a broad, working hypothesis that was intended to be the reference or starting point for future research. The intention was to fill a void given the absence of any simplified analytical methodology for ZEP-Wall applications up to that point in time (mid-1990s). Recall that an analytical methodology for REP-Wall applications (the original work of Partos and Kazaniwsky in the mid-1980s) had been proposed and used for over a decade at that point so there was a need to have a least a working hypothesis for the ZEP-Wall application. Other than the fact that the assumed trapezoidal shape of the lateral earth pressure envelope was based on research findings in the U.K., there was no attempt by the writer to vet this methodology further, e.g. by performing FE analyses, at that time.

The conceptual approach of basing a simplified ZEP-Wall analytical methodology on an empirically derived trapezoidal lateral pressure diagram was abandoned completely and with no further development by the writer a few years later (Horvath 2000) in order to bring this analytical methodology into conceptual consonance with what the writer had developed previously for the REP-Wall application. Specifically, the analytical model for the ZEP-Wall was modified to use lateral resultant forces rather than lateral pressures (surcharge effects were not considered in Horvath (2000) but could be added). This resultant force could either be absolute with dimensions or normalized/non-dimensionalized using the β variable, the latter being preferred in the writer's opinion. Otherwise, the basic elements of the analytical methodology remained the same, i.e. β was assumed to vary linearly between a maximum value and zero as yielding (lateral displacement) varied from zero to some assumed minimum value necessary to fully mobilize the inherent self-supporting potential of a geosynthetically reinforced soil mass.

Note that in developing this circa-2000 version of the writer's analytical methodology for full-depth ZEP-Wall applications, the stiffnesses of the compressible inclusion and reinforced retained soil, as shown qualitatively in Figure 5.3, were based on the <u>average</u> compressive strain of the compressible inclusion (which was assumed to be two-thirds of the maximum) but the <u>peak</u> extensive strain of the reinforced soil mass. Although it would have been more consistent to use the <u>average</u> strain for the latter, this information is simply not readily available. As noted previously, lateral displacement of MSEW facings is generally not a design issue so the only information readily available is the aforementioned empirical correlations involving maximum facing displacement (Δ_{max}), height of the MSEW (H), and relative stiffness of the reinforcements. The height-wise distribution of facing displacements relative to the maximum value is not something that is

174

consistently measured and reported in the published literature. Note that the height-wise variation in MSEW facing displacement would be expected to be strongly dependent of the nature and stiffness of the facing as well as various aspects related to the reinforcements themselves (e.g. composition, stiffness, spacing, time) and compaction of the retained soil (Bathurst et al. 2009).

In any event, based on a very limited comparison of this revised analytical methodology with FE analyses performed for Horvath (2000), the correlation between the simplified analytical method and FE results was not very good, with the simplified analytical method significantly underestimating the ZEP-Wall benefit in terms of reducing β for both the metallic and polymeric reinforcements considered in the FE analyses.

In retrospect, the problem studied and simulated in Horvath (2000) was not optimal for assessing the accuracy of this version of the writer's simplified analytical method for ZEP-Wall applications. This is because the IAB abutment modeled was not completely restrained against yielding (so did not represent a true non-yielding RERS) as the primary intent of the study reported in Horvath (2000) was to simulate IAB foundation behavior. Assessment of simplified analytical methodologies for both the REP- and ZEP-wall cases was a very secondary, incidental part of that overall study.

Independent of the outcomes presented in Horvath (2000), it is clear that a primary need for improving on a simplified analytical method for full-depth ZEP-Wall applications is developing a more-rational, theoretical way to quantify the lateral stiffness of a reinforced soil mass relative to the baseline unreinforced condition. This is because the current approach based on empirical Δ_{max}/H ratios for MSEW facings is clearly unsatisfactory. The lateral stiffness of a reinforced soil mass has much more variability than simply being either inextensible/metallic or extensible/polymeric reinforcement which are the only two choices for estimating facing displacement of MSEWs in most publications. Both the specific type of reinforcement and number of reinforcement layers need to be taken into account explicitly.

One possible approach for doing this is based on the work of Harrison and Gerrard (1972) which is one of the relatively few publications in the history of MSE and MSEW to approach the problem of soil reinforcement from the perspective of the theory of linear elasticity. A summary of their work is given in Appendix A.

Specifically, the concept is to quantify the dimensionless ratio of the Young's modulus of the reinforced soil mass in the horizontal direction, E_h, to the Young's modulus of the unreinforced soil, E_{soil}, using the following equation that is based on approximate relationships given by Harrison and Gerrard:

$$\frac{E_h}{E_{soil}} = K_r + 1 \qquad (5.21)$$

where the dimensionless variable K_r, which is defined as the *relative reinforcement stiffness*, reflects the aggregate stiffness of the reinforcement layers as follows:

$$K_r = \frac{\Sigma t_r \cdot E_r}{t_{total} \cdot E_{soil}} = \frac{\Sigma t_r \cdot E_r}{H \cdot E_{soil}} \qquad (5.22)$$

where E_r = operative Young's modulus of the geosynthetic reinforcement; Σt_r = aggregate (total) thickness of all reinforcement layers; and t_{total} = total thickness of the reinforced soil mass (taken to equal the geotechnical height of the retained soil, H, in this application).

Equation 5.22 can be generalized and improved in a number of ways. First, it can be generalized by allowing the thickness and modulus of each of the n number of layers of reinforcement, t_{ri} and E_{ri} respectively to be different. While this might not be important in most situations, there is no reason not to allow for it.

It can be improved by noting that the thickness and Young's modulus of geosynthetic tensile reinforcement are not commonly used explicitly in practice, especially for polymeric reinforcement (geotextiles and geogrids) where measuring thickness can either be problematic (geotextiles) or is ill-defined (geogrids). Rather, the parameter of *geosynthetic stiffness*, J, which is essentially $(t_r \cdot E_r)$, is what is reported in manufacturers' literature and used. Note that for polymeric reinforcements, which have both non-linear as well as time-dependent (creep) elements of force-displacement behavior, the value of J used is not unique for a given product as it varies with the assumed magnitude of maximum strain over some assumed period of time. Thus, the value of J used in analysis or design for polymeric reinforcements is typically a secant value that corresponds to an assumed specific strain level at an assumed specific point in time.

An additional improvement is to make use of a relatively new parameter called the *global reinforcement stiffness*, S_{global}, that has dimensions of force per unit length squared and is defined as

$$S_{global} = \frac{\sum_{i=1}^{n} J_i}{H}$$

(5.23)

where n = the number of reinforcement layers.

Putting all of these items together, Equation 5.22 becomes

$$K_r = \frac{\sum_{i=1}^{n}(t_{ri} \cdot E_{ri})}{H \cdot E_{soil}} = \frac{\sum_{i=1}^{n} J_i}{H \cdot E_{soil}} = \frac{S_{global}}{E_{soil}}.$$

(5.24)

Evaluating Equation 5.24 in a given application is straightforward enough for the geosynthetic component but not straightforward in terms of evaluating E_{soil}. The Young's modulus of soil is both stress and stress-level dependent, i.e. it will vary with depth (confining stress) and, for a given depth, how far along the non-linear stress-strain curve to failure (stress level). Therefore, it is necessary to develop a rational methodology for calculating a secant value of E_{soil} that reflects the average stiffness of the retained soil (without reinforcement) under some level of yielding in a given application.

In addition to the issue of evaluating Equation 5.24, there is also the issue of how to make use of the E_h/E_{soil} ratio calculated using Equation 5.21 once a value for K_r has been obtained from Equation 5.24. One possible approach would be use the empirical $(\Delta/H)_a$ ratios for mobilizing the active earth pressure state for an unreinforced soil mass that were discussed in detail and used extensively for the simplified REP-Wall analytical methodology. As this ratio reflects the minimum lateral displacement necessary to fully mobilize the shear strength of the retained soil, it indirectly reflects E_{soil} under the relatively large-strain conditions associated with the ULS for MSEWs. Therefore, it may be possible to develop a reasonably reliable methodology where the calculated E_h/E_{soil} ratio is used to empirically modify the $(\Delta/H)_a$ ratio to provide an estimate of Δ/H for the reinforced soil mass under consideration.

In conclusion, developing a reasonably accurate simplified analytical method for full-depth ZEP-Wall applications with non-yielding RERSs is still very much a work-in-progress. Quantifying the 'spring' stiffness of a reinforced-soil mass remains the primary need as the potential variability is so much greater than for unreinforced soils because the potential reinforcement choices in terms of type of reinforcement, number of reinforcement layers, etc. are infinite for all practical purposes.

5.5.2.1.2 Yielding Earth-Retaining Structures

The other gravity-loading topic addressed is the use of the ZEP-Wall concept with yielding ERSs, both rigid and flexible. As has been noted previously, one of the advantages of the ZEP-Wall concept relative to the REP-Wall concept is that there is always a potential advantage to using the ZEP-Wall concept even with a yielding ERS. This is because a yielding ERS, especially if it is a RERS, may displace laterally by an amount sufficient to mobilize the active earth pressure state within an unreinforced retained-soil mass but not necessarily enough to ensure full reinforcement mobilization within the same soil mass. The use of a ZEP-Wall application in conjunction with a yielding ERS can be designed so that sufficient lateral displacement of a reinforced soil mass can occur so that the full benefit of the reinforcements accrues and the lateral earth pressures on the ERS decrease to essentially zero.

5.5.2.2 Seismic Loading

As noted previously, there has been no known research, published or otherwise, to investigate how a ZEP-Wall application behaves under seismic loading. While the behavior of MSEWs under seismic loading has been studied extensively, this research has obviously focused on situations where the MSEW is a self-supportive, free-standing structure and not interacting with any other structural elements.

Nevertheless, seismic-related research of MSEWs is still useful in that it provides some broad, qualitative insight as to how a ZEP-Wall application might behave under seismic loading. Specifically, this research shows that a reinforced soil mass under seismic loading behaves more or less as a coherent block of 'rigid' material (and, in fact, is usually analyzed as such) that can undergo lateral displacement due to seismic shaking although the magnitude of these displacements is usually not quantified in routine practice.

This behavior suggests that in a ZEP-Wall application, even if the lateral pressures acting on the ERS were near-zero under gravity loading that some relatively significant lateral pressures and concomitant resultant forces could develop on the ERS as the coherent mass of reinforced soil displace under seismic inertia forces. In a sense, the benefit of the reinforcing, which was essential in reducing the gravity loads on the ERS to near zero, is lost completely and the retained soil behaves in a manner similar to a REP-Wall application.

There would, however, likely be some reduction of seismic-induced loads on the ERS due to further compression of the compressible inclusion but this reduction would likely be similar to that discussed previously for REP-Walls (or perhaps even less as the volume of the reinforced earth mass could be larger than the wedge of unreinforced soil that would otherwise develop). Lengthening the reinforcements to 'tie' the coherent mass of soil back into more-stable soil might help reduce seismic loads but since the reinforcements would have to strain further to resist the increased loads caused by seismic inertia this benefit could be reduced. In any event, it is clear that fundamental research into ZEP-Wall applications under seismic loading is much needed

5.5.2.3 Solution Assessment

5.5.2.3.1 Overview

The primary tool used by the writer to date to assess the basic full-depth ZEP-Wall application under gravity loading has been the FEM. Although numerous FE analyses for such applications have been performed by the writer since the late 1980s, a new suite of analyses was performed for this monograph using the same basic scenario used previously for the REP-Wall application, the basic details of which are given in Appendix E.

Figure 5.27 shows the FE mesh as specifically set up for analyses related to the ZEP-Wall application. As with the REP-Wall application, the yellow dotted lines indicate the portion of the mesh used to simulate the compressible inclusion. The actual thickness (horizontal dimension) of this zone was fixed at 1 metre so the stiffness of the 1-D bar (= axial spring) elements that comprised this zone was varied in order to simulate the desired variations in t_{ci}.

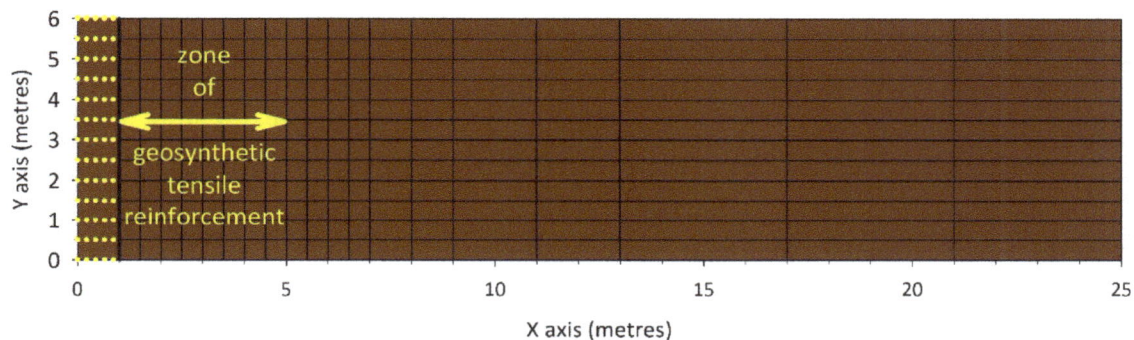

Figure 5.27. Finite-Element Mesh for ZEP-Wall/Full-Depth Placement Alternative Analyses.

The interface between the compressible inclusion and retained soil was fixed at $X = 1$ metre and the zone of reinforcement was assumed to extend horizontally 4 metres (13.1 ft) into the retained soil. This gave a ratio of reinforcement length to ERS height of 0.7 which is a reasonable value for MSEWs under gravity loads. For the basic full-depth placement case considered here, this zone of reinforcement was assumed to extend nominally for the full 6 metre (19.7 ft) height of the mesh (no reinforcement was assumed at the ground surface ($Y = 6$ metres).

The geosynthetic tensile reinforcement was not modeled explicitly in the mesh as the writer did in the earliest publications related to the ZEP-Wall concept (Horvath 1991a, 1991b, 1991c). Rather, a FE modeling technique involving an overlay of a second element type (in this case simple axial springs called *nodal links* in the FE software used by the writer) was utilized to impart the necessary lateral stiffness to the zone of reinforced soil without getting into details such as the actual vertical spacing of the reinforcements. The writer used this overlay technique previously in Horvath (2000)[82] and it has been used by others for a variety of applications involving the FEM, including a potential ZEP-Wall application (Ebeling et al. 1997).

---

[82] A detailed discussion of the overlay technique can be found in Section 3.2.4.4 of Horvath (2000).

The primary shortcoming of this overlay technique is that slippage or pullout of reinforcements relative to the surrounding soil cannot be modeled. In addition, the complex interaction between reinforcements and soil around them is not replicated. However, the benefit of the overlay technique is that it greatly simplifies the FE mesh and allows for analytical flexibility that would otherwise not be possible if each layer of reinforcement were modeled explicitly. Thus, for the purposes of this monograph this modeling approximation was considered reasonable and efficient.

The nodal links used as the overlay elements to simulate the geosynthetic tensile reinforcement in the analyses performed for this monograph were modeled stiffness-wise to approximate a generic high-stiffness polymeric product, the same as used by the writer in Horvath (2000). Such a product was judged to be a reasonable choice for the purposes of these analyses. However, this does not imply that this is the only type of reinforcement that should be considered or that it would be the optimal type of reinforcement to use in every application. Only a site- and project-specific design assessment by a qualified design professional can make such a determination.

5.5.2.3.2 Gravity Loading

To begin with, it is of at least academic interest to start the assessment of the basic full-depth application of the ZEP-Wall concept by considering the baseline case of geosynthetic tensile reinforcement of the retained soil alone with a non-yielding RERS, i.e. a full-depth ZEP-Wall application with a compressible-inclusion thickness, t_{ci}, equal to zero. The conventional wisdom has long been that under non-yielding conditions soil reinforcement has no benefit as the reinforcement cannot strain and develop the tensile forces that are fundamental to the concept of MSE. However, prior FE-based research by the writer (most recently Horvath (2000)) has shown that when soil placement is simulated adjacent to a non-yielding RERS without a compressible inclusion there can be some difference between the non-reinforced and reinforced cases due to secondary effects related to simulated construction as opposed to simply 'wishing' the retained soil, reinforced or not, in place. This is conceptually similar to the development of vertical shear stresses between a non-yielding RERS and unreinforced retained soil that do not exist with the simplistic wished-in-place approach, as was discussed in Chapter 2.

Figure 5.28 shows the calculated lateral earth pressures acting along the inside face of the ERS using results from both the 1-D interface and adjacent 2-D solid (soil) elements with a reinforced retained soil in both cases. The rationale for showing both results is discussed in detail in Appendix E and is not repeated here.

Overall, the agreement between the two elements types for the reinforced retained soil is good although the non-dimensional relative lateral resultant force, β, differs slightly between the two outcomes, with $\beta = 1.06$ for the 2-D elements and $\beta = 1.09$ for the 1-D elements.

Also shown in Figure 5.28 are the 2-D element results for the baseline/verification case of a non-yielding RERS with <u>unreinforced</u> retained soil (from Figure E.2 in Appendix E). There is broad agreement ($\beta = 1.09$) with the results for the reinforced soil case indicating that the effects of the reinforcement in the absence of yielding are essentially nil. This means that whatever differences in calculated behavior occur between the REP- and ZEP-Wall applications for the same compressible-inclusion thickness can be attributed solely to the presence of geosynthetic tensile reinforcement in the latter application and not other factors.

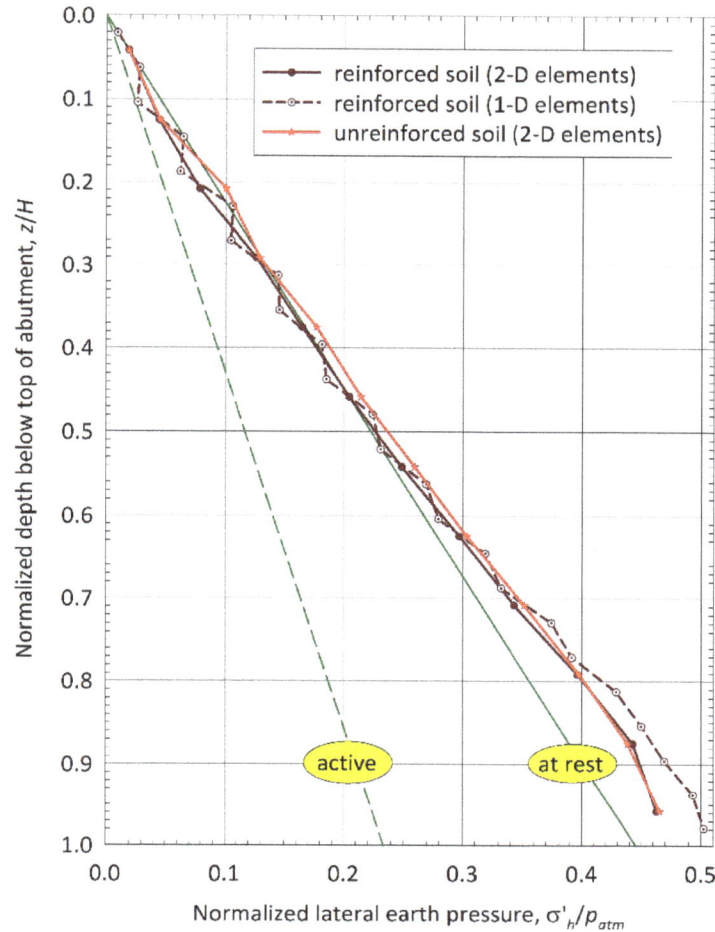

Figure 5.28. ZEP-Wall/Full-Depth Placement/Polymeric Reinforcements/Non-Yielding RERS Calculated Lateral Earth Pressures with No Compressible Inclusion.

As discussed earlier in this chapter and illustrated in Figure 5.9 for the REP-Wall case, the first issue to address when evaluating the effect of compressible-inclusion thickness, t_{ci}, on calculated results is to investigate the differences among the three sources (element types in the FE mesh) of calculated force/stress results that can be used to calculate lateral pressures acting on the ERS. This is shown in Figure 5.29 for the same t_{ci} = 175 mm (7 in) that was the 'design' thickness for the REP-Wall case shown in Figure 5.9.

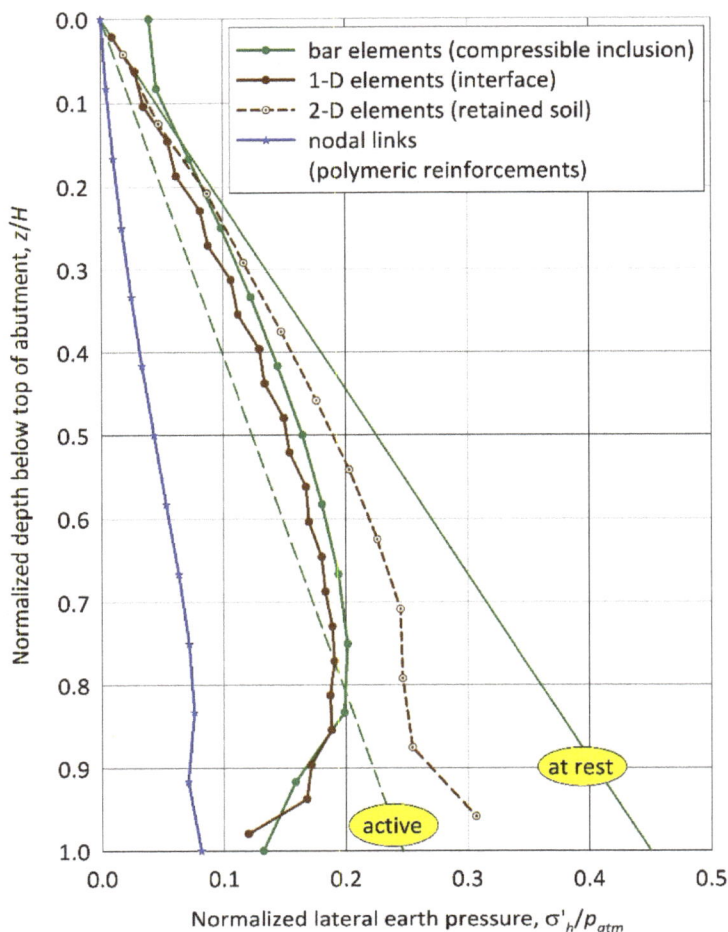

Figure 5.29. ZEP-Wall/Full-Depth Placement/Polymeric Reinforcements/Non-Yielding RERS Calculated Lateral Pressures with 175 mm (7 in) Compressible Inclusion.

This difference in results between and among these three element types can be attributed to the equivalent lateral pressure carried by the geosynthetic tensile reinforcement, the distribution of which is also shown in Figure 5.29. Note that because the reinforcement was modeled independently of the unreinforced retained soil in these analyses using an overlay of nodal links, the behaviors of the reinforcements and unreinforced retained soil are easily separated for independent review and assessment in the calculated results. In this case, the reinforcement represents an equivalent $\beta = 0.19$ which is exactly the difference between the 2-D soil elements ($\beta = 0.78$) and average of the compressible inclusion and 1-D interface elements ($\beta = 0.59$).

Continuing on, Figure 5.30 shows the calculated depth-wise distribution of tensile forces (normalized here to the total resultant at-rest force due to retained soil alone so the force at any depth i is labeled β_i) in the geosynthetic tensile reinforcement. Note that the sum of all these β_i adds up to $\beta = 0.19$ which is the aggregate benefit of the reinforcements as noted above. Again, the way in which the reinforcement was modeled using the element-overlay approach essentially 'smeared' the tensile resistance on top of the soil elements using nodal links placed within the FE mesh at the depths indicated by the red stars in this figure.

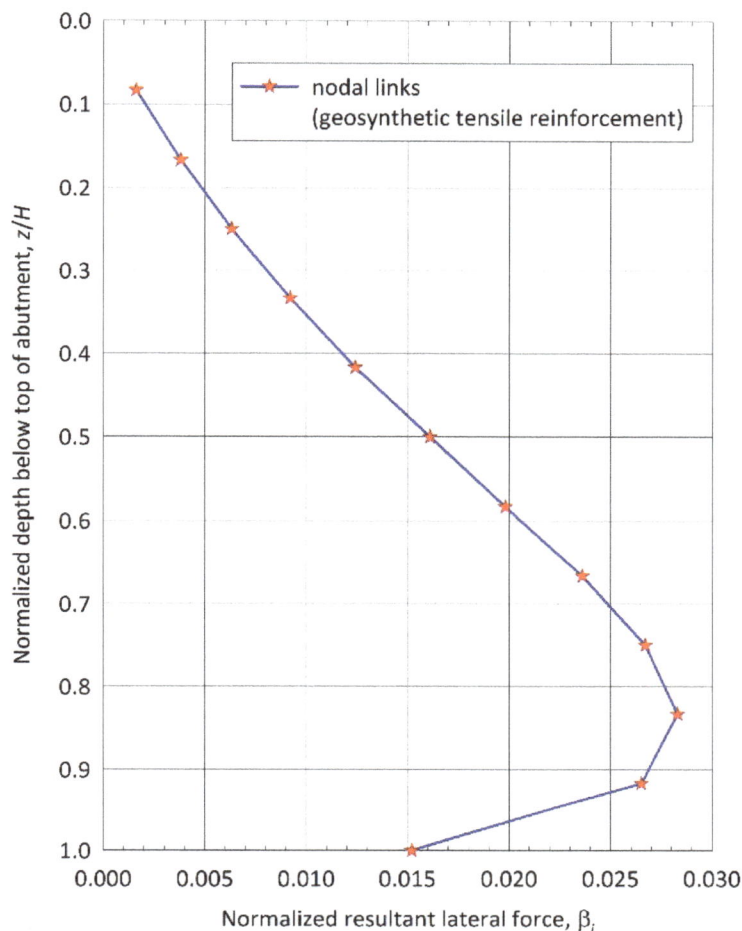

Figure 5.30. ZEP-Wall/Full-Depth Placement/Polymeric Reinforcements/Non-Yielding RERS Calculated Forces in Reinforcements with 175 mm (7 in) Compressible Inclusion.

The results shown in Figure 5.30 indicate that the forces in the reinforcements exhibit substantial depth-wise variation of almost one order of magnitude between the smallest and largest values. As will be seen subsequently, the calculated pattern of force mobilization in the reinforcements matches the pattern of compression within the compressible inclusion.

It is next of interest to examine in further quantitative detail what improvement the ZEP-Wall case offers relative to the REP-Wall case with the same t_{ci} (175 mm (7 in) in this example). Figure 5.31 shows the calculated lateral pressures acting on the compressible inclusion-retained soil interface using results for the 1-D interface elements. As can be seen, the ZEP-Wall case using simulated polymeric reinforcements ($\beta = 0.57$) provides about a 15% reduction in the overall resultant lateral force relative to the REP-Wall case ($\beta = 0.67$).

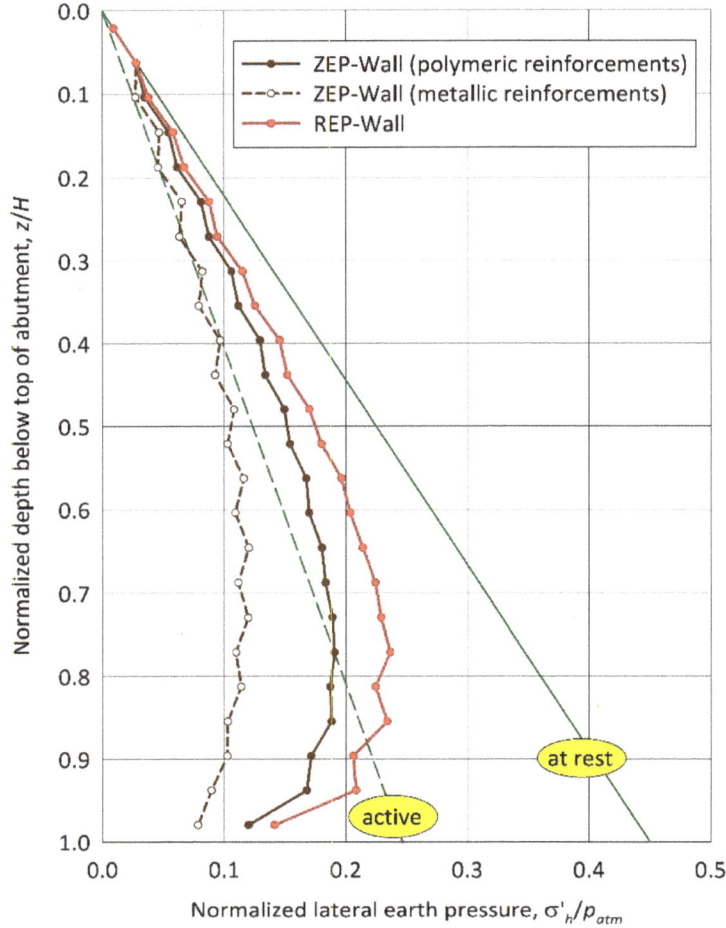

**Figure 5.31. ZEP-Wall/Full-Depth Placement/Non-Yielding RERS
Comparison of Lateral Pressures with 175 mm (7 in) Compressible Inclusion.**

To explore the influence of reinforcement stiffness for this particular example, the ZEP-Wall case was also analyzed using simulated metallic reinforcements (the same as used in Horvath (2000)) that are 50 times stiffer than the simulated polymeric reinforcements. The results from this analysis are also shown in Figure 5.31 and suggest that reinforcement stiffness is a significant variable in ZEP-Wall behavior (as would be expected) as the reduction in lateral pressures using metallic reinforcements ($\beta = 0.38$) was 43% relative to the REP-Wall case ($\beta = 0.67$) or almost three times as much as the reduction achieved with the ZEP-Wall case with polymeric reinforcements ($\beta = 0.57$).

Having examined and compared the force-related outcomes for both the ZEP- and REP-Wall cases with a compressible-inclusion thickness, t_{ci}, = 175 mm (7 in), it is of interest to examine and compare some lateral displacement-related parameters for the same cases. Figure 5.32 is comparable to Figure 5.18 presented previously for REP-Wall results and shows the depth-wise variation of compressive strains, ε_{hci}, within the compressible inclusion and the compressive displacement, Δ_{ci}, of the compressible inclusion-retained reinforced soil normalized to the geotechnical height of the ERS, H. Results are shown for both types of geosynthetic tensile reinforcement that were modeled in the ZEP-Wall analyses as well as for the REP-Wall results shown previously in Figure 5.18.

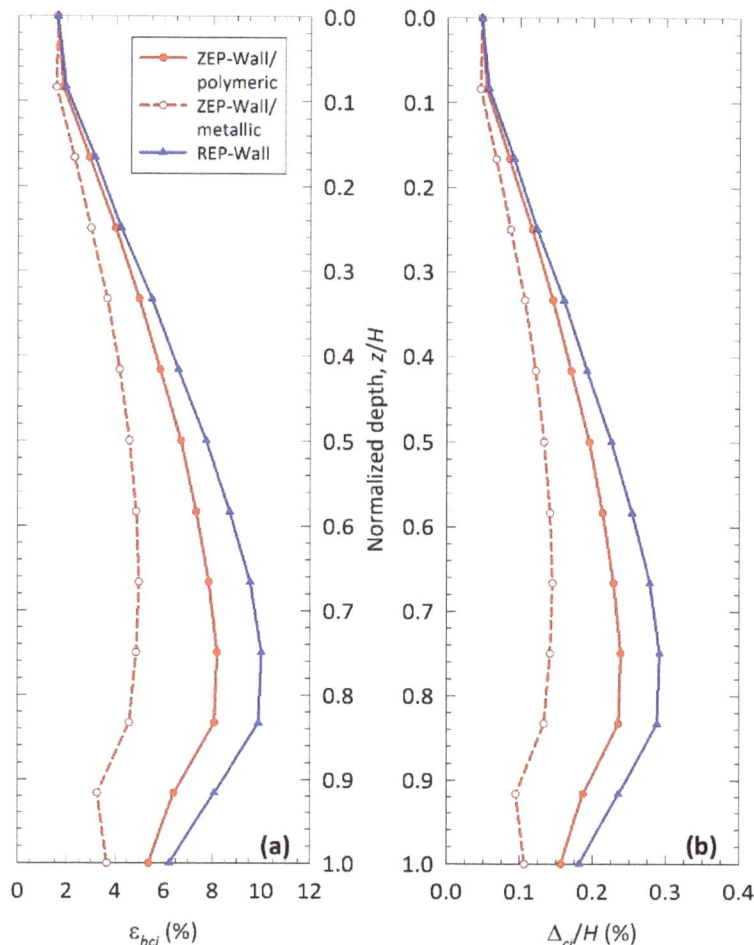

Figure 5.32. ZEP-Wall/Full-Depth Placement/Non-Yielding RERS Comparison of Lateral Strains and Displacements with 175 mm (7 in) Compressible Inclusion.

The results shown in this figure may seem somewhat counterintuitive in that lateral resultant force reduction <u>increases</u> with <u>decreasing</u> compression of the compressible inclusion. However, this is easily explained and understood when one realizes that the stiffness of the retained soil is <u>increasing</u> as well. Stated another way, as the stiffness of the retained soil goes from the unreinforced REP-Wall to ZEP-Wall with polymeric reinforcements to ZEP-Wall with metallic reinforcements it simply takes less lateral displacement and concomitant compression of the compressible inclusion to achieve lateral pressure and force reduction. This is consistent with the spring model shown in Figure 5.2 and behavior shown in Figure 5.3 that is used to visualize both REP- and ZEP-Wall applications.

Before proceeding with further investigation of the basic ZEP-Wall problem involving polymeric reinforcements to examine the effect of varying t_{ci}, it is of interest to view and discuss some additional results from the analysis made with simulated metallic reinforcements. Figure 5.33 is the same as portrayed previously in Figure 5.29 but with the results for the analysis assuming metallic reinforcements.

184

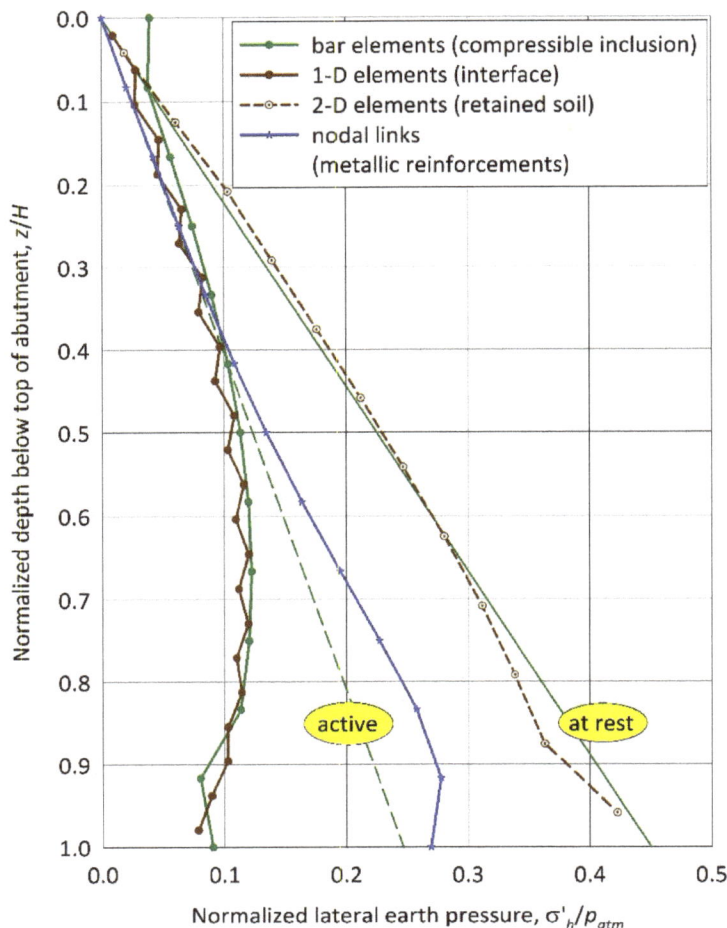

Figure 5.33. ZEP-Wall/Full-Depth Placement/Metallic Reinforcements/Non-Yielding RERS Calculated Lateral Pressures with 175 mm (7 in) Compressible Inclusion.

Once again, the bar and 1-D interface elements, both of which reflect the lateral earth pressure acting along the interior face of the ERS, are in overall good agreement and reflect a resultant lateral earth force that is substantially lower ($\beta = 0.39$ average) than the baseline reference case of the active state in unreinforced soil without a compressible inclusion ($\beta = 0.52$). The interesting thing here is that the lateral earth pressures within the unreinforced retained-soil component are virtually unchanged from the initial at-rest state with $\beta = 1.00$. Essentially all of the reduction of lateral earth pressure acting on the ERS is the result of the relatively stiff ('inextensible') metallic reinforcements which in this case support an equivalent resultant lateral earth force of $\beta = 0.64$. Note that this is a substantial difference from the same problem with polymeric reinforcement that was shown previously in Figure 5.29. In that case, the reinforcement only provided $\beta = 0.19$ and the retained soil mobilized a shearing resistance that was almost exactly equidistant between the at-rest and active states with a $\beta = 0.78$.

Figure 5.34 portrays the same information as shown in Figure 5.30 concerning the calculated depth-wise variation of forces within the geosynthetic tensile reinforcement but in this case compares results for both the polymeric and metallic reinforcements. While the two types of reinforcements have the same qualitative distribution of lateral forces, the magnitudes in the metallic reinforcements, with an equivalent overall β = 0.64, reflect the fact that these reinforcements are collectively carrying more than three times the load of the polymeric reinforcements (overall β = 0.19). But note that this three-fold increase in load bearing comes as a result of a fifty-fold increase in stiffness of the metallic reinforcements relative to the polymeric reinforcements.

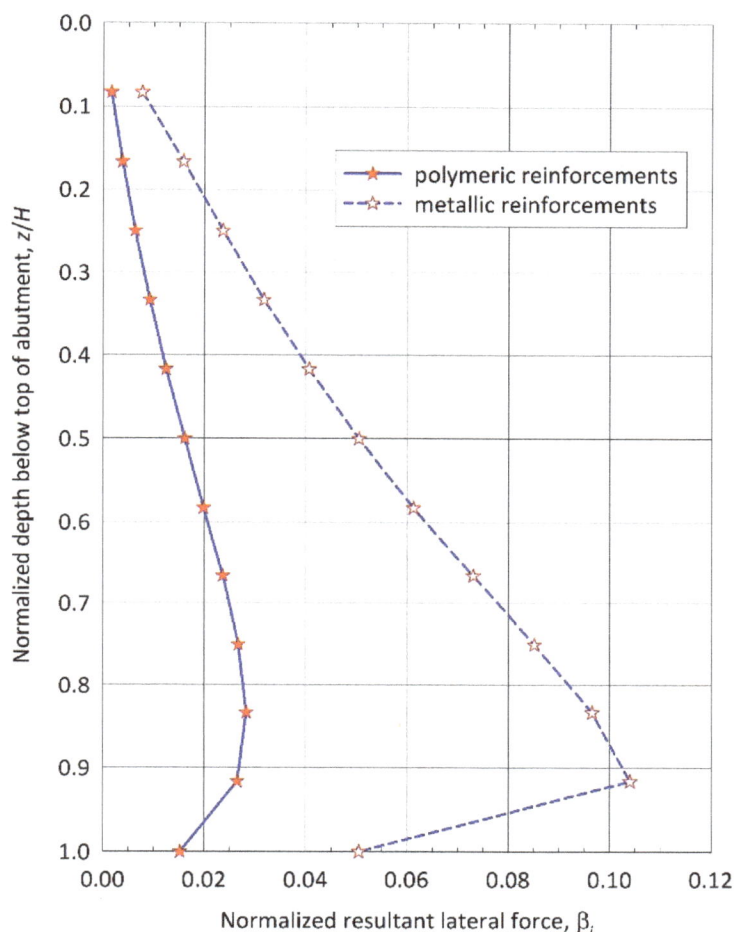

Figure 5.34. ZEP-Wall/Full-Depth Placement/Non-Yielding RERS Comparison of Lateral Forces in Reinforcements with 175 mm (7 in) Compressible Inclusion.

Next to present and discuss are the results of varying t_{ci} for the full-depth ZEP-Wall application. The same suite of plots as presented earlier in this chapter for the full-depth REP-Wall application will be shown with the exception of the contoured stress levels, S, within the retained soil. Because of the approximate way in which soil reinforcement was modeled using the element-overlay technique, it was felt that the calculated stress levels within the retained-soil elements were unrepresentative of the actual complex stress pattern that develops with a reinforced soil mass.

As will be seen, results are shown in the subsequent plots for both the simulated polymeric and metallic reinforcements (the latter being 50 times stiffer than the former as noted previously) as well as the REP-Wall results, where appropriate, that were shown and discussed previously. The intention here is to provide some sense of not only how the compressible-inclusion stiffness (as reflected in t_{ci}) affects results but the influence of retained-soil stiffness as well when geosynthetic tensile reinforcement is used. With regard to the latter, some of the concepts introduced earlier in this chapter, such as use of the dimensionless reinforcement stiffness factor, K_r, will be explored in the following discussion.

To begin with, Figures 5.35 and 5.36 show the calculated variation in lateral earth pressure acting on the inside face of the simulated non-yielding RERS for full-depth ZEP-Wall applications using polymeric and metallic reinforcement respectively. These figures are conceptually identical to Figure 5.10 for the full-depth REP-Wall application and also utilize the calculated results from the 1-D elements along the interface between the compressible inclusion and reinforced retained soil. Note that the baseline no-geofoam (t_{ci} = 0) cases are shown as well so that the relative reduction in lateral earth pressures benefitting from a compressible inclusion are more apparent.

More illustrative and useful is to examine the variation of several dimensionless parameters that have been defined and used throughout this monograph. To begin with, Figure 5.37 shows the variation, for the polymeric reinforcement case, of the relative lateral resultant force, β, with compressible-inclusion thickness non-dimensionalized to the geotechnical height of the ERS, H, and expressed as a percent. This is qualitatively identical to Figure 5.11 for the REP-Wall case with the addition of results from the nodal links used to model the polymeric reinforcement and results from all four element types are shown.

The results shown in Figure 5.37 are qualitatively very different from those for the REP-Wall case shown in Figure 5.11. In the latter, the results from all three element types (bar/1-D/2-D) were for all intents and purposes identical. Here, the results for the bar elements (reflecting the forces within the compressible inclusion) and 1-D elements (reflecting the forces along the interface between the compressible inclusion and reinforced soil mass) are essentially the same. However, the forces within the 2-D element (reflecting the forces in the unreinforced soil adjacent to this interface) are substantially different, reaching a plateau roughly between the at-rest and active earth pressure states.

The reason for this divergence can be seen in the increasing contribution of the geosynthetic tensile reinforcement (modeled here as an overlay element using nodal links) with increasing (t_{ci}/H). In ZEP-Wall applications, it is the combined effect of mobilized lateral forces within the unreinforced retained soil plus reinforcement (which are physically separate in the FE model used for this monograph but in reality function as a single composite material) that essentially match the lateral forces in the compressible inclusion and along the compressible inclusion-reinforced retained soil interface.

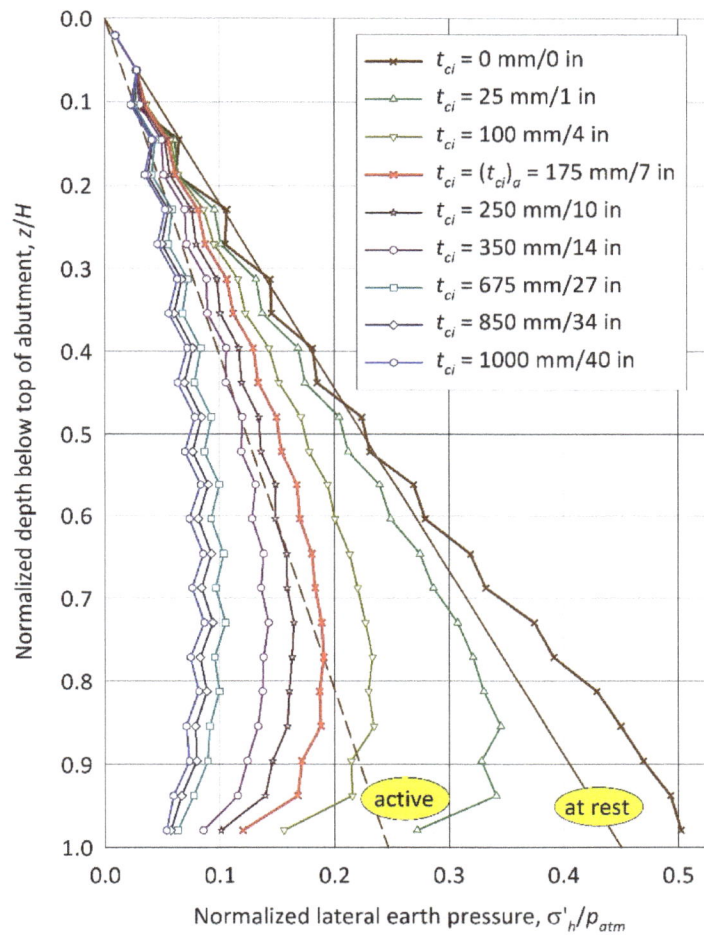

Figure 5.35. ZEP-Wall/Full-Depth Placement/Polymeric Reinforcements/Non-Yielding RERS Calculated Lateral Pressures as a Function of Compressible-Inclusion Thickness.

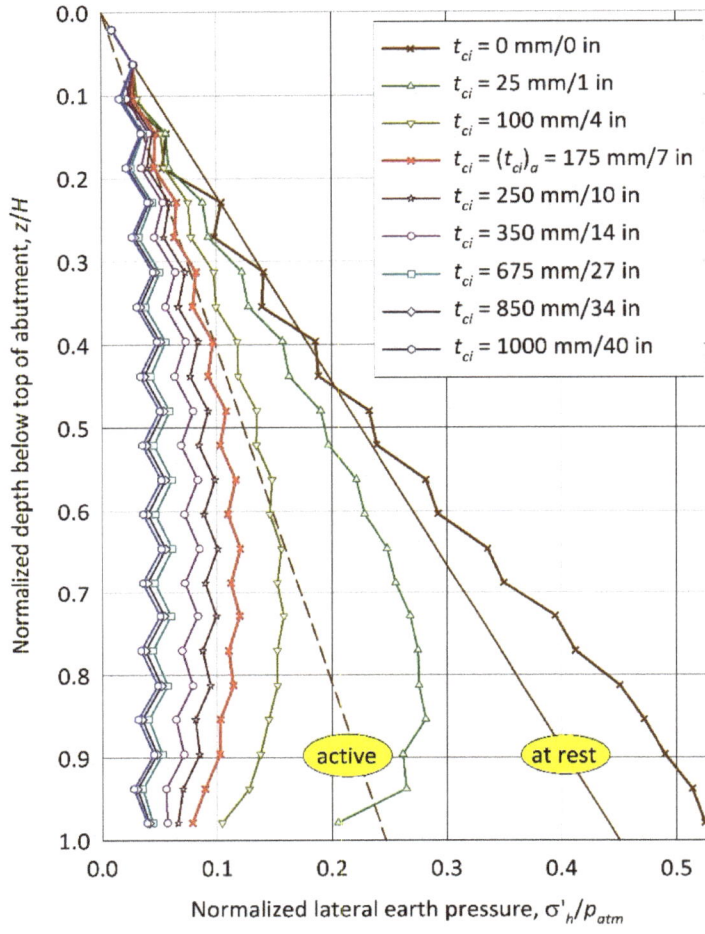

Figure 5.36. ZEP-Wall/Full-Depth Placement/Metallic Reinforcements/Non-Yielding RERS Calculated Lateral Pressures as a Function of Compressible-Inclusion Thickness.

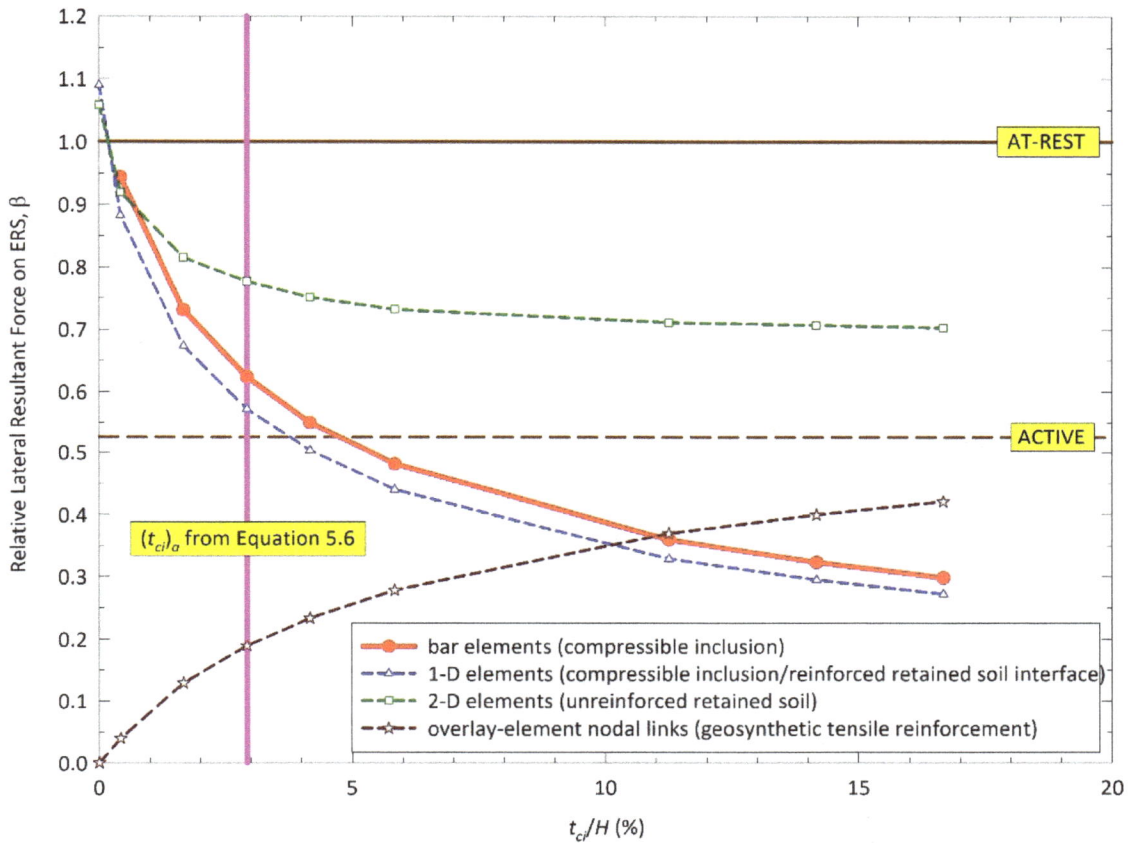

Figure 5.37. ZEP-Wall/Full-Depth Placement/Polymeric Reinforcements/Non-Yielding RERS Calculated Relationship between β and t_{ci}/H.

Figure 5.38 is qualitatively identical to Figure 5.37 but this time for the case of metallic reinforcements which, as noted previously, are 50 times stiffer than the polymeric reinforcements. The results are qualitatively the same as for polymeric reinforcements although the overall reductions in lateral resultant forces are greater and the contribution of the geosynthetic tensile reinforcement is significantly greater. It is also of interest to note that the contribution of the unreinforced retained soil remains essentially constant at the at-rest earth pressure state. This implies that the geosynthetic tensile reinforcement is so much stiffer laterally than the retained soil that, as a result, it picks up all the load with increasing lateral displacement with essentially zero shear-strength mobilization within and concomitant load-carrying support from the retained soil. This is admittedly a simplistic outcome that is a consequence of the approximate overlay modeling technique used in the FE model but it does illustrate how relatively stiff ('inextensible') reinforcement functions in a MSE applications.

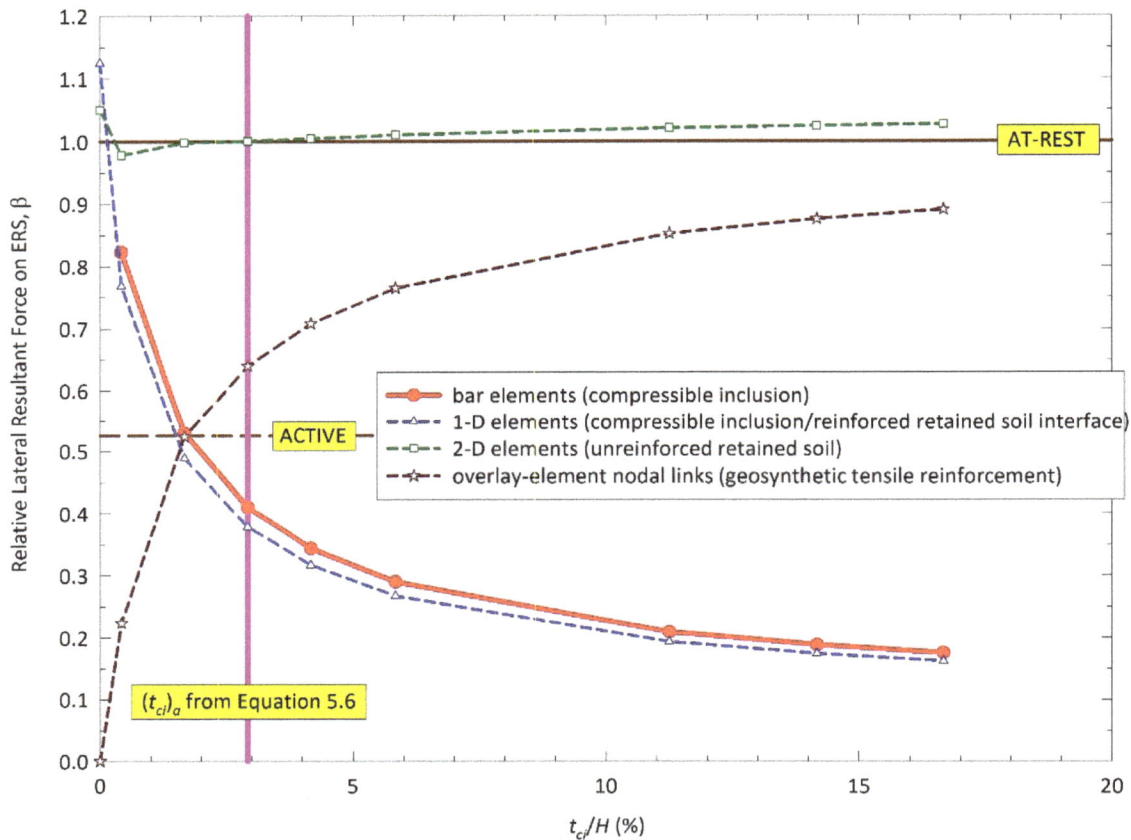

Figure 5.38. ZEP-Wall/Full-Depth Placement/Metallic Reinforcements/Non-Yielding RERS Calculated Relationship between β and t_{ci}/H.

Figure 5.39 compares the results for the REP-Wall and both variations of ZEP-Wall using results from the 1-D interface elements only as they provide results for the entire range of t_{ci} considered, including the baseline no-geofoam (t_{ci} = 0) case. The primary take-away from this figure is that the relationship between lateral earth pressure/force reduction and stiffness of the compressible inclusion (as reflected here by t_{ci}) is highly non-linear as the writer has conjectured in the past (e.g. Horvath 2000) and thus very different from the linear relationship assumed in simple analytical models presented both in this monograph as well as the past (Horvath 2000).

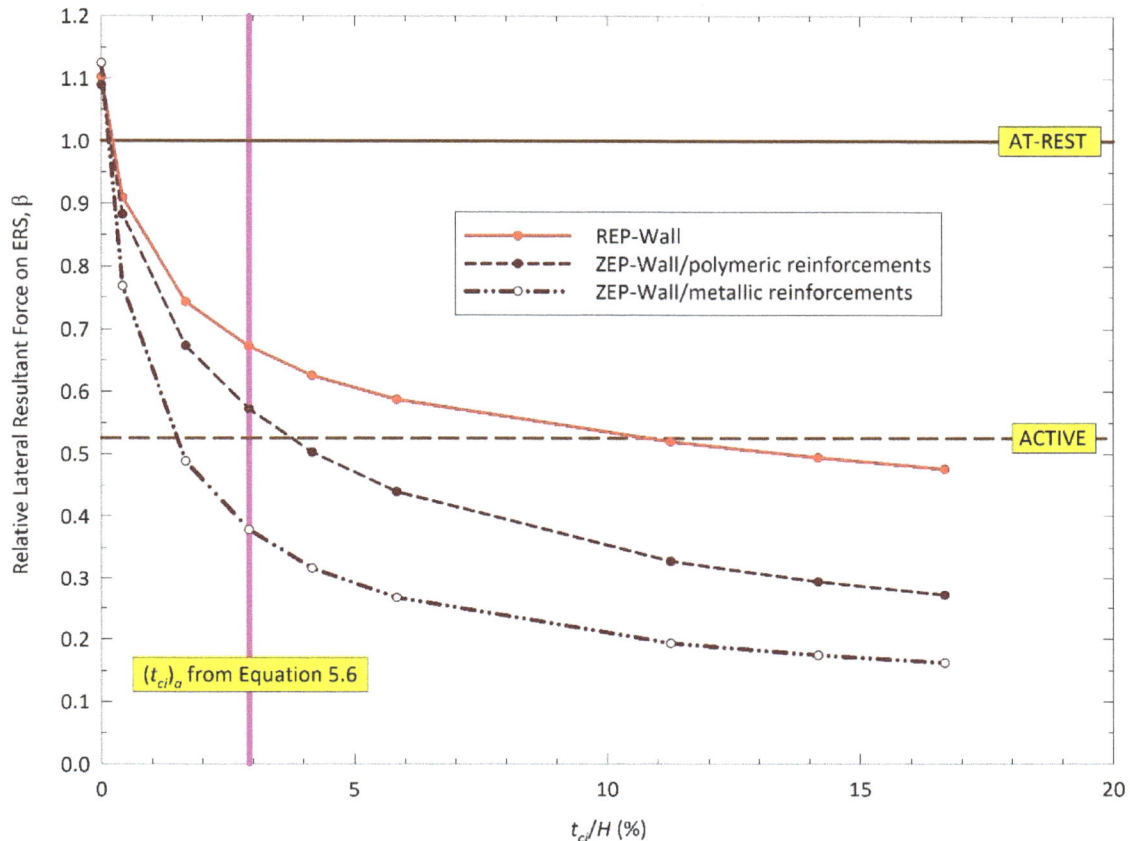

Figure 5.39. Comparison of Relationships between β and t_{ci}/H for Full-Depth REP- and ZEP-Wall Applications (Linear Scale).

Figure 5.40 shows the same results as Figure 5.39 but with semi-log plotting as was done previously in Figure 5.12 for the REP-Wall results alone. It is of interest to note that the ZEP-Wall results do not display the same near-perfect linearity in this type of plot as the REP-Wall results did although the deviation from linear is relatively small.

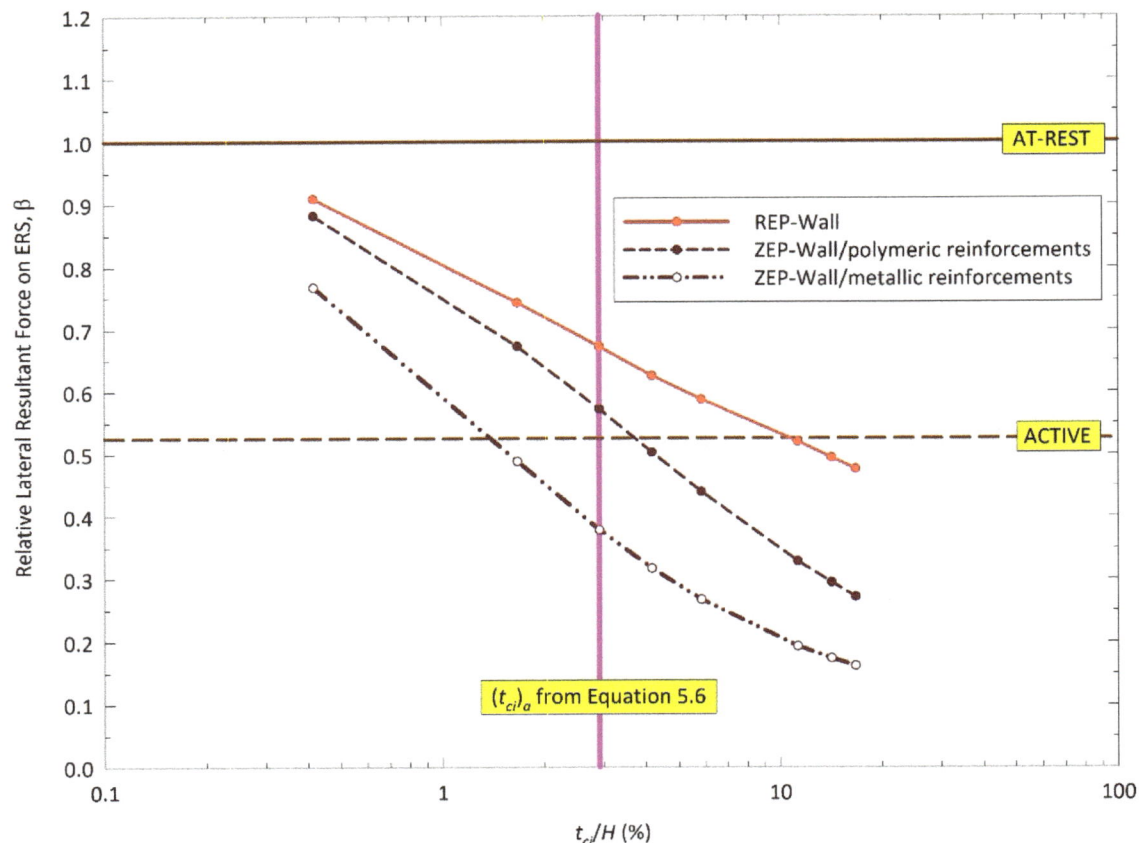

Figure 5.40. Comparison of Relationships between β and t_{ci}/H for Full-Depth REP- and ZEP-Wall Applications (Log Scale).

As discussed previously with the REP-Wall case, while correlations involving t_{ci} are of use when investigating a specific problem, it is of broader potential use in both research and practice to express the overall system stiffness using the dimensionless parameters K_{ci} and λ_{ci} that are, respectively, absolute and relative (to the geotechnical height of the ERS, H) measures of compressible-inclusion stiffness. This is because these two parameters are completely transportable and interchangeable across the entire spectrum of ERS applications that involve a compressible inclusion: different types of ERSs (rigid vs. flexible); yielding conditions (including self-yielding) and modes; materials; and loading (gravity, seismic, expansive soils, etc.). In addition, the non-dimensional nature of these parameters makes them independent of systems of units and thus transportable and interchangeable globally as well.

Figures 5.41 and 5.42 show the relationship between β, and K_{ci}, with the latter plotted to linear and $\log_{10}$ scales respectively. Results from both ZEP-Wall cases as well as the REP-Wall results shown previously in Figures 5.13 and 5.14 are shown. The overall trends are identical both with and without reinforcements. It is also clear that the lateral stiffness of the retained soil is a significant variable influencing the lateral resultant force on the ERS. This is consistent with the published literature as well as assumptions and hypotheses made throughout this monograph.

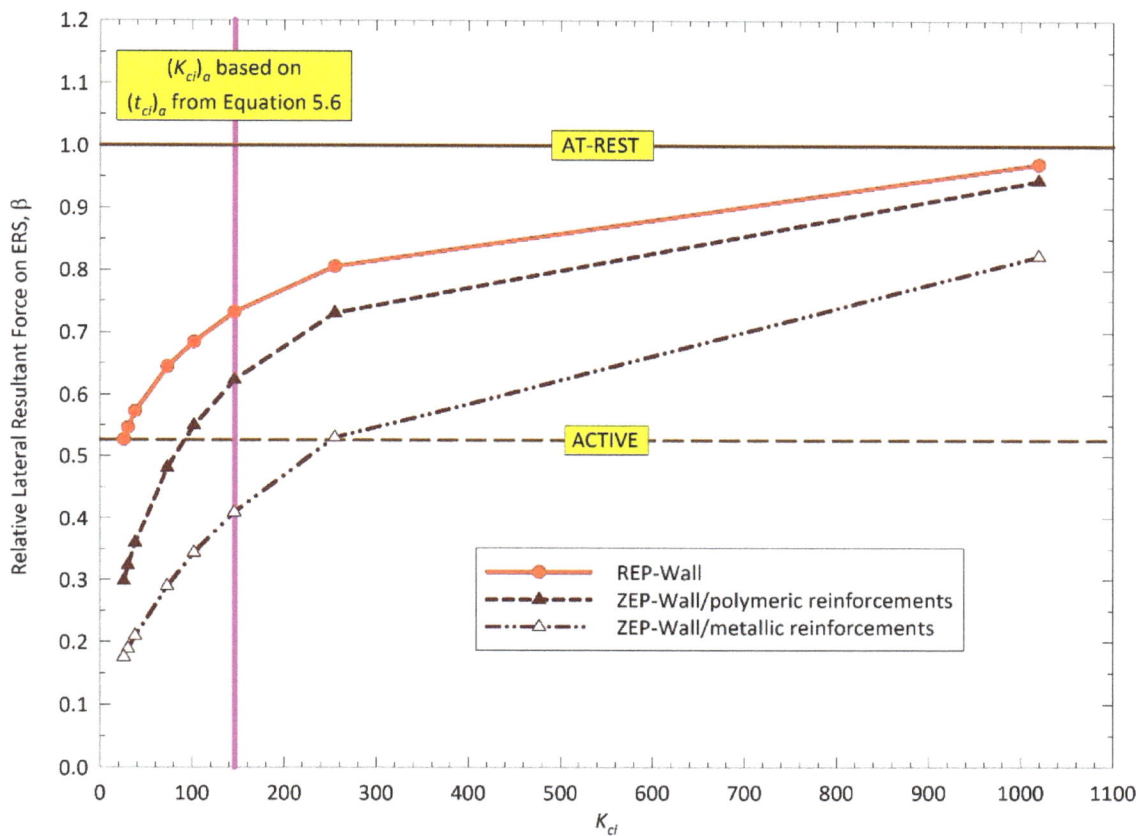

Figure 5.41. Comparison of Relationships between β and K_{ci} for Full-Depth REP- and ZEP-Wall Applications (Linear Scale).

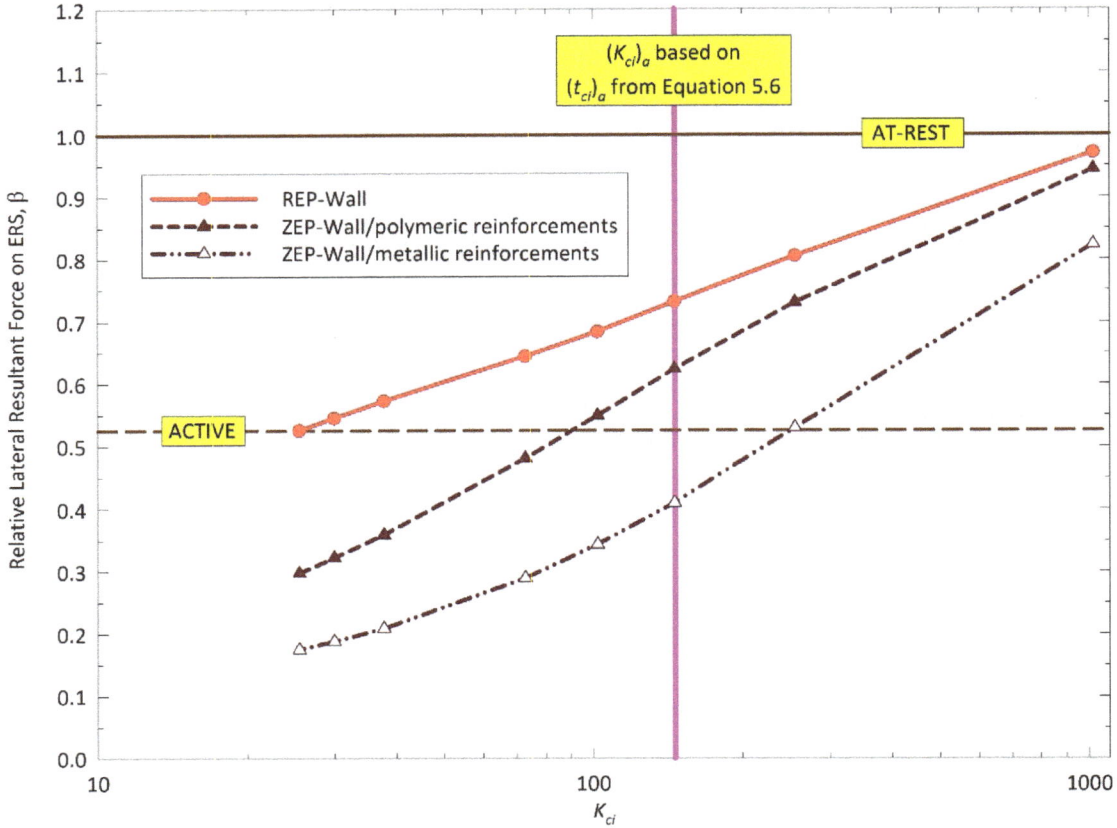

Figure 5.42. Comparison of Relationships between β and K_{ci} for Full-Depth REP- and ZEP-Wall Applications (Log₁₀ Scale).

Figures 5.43 and 5.44 show the relationship between β, and λ_{ci}, with the latter plotted to linear and $\log_{10}$ scales respectively. Results from both ZEP-Wall cases as well as the REP-Wall results shown previously in Figures 5.15 and 5.16 are shown. Given the linear relationship between λ_{ci} and K_{ci}, the observations expressed above for the latter parameter apply to Figures 5.43 and 5.44 as well.

Figure 5.45 is the final force-based figure presented in this series and depicts the calculated relationship between β and lateral displacement, Δ_{ci}, of the compressible inclusion-retained soil interface, with the latter parameter non-dimensionalized to the geotechnical height of the ERS, H, as was done previously in the REP-Wall case. As discussed in detail earlier in this monograph, especially with regard to the REP-Wall case, the Δ/H ratio in general has frequently been used in both the literature and practice as a metric for defining the magnitude of displacement necessary to mobilize the two lateral earth pressure limiting states, i.e. active and passive. This ratio has also been used to provide an approximate estimate of the maximum lateral displacement of the facing of MSEWs, a fact also noted earlier in this chapter. Thus, it is of interest to present this ratio for the ZEP-Wall analyses presented here.

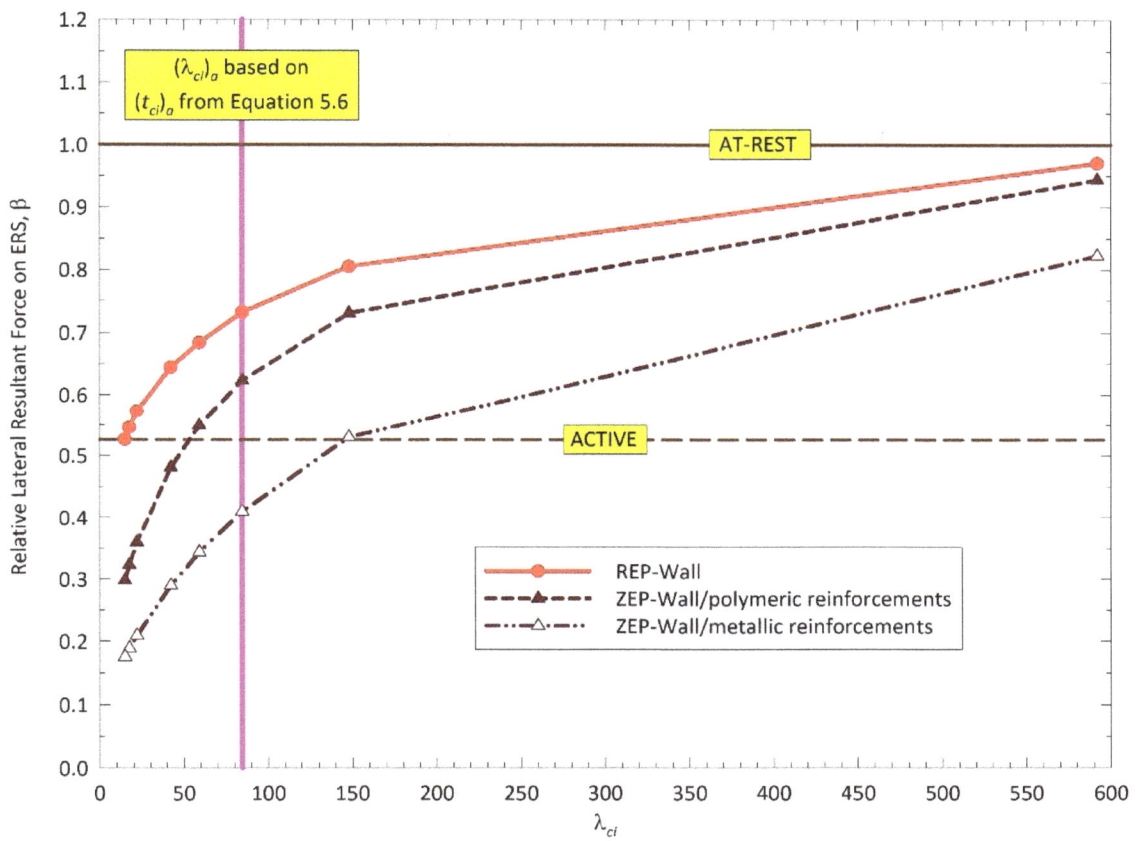

Figure 5.43. Comparison of Relationships between β and λ$_{ci}$ for Full-Depth REP- and ZEP-Wall Applications (Linear Scale).

196

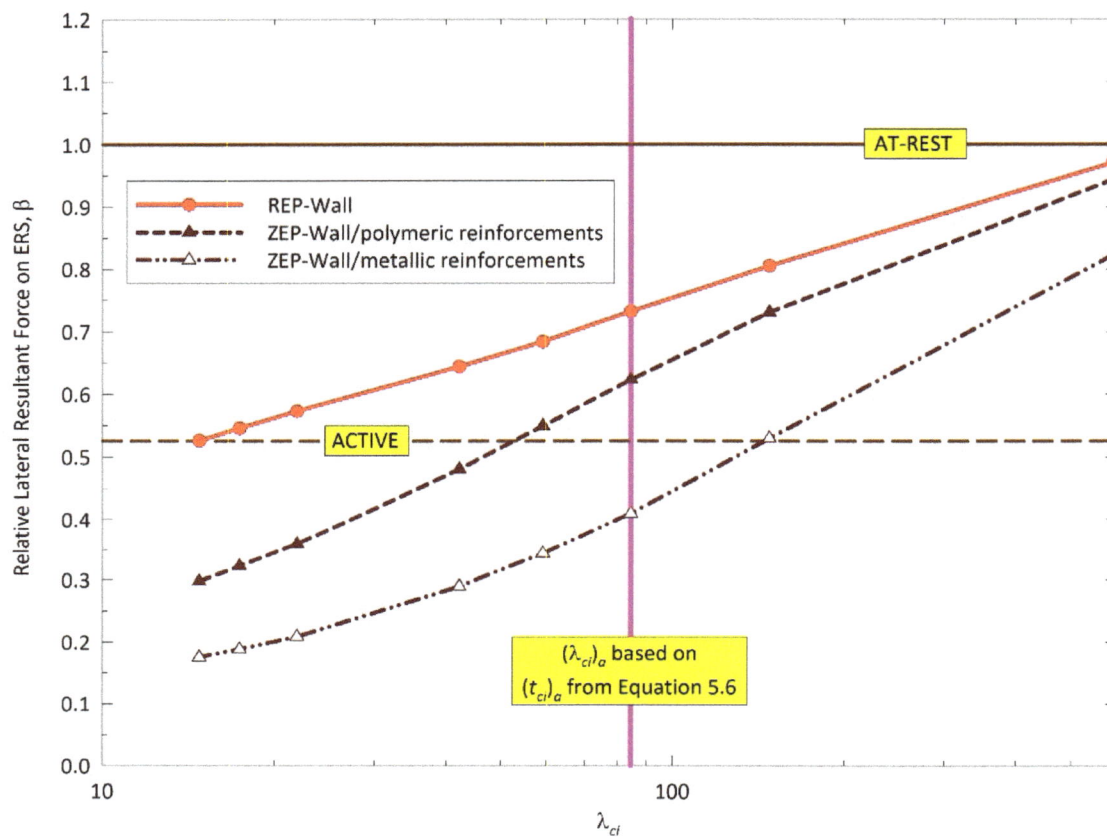

Figure 5.44. Comparison of Relationships between β and λ_{ci} for Full-Depth REP- and ZEP-Wall Applications (Log₁₀ Scale).

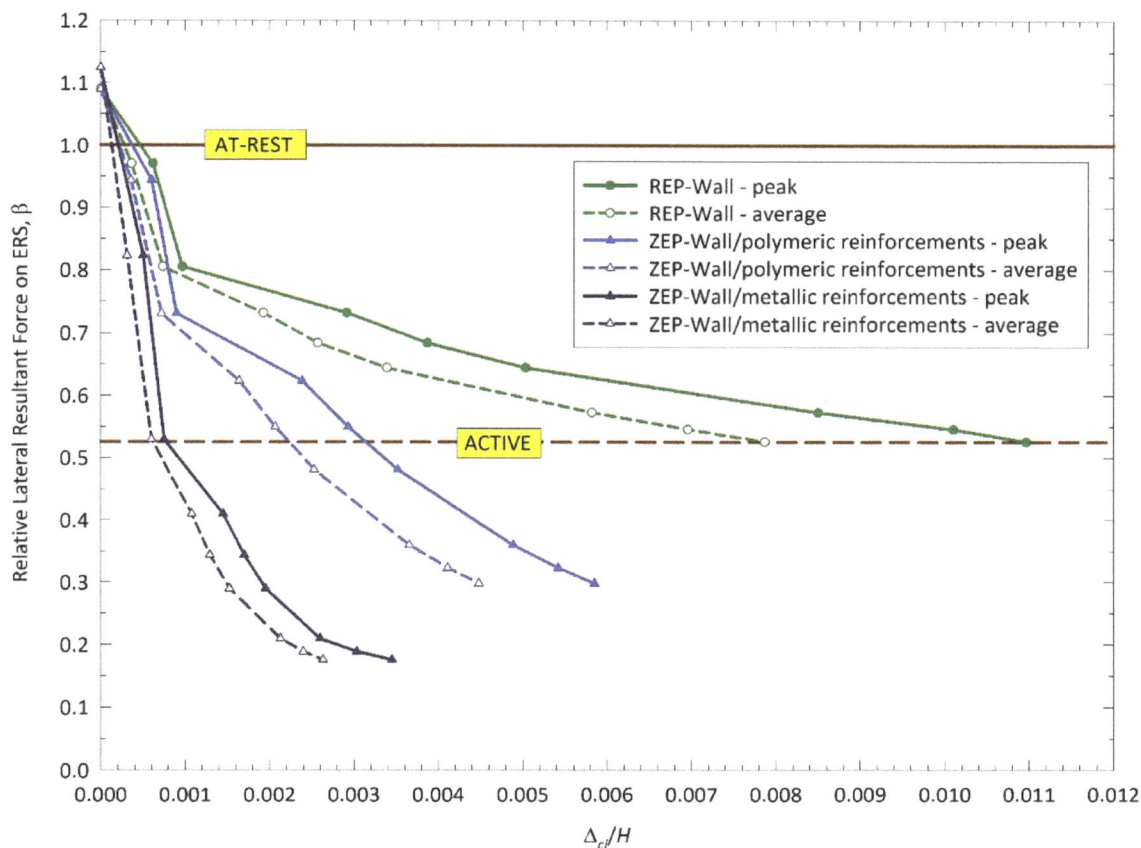

Figure 5.45. Relationships between β and Δ_{ci}/H for Finite-Element Model Used.

The Δ_{ci}/H ratio in Figure 5.45 is presented in decimal form as this appears to be the more common form used in practice for ERS applications. In some published works (including this monograph when convenient for plotting purposes), it is alternatively expressed as a percent. Also, the REP-Wall results shown previously in Figure 5.19 are shown in addition to the results for the two ZEP-Wall cases to again qualitatively illustrate the effect that lateral stiffness of the retained soil, reinforced or not, has on the calculated outcomes. Finally, for all three cases both the peak as well as average Δ_{ci}/H ratios are shown.

The ratio of average-to-peak Δ_{ci} tends to increase slightly with increasing lateral stiffness of the retained soil. For the REP-Wall case, this ratio is approximately ⅔ (= 0.67) as noted earlier in this chapter. For the ZEP-Wall with polymeric reinforcements, it is approximately 0.71 while for the metallic reinforcements it is approximately 0.75. A reasonable overall average for all compressible-inclusion applications is 0.7.

However, the larger, more significant issue is that whenever the compressible-inclusion function is used to reduce lateral pressures acting on a RERS, the pattern of lateral displacements, Δ, acting along the compressible inclusion-retained soil interface is depth-wise non-uniform by a considerable margin. This differs significantly from uniformity assumptions implied in $(\Delta/H)_a$ ratios found in the published literature for mobilizing the active earth pressure state in traditional yielding-RERS applications. Consequently, care must be taken when using such published correlations with compressible-inclusion applications. This issue has been noted and discussed in detail earlier in this chapter for the REP-Wall application but merits restating here as it is equally applicable with ZEP-Wall applications as well.

This discussion of lateral displacement provides a natural segue to discuss the depth-wise variation of lateral compressive strain within the compressive inclusion, ε_{hci}, and lateral displacement at the compressible inclusion-retained soil interface, Δ_{ci}, for different compressible-inclusion thicknesses, t_{ci}, as was done for the REP-Wall case in Figure 5.18. Figures 5.46 and 5.47 show these results for the ZEP-Wall with polymeric and metallic reinforcements respectively. Note that the Δ_{ci}/H ratios presented in these figures are expressed in percent as opposed to decimal form solely as a plotting convenience. Note also that as the lateral stiffness of the retained soil increases from no reinforcement (Figure 5.18) to polymeric reinforcement (Figure 5.46) to metallic reinforcement (Figure 5.47) the strain and displacement profiles both trend toward being more uniform with depth, especially for the greater thicknesses of the compressible inclusion.

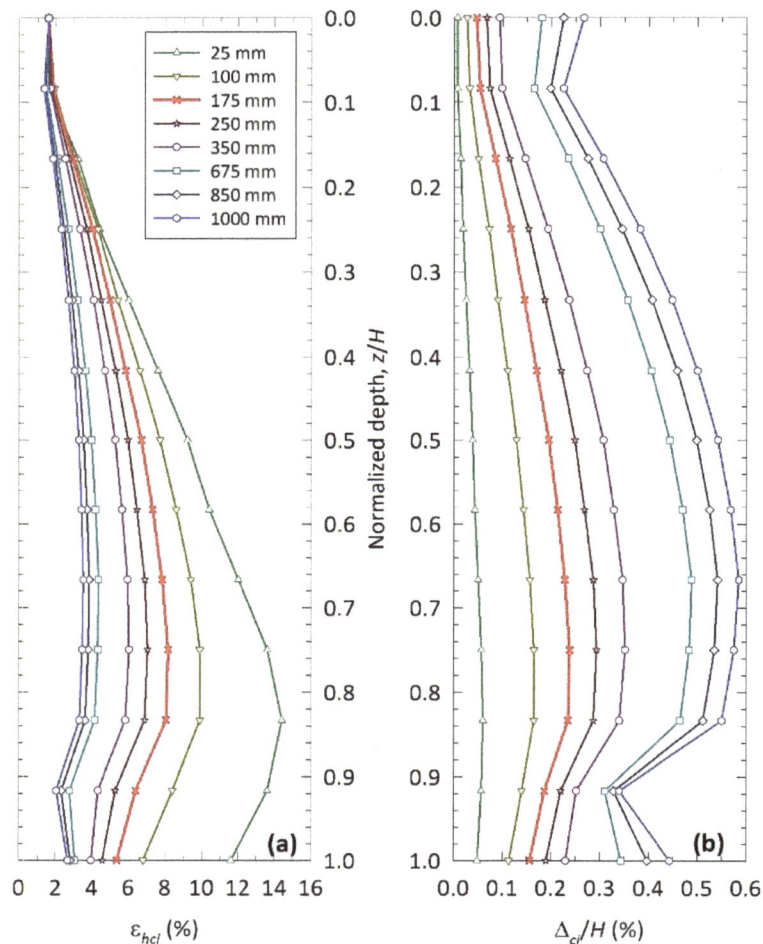

Figure 5.46. ZEP-Wall/Full-Depth Placement/Polymeric Reinforcements/Non-Yielding RERS Compressible-Inclusion Lateral Strains and Displacements as a Function of Compressible-Inclusion Thickness.

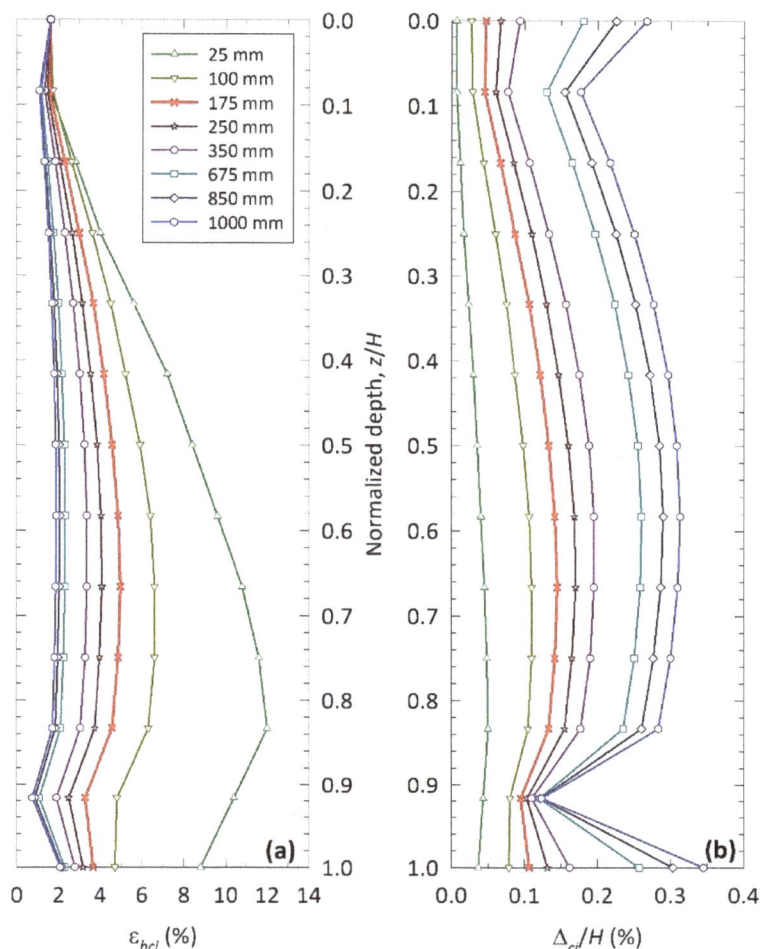

Figure 5.47. ZEP-Wall/Full-Depth Placement/Metallic Reinforcements/Non-Yielding RERS Compressible-Inclusion Lateral Strains and Displacements as a Function of Compressible-Inclusion Thickness.

Note that the sharp inflection of the Δ_{ci}/H curves in particular at depth ratios, z/H, of approximately 0.9 and 0.1 for the larger thicknesses of the compressible inclusion are the result of the way in which the reinforcements were modeled. No reinforcements were assumed to exist at the very bottom ($z/H = 1.0$) or very top ($z/H = 0$) of the retained soil.

The final issue explored in this discussion of the ZEP-Wall case with full-depth placement is to revisit the concepts presented earlier in this chapter involving potential use of the theory of linear elasticity in general and the solution of Harrison and Gerard (1972) in particular, as expressed in Equations 5.21 through 5.24, as a methodology for quantifying the lateral stiffness that geosynthetic tensile reinforcement adds to a retained-soil mass. As noted previously, the primary difficulty with using this conceptual approach is quantifying the Young's modulus of the unreinforced soil in lateral compression, E_{soil}.

To begin with, it is clear (and no surprise) that the lateral stiffness of the retained soil, whether reinforced or not, is non-linear as is apparent from Figure 5.45. Therefore, the Young's moduli are variable. The typical way in which this is dealt with in a solution process is to work with either tangent moduli applied incrementally or a secant modulus applied over the operative range of stress or strain in a given application. At this point, it is not obvious which is the better approach to use here so both tangent and secant moduli will be developed.

To better understand and interpret the following graphics and accompanying discussion, it is useful to remember that a secant-based assessment is essentially an overall (i.e. end-point to end-point) linear approximation of some non-linear behavior. Most importantly, a secant-based assessment reflects an approximation over some variable range that is often only a portion of the entire range of the problem. Frequently, this portion is referred to as an 'operative' range because it reflects that portion of the overall problem range that is expected to occur or develop in some particular application and is thus the primary, if not exclusive, range of interest in that application.

On the other hand, a tangent-based assessment is an incremental (usually linear), sequential assessment of some non-linear behavior such that all of the sequential assessments combined cover the same operative range of a problem. Stated another way, a tangent-based assessment can be viewed as approximating some non-linear behavior with a finite number of linear segments that, in general, may be of varying length. It is obvious that the greater the number of (linear) segments the better the approximation of the actual behavior as in the limit an infinite number of linear segments would converge to the actual non-linear behavior.

As is well known in both practice and research, both the secant and tangent approaches have their benefits and limitations. As such, these approaches tend to be complementary as opposed to competitive. The secant approach is most useful when a single-value estimate is required of some variable that is actually non-single-valued. This typically occurs when some relatively simple closed-form solution is being used that requires a single value of an input variable. A relevant example is the Harrison and Gerard (1972) theory of elasticity solution presented in this chapter. It requires single values of Young's modulus for both the retained soil and geosynthetic tensile reinforcement. So, an average value of each of these moduli over the operative range of a particular application needs to be developed. The down side of the secant approach is that while the overall behavior over the operative range may be captured correctly, the intermediate values within that range will differ from the actual values.

On the other hand, the tangent approach is best suited to numerical analyses such as the FE analyses performed for this monograph. The program that was used essentially applies the theory of linear elasticity for stress-strain behavior to each element within the FE mesh. By applying this linear theory in steps or increments and performing more than one analytical iteration for each increment using a 'predictor-corrector' algorithm, the non-linear, inelastic behavior of real soil can be approximated. Clearly, the more elements used in a FE mesh and the more construction or loading steps simulated the more accurate the final answer.

With this broad understanding of secant vs. tangent approximations in mind, as a first-order approximation to investigate this problem in at least a preliminary manner the starting point was to assume that the lateral displacement could be modeled as a problem in uniaxial stress (in the horizontal direction in this case) in which case Hooke's Law becomes

$$E = \frac{\sigma}{\varepsilon} \qquad (5.25)$$

where E = either E_{soil} or E_h depending on whether the soil is unreinforced (REP-Wall case) or reinforced (ZEP-Wall case); σ = average lateral stress imposed on the compressible inclusion-retained soil interface calculated using the β value for a given loading increment and the geotechnical height of the ERS, *H*; and ε = average lateral extensive strain of the retained soil for the corresponding loading increment under consideration. For the ZEP-Wall case, this last parameter was calculated using the lateral displacements of the compressible inclusion-retained soil interface, Δ_{ci}, and using the length of the geosynthetic tensile reinforcements (4 metres (13.1 ft) in this case) as the original length.

Figure 5.48 shows the back-calculated secant Young's moduli (non-dimensionalized solely for generality using atmospheric pressure) as a function of the average Δ_{ci}/H ratio for both ZEP-Wall cases as well as the REP-Wall case. Note that these moduli correspond to E_h (for the ZEP-Wall cases) and E_{soil} (for the REP-Wall case) using the notation defined previously in Equation 5.21. Note also that the Δ_{ci}/H ratio is the value at the <u>end</u> of the range for which a particular modulus was calculated.

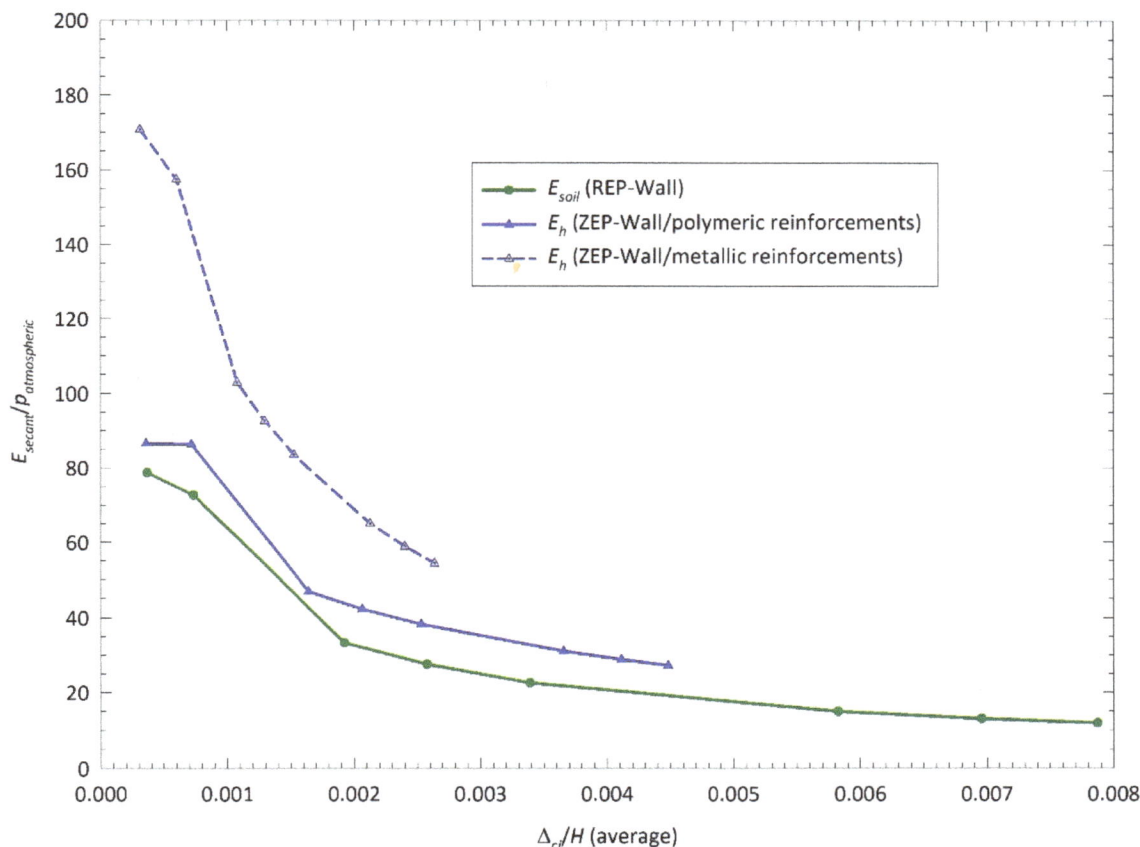

Figure 5.48. Relationships between E_{secant} and Δ_{ci}/H for Finite-Element Model Used.

It is no surprise that these moduli are highly non-linear and each decreases by approximately one order of magnitude with increasing displacement. It is of interest to note that the ZEP-Wall case with polymeric reinforcements is only marginally stiffer than the unreinforced soil of the REP-Wall case. The ZEP-Wall case with metallic reinforcements is initially substantially stiffer than the other two cases but this differential decreases with increasing lateral displacement.

Figure 5.49 shows the same results as Figure 5.48 but in a slightly different format using $\log_{10}$ scaling for Δ_{ci}/H and depicting the calculated moduli as dimensionless ratios of their maximum 'small-strain' value (the initial value was used for this in this case). The reason is that this format is the one typically used to show degradation (reduction) of soil modulus (usually shear but Young's as well) as a function of soil strain (again, usually shear but others as well) in problems involving soil dynamics. More recently, this concept has been extended to applications involving gravity loading of foundations, especially deep foundations. The purpose of such plotting is to develop mathematical relationships for modulus-degradation that can be used in analytical models and methods in routine practice.

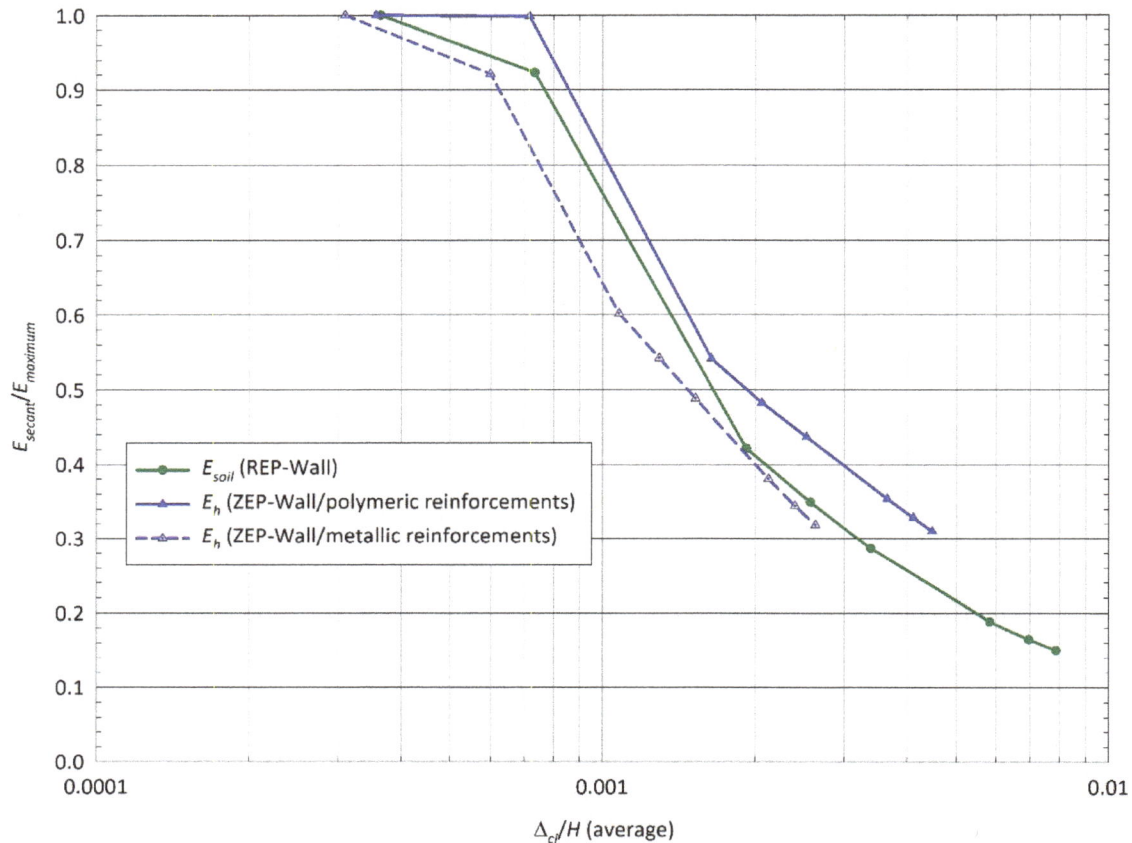

Figure 5.49. Relationships between $E_{secant}/E_{maximum}$ and Δ_{ci}/H ($\log_{10}$ scale) for Finite-Element Model Used.

As can be seen in this figure, this plotting format is very enlightening and potentially useful. It minimizes the absolute differences between the various cases and shows that they all have similar qualitative behavior. Furthermore, this behavior is broadly similar to that of a wide range of geotechnical engineering problems where modulus degradation occurs. This suggests that one or more of the empirical modulus-degradation models (e.g. Fahey and Carter 1993) that have been developed over the years for various applications may have applicability and utility for developing simplified analytical methodologies for both the REP- and ZEP-wall full-depth applications with non-yielding RERSs.

Figure 5.50 is similar to Figure 5.48 but shows the non-dimensionalized <u>tangent</u> Young's moduli. Note that here the Δ_{ci}/H ratio is the value at the <u>beginning</u> of the load increment for which a particular modulus was calculated.

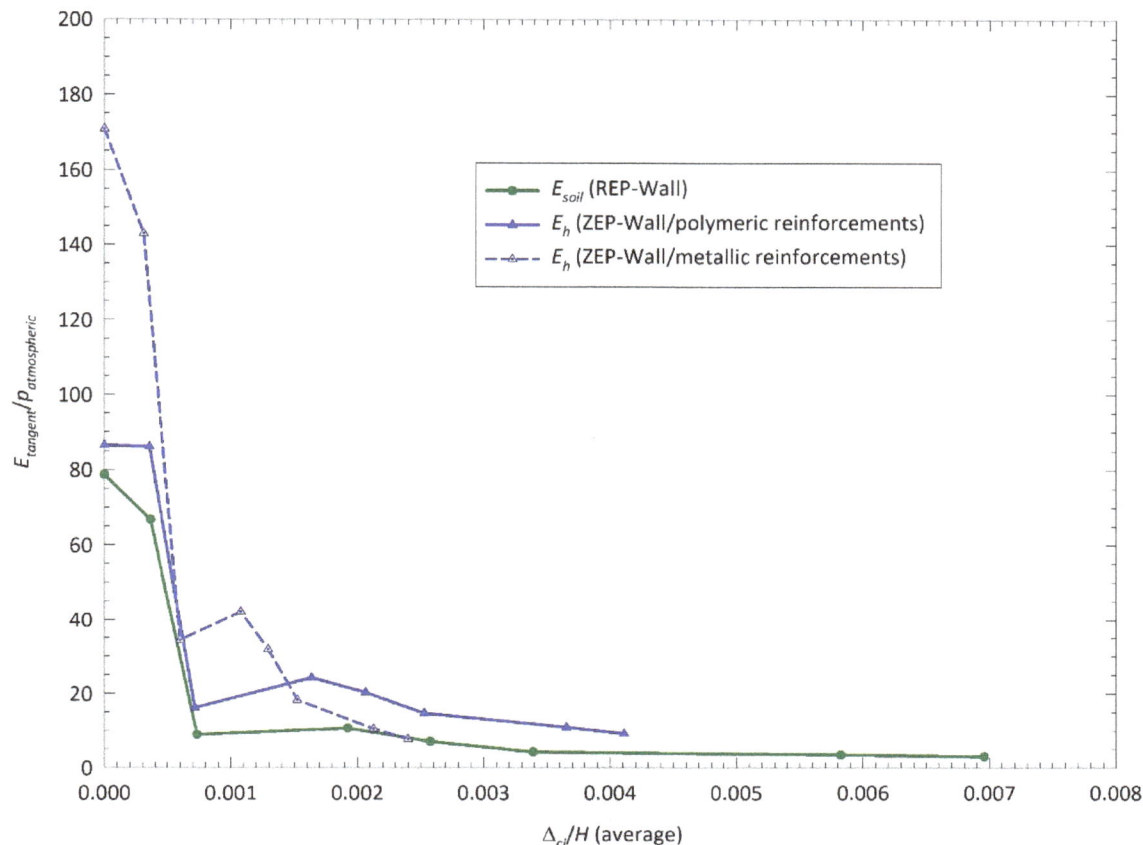

Figure 5.50. Relationships between $E_{tangent}$ and Δ_{ci}/H for Finite Element Model Used.

The results in this figure are quite interesting. The ZEP-Wall with metallic reinforcements is initially substantially stiffer than both the ZEP-Wall with polymeric reinforcements and unreinforced soil (REP-Wall). This would correspond to what is usually referred to as the *initial tangent Young's modulus*, E_{ti}. However, this initial stiffness decreases rapidly with increasing lateral displacement and by the time a Δ_{ci}/H of approximately 0.002 is reached the <u>incremental</u> stiffness of all three cases is very similar. However, this is deceptive in its implications as by that displacement ratio the ZEP-Wall with metallic reinforcements has reduced the lateral resultant force on the ERS by approximately 80% to a $\beta \cong 0.2$ whereas the reductions for the ZEP-Wall with polymeric reinforcements and the REP-Wall are only about 50% and 40% respectively ($\beta \cong 0.5$ and 0.65 respectively).

As discussed earlier in this chapter, the stiffness benefit of placing geosynthetic tensile reinforcement within the retained soil relative to the baseline unreinforced-soil (REP-Wall) case can be quantified using the parameter of relative reinforcement stiffness,

K_r, that was introduced by Harrison and Gerrard. Equation 5.21 can be rearranged and written as

$$K_r = \frac{E_h}{E_{soil}} - 1 \qquad (5.26)$$

where the Young's moduli in this equation can be defined using the aforementioned secant or tangent moduli. By definition, $K_r = 0$ for the baseline case of an unreinforced retained soil mass (i.e. the REP-Wall case) so K_r values greater than 0 reflect the fractional (or percentage if multiplied by 100) improvement in lateral stiffness provided by reinforced retained soil of a ZEP-Wall case.

Figure 5.51 shows the calculated values of K_r (expressed in decimal form) for the two ZEP-Wall cases studied. Note that results for assessments based on both secant and tangent Young's moduli perspectives are shown.

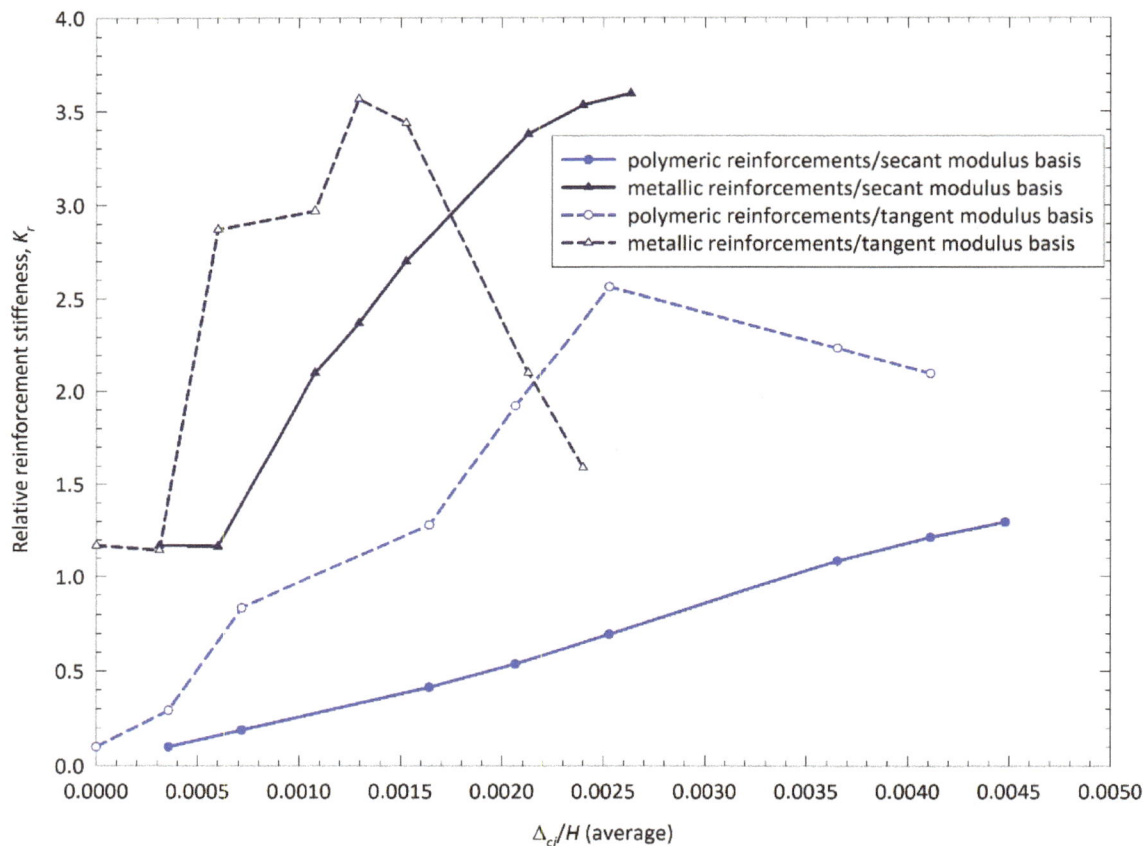

Figure 5.51. Relationships between Relative Reinforcement Stiffness, K_r, and Δ_{ci}/H.

First considering the secant modulus-based cases (shown with solid-line curves) that reflect the <u>overall</u> benefit of soil reinforcement, both types of reinforcements show increasing K_r with increasing lateral displacement. This means that the benefit of the reinforcements relative to an unreinforced soil mass ($K_r = 0$) is increasing monotonically with increasing lateral displacement of the compressible inclusion-retained soil interface. This is not surprising as increasing lateral displacement means increasing tensile strain within the reinforcements with concomitant increases in resisting forces provided by the reinforcements.

However, within this (expected) overall observation that geosynthetic tensile reinforcement provides for increased lateral stiffness of the retained soil relative to the baseline no-reinforcement (REP-Wall) case there are some significant differences. The retained soil with polymeric reinforcements shows vary little improvement relative to unreinforced soil at small displacements but then the improvement increases essentially linearly with increasing displacement. On the other hand, the retained soil with metallic reinforcements shows a significant benefit of the reinforcement almost immediately but then the rate at which that benefit increases tends to diminish slightly with increasing displacement as evidenced by the concave-downward nature of the curve.

Next considering the tangent modulus-based cases (shown with dashed-line curves) that reflect the rate-of-change of the stiffness benefit of reinforced vs. unreinforced soil, one implication of the observed results is that the metallic reinforcements offer significant benefit from the start with negligibly small displacements whereas the polymeric reinforcements offer relatively little benefit initially but then the rate of improvement increases to levels comparable to the metallic reinforcements. Again, this is no surprise given the relative stiffness between the two reinforcement types, with the simulated metallic reinforcements being 50 time stiffer than the simulated polymeric reinforcements. In addition, it is interesting to note that both types of reinforcements reach a point where the rate of improvement peaks and then decreases. This means that while both types of reinforcement continue to provide increasing stiffness relative to unreinforced soil (as reflected in the secant-based curves which continue to increase in magnitude), the rate at which this benefit is increasing decreases, dramatically so for the metallic reinforcements. But once again, it is important to note that by the time this dramatic decrease in the rate-of-improvement occurs the metallic reinforcements have already reduced the lateral resultant force on the ERS by over 80%.

In conclusion, there is still a lot of R&D work to be done to develop an acceptably accurate simplified analytical model for the basic full-depth ZEP-Wall application with non-yielding RERSs. It is clear that the simplistic assumption of a linear relationship between lateral displacement, Δ_{ci}, of the compressible inclusion-retained soil interface that the writer made when developing a simplified analytical model for the basic full-depth REP-Wall application with non-yielding RERSs is inadequate for ZEP-Wall applications. The non-linearity of Δ_{ci} as well as a way to quantify the stiffness of geosynthetic tensile reinforcement are two important behavioral components that need to be incorporated into even a relatively simple-yet-acceptable analytical methodology for ZEP-Wall applications. It is likely that whatever methodology is used to incorporate non-linear behavior would improve the existing simplified analytical methodology for REP-Wall applications as well.

5.5.3 Partial-Depth Placement Alternative

There is no technical reason why the ZEP-Wall concept could not be used for a partial-depth application although such usage is not known to have either occurred in practice or been researched in academia. Given all the variables involved (the relative portion of an ERS covered by the compressible inclusion; where along the inside face of the ERS the compressible inclusion is placed; type and length(s) of reinforcements), it would be

206

a substantial task to develop a generalized, simplified analytical methodology for such an application and thus well beyond the scope of this monograph. Therefore, for the foreseeable future any contemplated partial-depth ZEP-Wall application would need to be assessed using an advanced numerical-analysis methodology such as the FEM.

5.6 APPLICATION TO FLEXIBLE EARTH-RETAINING STRUCTURES

There has been no known research or usage to date of the compressible-inclusion function in either its REP- or ZEP-Wall variant with FERSs. While there is nothing preventing this in principle, there are several pragmatic reasons why this is not likely to be a significant potential application of this function:

- By definition, FERSs allow both displacement and deformation of the ERS-retained soil interface so that the active earth pressure state is inherently mobilized. Thus, there would be no benefit to using the REP-Wall concept and only marginal benefit to using the ZEP-Wall concept.

- Many FERSs are intended only for relatively short-term usage, e.g. for support-of-excavation (SOE) applications, so there is not likely to be a cost benefit to using geosynthetics in such applications.

- Many FERSs, especially in SOE applications, are installed in whole or in part from the ground surface with no excavation of the retained soil. Thus, installation of geosynthetics to reduce lateral earth pressures would require an additional construction step.

- The inherent flexible nature of FERSs raises issues concerning the integrity of the geofoam compressible-inclusion panels placed against such ERSs, i.e. the geofoam panels might crack due to ERS deformation.

- Many permanent FERS applications involve bulkheads in a marine environment where placing the compressible inclusion and, especially, layers of geosynthetic tensile reinforcement would be impractical.

Chapter 6

Self-Yielding Rigid Earth-Retaining Structures

6.1 INTRODUCTION AND OVERVIEW

The concept of a self-yielding RERS was defined in Chapter 2. Although such ERSs have become common in the U.S. and elsewhere for certain applications (primarily in the form of IAB abutments), the geotechnical design issues of self-yielding RERSs do not seem to be fully appreciated by many design professionals. This is likely because structural (bridge) engineers are typically the lead design professionals on projects involving IABs. Within the organizational structure of many state DOTs and private consulting firms in the U.S., geotechnical engineering input is not always provided to the extent that it could be useful whenever bridge designers control a project. This is because the focus tends to be on the economics and concomitant performance of the bridge superstructure and AASHTO code compliance without full appreciation of the fact that structural design decisions can have unanticipated geotechnical consequences. This is compounded by the fact that even many geotechnical engineers do not fully appreciate and understand self-yielding RERSs (typical undergraduate textbooks tend not to even mention the topic) and the very complex nature of their geomechanics behavior.

As an aside, one of the significant exceptions to these general comments is the Commonwealth of Virginia DOT in the U.S. They have not only been aware of the geotechnical issues involving self-yielding RERSs (most of their work to date appears to be with SIABs which they refer to using the alternative 'integral backwall bridge' terminology noted in Chapter 2) but pioneered the use of geofoam compressible inclusions more than 20 years ago. Some of their work included long-term monitoring of instrumented projects and as such represents a valuable contribution to the use of compressible inclusions with self-yielding RERSs. Fortunately, they have documented at least some of their project experience in a series of publicly available research reports (Hoppe and Gomez 1996; Hoppe 2005a, 2005b, 2006; Hoppe and Bagnal 2008; Hoppe and Eichenthal 2012).

At the present time, there are no relatively simple analytical methodologies for use with self-yielding RERSs in general as these are very diverse and complex systems with earth forces and ERS displacements that vary continuously over the life of the structure. Consequently, designs are best done on a project-specific basis using advanced numerical methods such as the FEM. Thus, the goal of this chapter is limited to illustrating fundamental design considerations that apply to all self-yielding RERSs and how geosynthetics can be used conceptually to improve the performance of structures that contain a self-yielding RERS as part of the overall structural system. How these concepts are implemented for a specific type and application of a self-yielding RERS is left to the design professionals responsible for that application.

6.2 FUNDAMENTAL DESIGN CONSIDERATIONS

The references cited in Chapter 2 illustrate that structural systems that incorporate one or more self-yielding RERSs as part of the overall system can encompass a wide variety of applications, from the IAB abutments to the walls of navigation locks and

208

water/wastewater treatment tanks. While each application has its unique features, there are fundamental design elements and considerations that they share. Understanding these elements and, more importantly, how they cause the ERSs involved to behave in an atypical (for RERSs) fashion is crucial to understanding the design concepts presented subsequently. These common design elements are:

- The overall structural system is subjected to natural (usually seasonal) variations in ambient air temperature in a way that significantly affects the performance of the structure because it results in the structure being subjected to significant thermal-induced displacements and/or stresses depending on the displacement-constraint conditions inherent in the structural system. Furthermore, these temperature variations exist for the life of the structure. This is a simple fact of nature that can never be eliminated completely although in some cases the effects can be moved from one part of the overall structural system to another. However, they cannot be made to disappear so must ultimately be dealt with. Overlooking, wishing away, or being ignorant of all the potential impacts of these thermal effects appears to be the single biggest, most consistent mistake made by practitioners.

- The overall structural system is intentionally monolithic in its structural design and performance. Thus, the ERS component(s) are fully connected to the rest of the structure in the sense that there is full structural continuity with respect to both bending moments and displacements. This means that the ERS component(s) are affected by the aforementioned thermal effects and the force-displacement behavior of the overall structural system 'drives' the behavior of the ERS component(s).

- Because the force-displacement behavior of the ERS component(s) is controlled by the thermal behavior of the overall structural system, the retained soil adjacent to the ERS component(s) is put in the position of reacting to the behavior of the ERS component(s). This is a radical paradigm shift and role reversal from more traditional RERS applications where it is the ERS that reacts to forces imposed on it by the retained soil.

- Another significant paradigm shift is that the lateral earth pressures acting on the ERS component(s) not only vary throughout the year because of thermal effects on the overall structural system but also vary over the entire life of the structure. This is because soil is inherently inelastic in its behavior so that at the end of an annual thermal cycle the lateral earth pressures do not, in general, return to what they were at the start of the cycle. They are typically somewhat higher at the end of an annual cycle than at the beginning due to a geomechanical behavioral phenomenon that England (1994) called *ratcheting*. More importantly, this ratcheting is cumulative so that the net lateral earth pressures can progressively and irreversibly continue to increase over time with each annual thermal cycle. If such pressures are not addressed in one way or another during design, failure in the form of structural distress (SLS) or even collapse (ULS) can occur.

- Another geotechnical outcome of the inelastic cyclic load-displacement response of the ERS component(s) is the development and progression of a settlement 'bowl' on the ground surface adjacent to the ERS component(s). This is because in traditional RERS applications the ERS is, for all intents and purposes, relatively fixed in space and time (at least under gravity loads). This is true even with yielding RERSs because the magnitude of yielding (lateral displacement) for a properly designed yielding RERS is negligibly small. This means that the retained soil is equally fixed in space and time.

However, the retained soil adjacent to a self-yielding RERS (which can undergo relatively large lateral displacements during the course of an annual thermal cycle) no longer has any restraint against displacement and will 'follow' the self-yielding RERS as it displaces. However, the aforementioned inelastic nature of soil behavior results in the retained soil not returning to its original position at the end of an annual thermal cycle. This net lateral displacement of the retained soil increases over time due to the aforementioned ratcheting phenomenon and translates into surficial settlements that progress vertically and horizontally in their extent over time. Depending on the particular nature of the self-yielding RERS, these surface settlements can be problematic to the overall performance of the structural system of which the self-yielding RERS(s) is (are) a component.

6.3 BASIC DESIGN ELEMENTS

To illustrate how all these design issues come together and can potentially be addressed in an actual application, the example of an IAB abutment will be used. This problem was addressed in considerable detail in Horvath (2000), with summaries presented in Horvath (2004c, 2005). The figures presented below are modified versions of ones that appeared originally in Horvath (2000).

To begin with, Figure 6.1 illustrates the primary design elements of a traditional single-span bridge where the superstructure is designed as a simple span. Such a design allows for seasonal thermal changes to be accommodated by (theoretically) unrestrained lateral displacement of the superstructure so that no thermally induced stresses develop in the system.

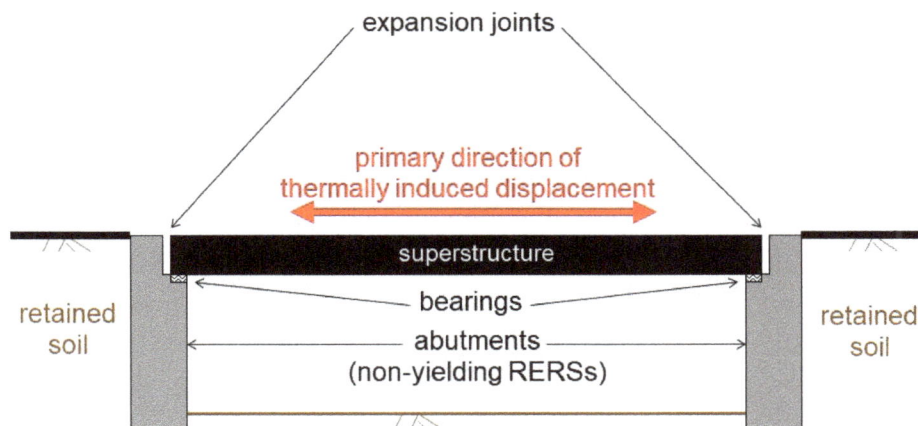

Figure 6.1. Traditional Design Concept to Accommodate Unrestrained Thermally Induced Displacement of a Bridge Superstructure.

The key takeaways concerning this traditional design solution that are relevant to the present discussion is that the overall system consists of three elements:

- A component that laterally displaces continuously for the life of the structure due to natural thermal effects. In this case, this is limited to the bridge superstructure.

- A component that is fixed in space and time to provide a stable approach to the bridge superstructure. In this case this consists of abutments on either end that are non-yielding RERSs and thus provide non-displacing support to the retained soil. The earth loads from the retained soil are for design purposes unchanging over time and are those from the at-rest earth pressure state.

- A combination of structural components (expansion joints and bearings[83]) to accommodate the relative lateral displacement between the aforementioned displacing and non-displacing elements of the overall system. These components ensure that the overall bridge system is statically determinate and thus 'immune' stress-wise from thermal effects.

As is well known, the expansion joint and bearing components tend to be significant structural-maintenance issues for the entire life of a conventional bridge. Consequently, it is no surprise that alternative designs were investigated as far back as the 1930s (Card and Carter 1993) although such efforts did not gain much traction and momentum in the U.S. at least until much later in the 20th century, well after the post-World War Two road-building boom had ended. The most common alternative used to date is the IAB, the basic elements of which are shown in Figure 6.2 (the approach slabs are optional, their use is explained subsequently). Note that the IAB design concept is driven totally by structural considerations to eliminate the troublesome expansion joint + bearing detail completely.

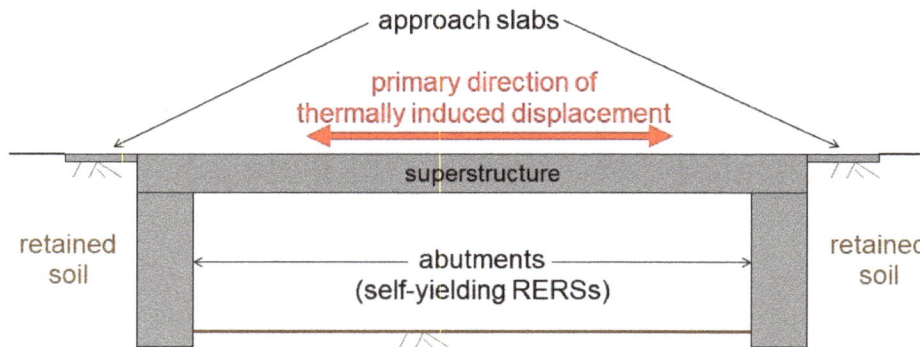

Figure 6.2. Basic Elements of an Integral-Abutment Bridge.

Relevant to the present discussion is the fact that because of the unchangeable environmental and physical elements that drive the overall problem shown in Figures 6.1 and 6.2, any alternative to the traditional design concept shown in Figure 6.1 will always have some structural component or components that will be subjected to environmental thermal effects and consequently expand and contract seasonally in the horizontal direction. Furthermore, there will always be soil at either end of a bridge that interacts with the overall bridge structure in some way through the abutments and thus has to be considered.

---

[83] In this simple example, one bearing would typically be of the hinge type and the other of the roller type. The specific details are irrelevant to the present discussion.

The IAB concept shown in Figure 6.2 is no exception to these precepts. To begin with, Figure 6.3 illustrates the thermal effects on IAB abutments which are now self-yielding instead of non-yielding as in the traditional bridge design (Figure 6.1). This is because the abutments are now part of a structural monolith that has the shape of an inverted letter 'U'. As a result, the tops of the abutments are displaced horizontally by the thermally induced displacements of the bridge superstructure. This results in a complex displacement mode of the abutments that has elements of both rotation-about-bottom and translation. More importantly, the peak magnitude of these displacements, which are primarily a function of the length of the bridge span, are relatively large. They are typically an order of magnitude larger than those necessary to mobilize the active earth pressure state and thus of the same order of magnitude necessary to mobilize the passive earth pressure state.

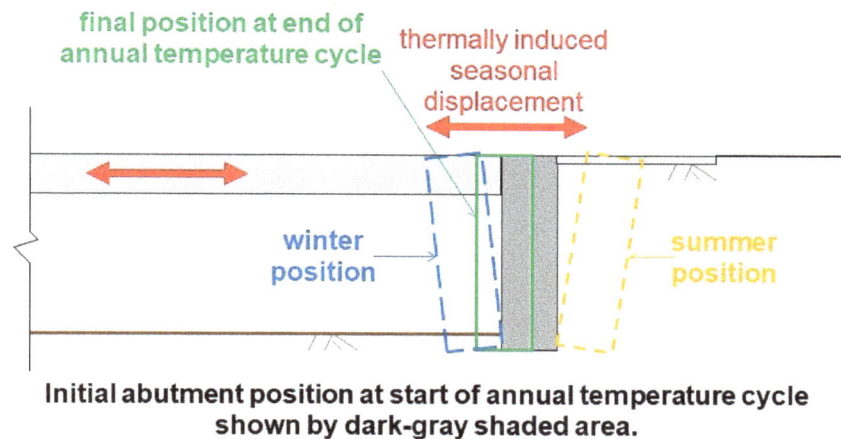

Initial abutment position at start of annual temperature cycle shown by dark-gray shaded area.

Figure 6.3. Thermally Induced IAB Abutment Displacement Pattern.

Historically, the primary (and often sole) geotechnical concern in routine design practice related to IAB abutments has been the relatively large lateral earth pressures generated when the abutments are pushed into the retained soil, i.e. the "summer position" shown in Figure 6.3. This has led to a number of proposed design lateral earth pressure diagrams ('envelopes') for use in routine practice although both field measurements of actual IABs as well as the writer's FE analyses indicate broad disagreement with these design guidelines (Horvath 2000).

However, this narrow geotechnical concern over summer lateral earth pressures misses the 'big picture' that involves multiple geotechnical concerns. The root cause of these larger geotechnical concerns is the consequence of the inability of IAB abutments to return to their original position at the conclusion of an annual thermal cycle as depicted qualitatively in Figure 6.3.

To begin with, this net inward displacement of the abutments tends to increase over time with each thermal cycle due to the phenomenon of ratcheting that was noted previously[84]. This means that the increased lateral earth pressures that occur during each

---

[84] Although not addressed in this monograph, this cumulative inward migration of IAB abutments toward each other suggests that this would be accompanied by a net increase in compressive forces acting on the bridge superstructure.

summer cycle are larger with each successive summer. Given that an IAB may have a design life of 100 years or more, it is obvious that in the long term structural distress (SLS) or even collapse (ULS) could develop due to lateral earth pressure magnitudes that were not considered during design.

Another geotechnical consequence is that during the winter cycle the active earth pressure state would be expected to develop within the retained soil. This would be accompanied by development of a failure wedge of soil that would tend to slump inward (toward the abutment) and downward. This displacement would be exacerbated by the fact that the inward displacement of an IAB abutment during the winter cycle is typically much larger than that required to just mobilize the active state, often by a factor of 10 as noted above. This excess of actual displacement over that necessary to simply mobilize the active earth pressure state would thus cause the failure wedge to displace excessively as the soil, lacking any lateral restraint, would simply follow the displacement of the abutment.

This inward and downward displacement of the soil comprising the active failure wedge results in settlement of the overlying ground surface adjacent to the abutments. Because of the inelastic nature of soil, this settlement would remain even during the summer expansion cycle. In fact, due to the cumulative inward displacement of the abutments over time as noted above, this settlement bowl will tend to continuously expand laterally and vertically over time as illustrated qualitatively in Figure 6.4.

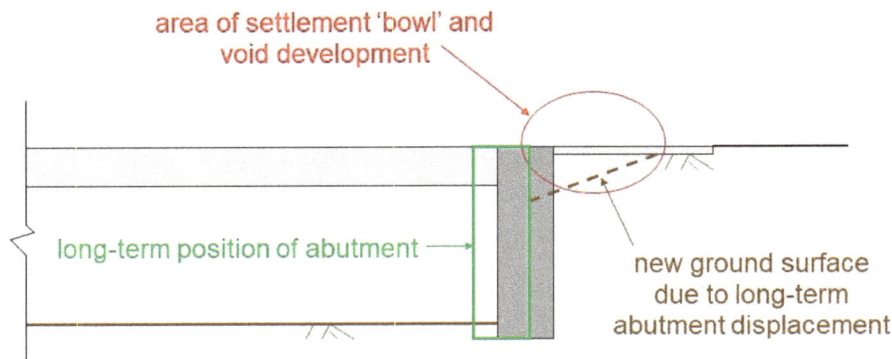

Figure 6.4. Settlement 'Bowl' and Void Development Adjacent to IAB Abutments.

It is relevant to note that development of this settlement bowl is not just a theoretical or hypothetical exercise but something that has been found to develop with actual IABs. In fact, it is the reason why approach slabs as shown in Figures 6.2 through 6.4 have become a standard part of IAB design. This is to avoid the infamous 'bump-at-the-end-of-the-bridge' that would occur otherwise with the roadway pavement on the landside of the abutments.

However, as noted in Horvath (2000, 2004c, 2005), this approach-slab solution has not been entirely successful. This is because numerous approach-slab failures have occurred with actual IABs. This indicates that not only does ground-surface subsidence adjacent to IAB abutments develop but it develops to an extent that is not properly addressed, at least in some cases, by current approach-slab guidelines.

The conclusion drawn by the writer circa 2000 was that IAB design methodologies were insufficient and required modification. This is because IAB methodologies failed to recognize that in replacing the traditional design shown in Figure 6.1 with the one shown in Figure 6.2:

- the basic problem environmental factors and concomitant physics related to seasonal temperature changes remained unchanged and

- the requirements for addressing these unchanging technical issues that were inherent in traditional bridge designs remained unchanged.

Specifically, with regard to the second item, there are two design objectives that simply must be met every time when designing an IAB as they were inherent in traditional bridge designs:

- The retained soil <u>must</u> be stabilized year-'round with respect to displacements in some fashion. In traditional bridge design, this was accomplished by standalone abutments that were non-yielding RERSs that could be counted on to remain spatially stable year-'round indefinitely. Experience indicates that, in general, traditional abutments fulfilled this role and need adequately. With the abutments of an IAB made part of a constantly deforming and displacing structural system, this means that the retained soil itself must now be made self-stable in some way as the retained soil can no longer count on the abutments for this. In essence, the retained soil has to be its own abutment.

- An engineered equivalent to an expansion joint is required between the inside face of the IAB abutments and the self-stabilized retained soil to accommodate, in a controlled and predictable fashion, the relative displacement that occurs between the two. An important consideration in the design of such an equivalent expansion joint is that it should also be able to reduce, at least to some degree, the relatively large lateral earth pressures that develop during each summer cycle of abutment displacement outward toward the retained soil. This would have potentially significant economic benefits relative to the structural design of the abutments.

The writer developed two geosynthetics-based alternatives that satisfy both of these requirements for addressing these serious geotechnical performance issues (Horvath 2000, 2004c, 2005). These are illustrated in Figure 6.5.

The common element in each alternative is a geofoam compressible inclusion that functions as the requisite expansion joint. The difference between the alternatives is the manner in which the retained soil is rendered self-supporting. In what would likely be the more common alternative used in practice (Figure 6.5a), the retained soil adjacent to the compressible inclusion is turned into a MSEW. Therefore, this overall alternative is essentially a full-depth ZEP-Wall application as discussed at length in Chapter 5.

In the second alternative (Figure 6.5b), a portion of the retained soil adjacent to the compressible inclusion would be replaced by EPS-block geofoam. This is essentially the lightweight-fill application that was discussed at length in Chapter 4. This alternative would likely be cost-effective only for applications where the overall bridge site was underlain by compressible soil. In such a case, the IAB abutments would likely be supported on deep foundations so it would be desirable to minimize long-term settlement of the approach earthworks on either side of the bridge so the use of a lightweight-fill material would be justified economically. In any event, the EPS blocks would satisfy the need to make the retained soil self-supporting.

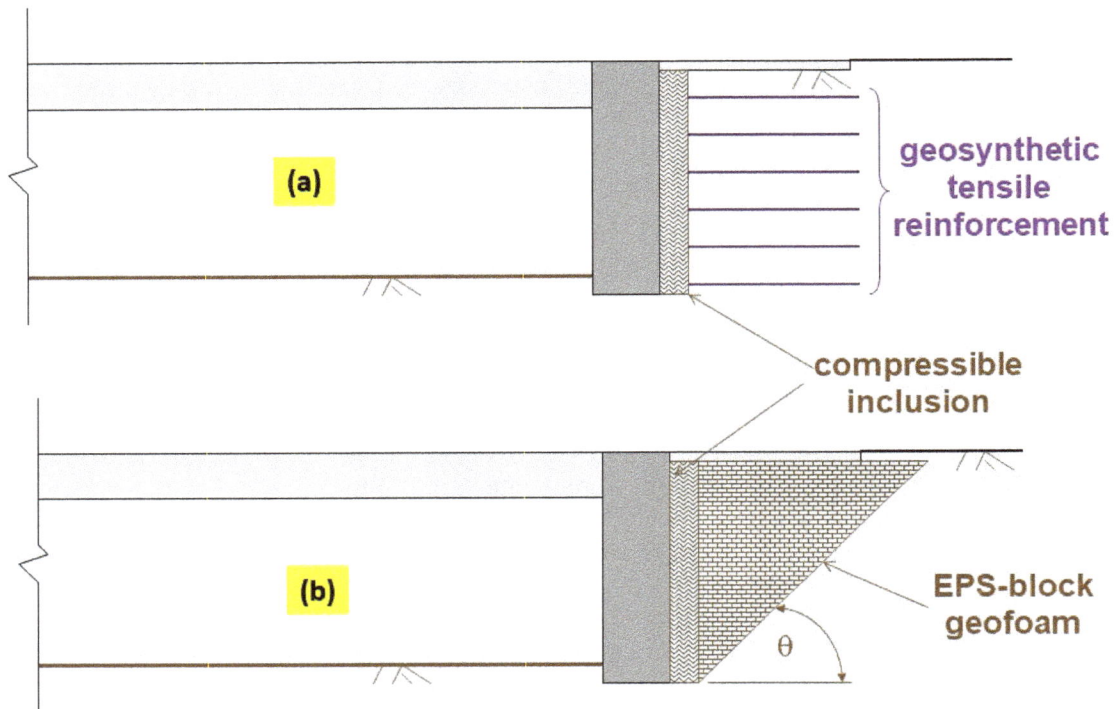

Figure 6.5. IAB Design Alternatives to Address Geotechnical Performance Issues.

For the sake of completeness, it is noted that other alternatives in addition to the ones shown in Figure 6.5 could be developed as there are other ways to render the retained soil self-supporting other than to use geosynthetic tensile reinforcement or replacement with EPS-block geofoam. However, the geofoam compressible inclusion would still be a necessary component of any other alternative.

As a final comment, it is relevant to note that the design alternatives illustrated in Figure 6.5 are equally applicable to retroactive installation with existing IABs as they are in new construction. Given the substantial inventory of existing IABs, many if not most of which likely exhibit at least settlements (in the absence of approach slabs) and settlement voids beneath approach slabs if not structural distress to the abutments due to excessive lateral earth pressures, there is clearly a need for rehabilitation strategies that will improve the long-term performance of these IABs. As evidenced by the paper by Edgar et al. (1989), there was a problem with existing IAB performance over 30 years ago. It is likely that the problem is even larger now. Note that Edgar et alia used the solution shown in Figure 6.5a although in lieu of a geofoam compressible inclusion they simply left a physical gap between the inside face of the abutments and geosynthetically-reinforced retained soil.

This page intentionally left blank.

Chapter 7

Additional Geosynthetic-Function Design Considerations

7.1 INTRODUCTION

The goal of this chapter is to discuss in broad terms some of the other key technical considerations in terms of geosynthetic functions that usually accompany a desire to use geofoam to reduce lateral pressures on ERSs. There is a focus on the inherent multi-functionality of geofoam materials as the various functions need to be considered in every design whether they are explicitly desired or not on a given project.

7.2 OVERVIEW OF GEOFOAM MULTI-FUNCTIONALITY

Geofoam materials are inherently multi-functional, with most of the functions they offer unique compared to other categories of geosynthetics. Because most design professionals at present are much more familiar with the functions of planar geosynthetics such as geotextiles and geogrids, they may be unaware of both the breadth and novelty of geofoam functions.

When desired, the inherent multi-functionality of a geofoam material can be extended even further by creating geocomposite products that consist of more than one geosynthetic material, including traditional planar products. Alternatively, geofoam can be used in conjunction with other geosynthetics in a complementary, synergistic fashion to provide end results that neither material or product could produce alone. The ZEP-Wall application with non-yielding RERSs is an example of this where neither the geofoam nor geosynthetic tensile reinforcement alone could accomplish what both can do together.

The multi-functionality of geofoam, whether provided by a single material/product, geocomposite product, or combined application with other geosynthetic(s), can provide both a technical and economic benefit that enhances the cost-effectiveness of using geofoam in many applications, including for reducing lateral pressures on ERSs as is the focus of this monograph. However, there is <u>always</u> the 'other side of the coin' that needs to be considered in all cases. This is the fact that geosynthetic functions cannot, in general, simply be 'turned off' when they are not needed or desired. This is important because in certain specific applications a geofoam function may not only be unneeded but unwanted as it can have potentially negative consequences that must be addressed in design.

It is important to note that such negative consequences are not unique to geofoam as it is no different in this regard than any other materials used in engineered construction. All materials...steel, Portland-cement concrete, wood, polymerics, composites...have performance issues and downsides that do not inhibit their use but simply need to be addressed during design, construction, and long-term maintenance.

7.3 DRAINAGE

Unless an ERS is designed to be permanently submerged below groundwater or within a body of open water, there is always a need to provide for positive drainage of groundwater. This is true whether or not geofoam is used to reduce lateral pressures acting on the ERS. In some applications, such as building basement walls, there may also be a need to vent ground-borne gases such as radon or naturally occurring methane.

In either case, the geosynthetic functions of fluid collection and transmission that are the core requirements of a positive drainage system are usually accomplished by placing a chimney drain along the inside face of the ERS. Nowadays, a panel-shaped geocomposite product for this would generally be preferable compared to natural or synthetic aggregate. Such geocomposite products consist of a high-permeability core with a factory-laminated geotextile on one face of the panel to provide the geosynthetic functions of separation and filtration. There may or may not be a geomembrane factory-laminated to the other face of the panel to act as a vapor barrier.

Selection of the appropriate geosynthetic product to use as part of a positive drainage system depends primarily on which geofoam function (lightweight fill or compressible inclusion) is used to reduce lateral pressures on the ERS, keeping in mind that geofoam materials have negligible inherent fluid permeability unless the material is intentionally manufactured to be otherwise (as in the case of GPS-PB) or the final product is shaped during the manufacturing process to have channels or grooves in at least one face.

In general, when the lightweight-fill function is used either a traditional geocomposite sheet drain or a geocomposite based on GPS-PB would be the drainage product of choice. The latter is arguably preferable in that it also functions as a compressible inclusion to further reduce the magnitude of lateral pressures that reach the ERS.

On the other hand, when the compressible-inclusion function is used, a drainage geocomposite based on GPS-PB would always be preferable. This is because GPS-PB has stiffness properties that are comparable to the R-EPS that would be the material-of-choice for the compressible inclusion. Thus, economy could be achieved by using a three- or four-component geocomposite consisting of R-EPS, GPS-PB, a geotextile (for filtration), and (optionally) a geomembrane vapor barrier as the GPS-PB could function for both drainage as well as contributing to the required stiffness of the compressible inclusion.

7.4 THERMAL INSULATION

The geosynthetic function of thermal insulation is unique to geofoam and appears to have been the first geofoam function that was investigated and used in practice beginning circa 1960 (Horvath 1995b). This function is also a classic example of the point made earlier in this chapter that geosynthetic functions cannot be ignored even when they are not required or even desired in a particular project application.

Discussing first the benefits of the thermal-insulation function, applications involving the below-ground space of buildings benefit thermally from both the lightweight-fill and compressible-inclusion functions as both functions will significantly retard the flow of heat either into or out of the below-ground space. This can improve the energy efficiency of this space with regard to both seasonal heating and cooling. An additional benefit is improving the relative humidity of the space in warm weather as the thermal profile through the ERS is shifted significantly by the external presence of the geofoam.

A much broader benefit occurs when a drainage geocomposite that uses GPS-PB as its high-permeability core is used as the primary component of a positive-drainage system. Such a geocomposite provides what is called *insulated drainage* that continues to perform even when the outside face of the ERS is exposed to air temperature that is below the freezing point of water.

Next considered is a situation where thermal insulation is not only not required, it is not desired as it can have negative implications. This involves the lightweight-fill function.

An inherent aspect of this function is that the uppermost layer of EPS blocks will always extend the farthest out from the inside face of the ERS and into the retained soil as shown conceptually in Figure 3.2. It is likely in many applications that the ground surface overlying these EPS blocks will be paved for either vehicular or pedestrian traffic. If the upper surface of the EPS blocks is too close to the pavement surface, the phenomenon of *differential icing* can occur in climates where seasonal freezing occurs (Horvath 1995b, 2001b). This is a potential safety issue for the vehicles or persons traversing the pavement, not for the EPS blocks themselves. As this problem has been known since circa 1970, the need to design for this by providing adequate cover above the EPS blocks is now a matter of routine in practice (Stark et al. 2004a, 2004b). It is certainly not a reason not to use geofoam.

7.5 NOISE AND SMALL-AMPLITUDE VIBRATION DAMPING

Another geosynthetic function that is unique to geofoam is its inherent ability to provide damping of noise and small-amplitude vibrations because of the relatively low density of geofoam materials. There are applications where this is the primary functional reason for using geofoam although they appear to be relatively rare to date. Even explicitly citing this function as a secondary reason for using geofoam appears to be relatively uncommon to date.

The benefit of this function in applications where lateral pressure reduction for an ERS is the primary reason for using geofoam would appear to be for below-ground walls of buildings. Whether the lightweight-fill or compressible-inclusion function is used in such applications it is obvious that the geofoam will act as a barrier between an external source of ground-borne vibrations such as from motor vehicles or trains and the below-ground interior of the structure.

At the present time, there is no known simplified analytical approach for assessing the vibration-damping benefit of a geofoam application. This is likely one of the primary reasons why this functional usage of geofoams has been very limited to date.

This page intentionally left blank.

Chapter 8

Geofoam Material and Product Selection

8.1 INTRODUCTION

The purpose of this chapter is to concisely organize and summarize comments and observations made throughout this monograph in order to provide broad guidance to design professionals to assist them in making informed, rational decisions about geofoam materials and products that are used for the applications that are the focus of this monograph, i.e. reducing lateral pressures on ERSs. This includes offering guidance about material specifications where choosing specifications is part of the design process.

It cannot be emphasized too strongly that the writer's intent in this chapter is to offer generic, objective information and <u>not</u> to promote a particular product or manufacturer. However, in some cases it is necessary to mention specific names for informational purposes but this is neither an endorsement nor a recommendation.

8.2 OVERVIEW

It is important for design professional to recognize that in the current state of technological evolution, the material and product selection process for the lightweight-fill and compressible-inclusion functional applications are markedly different with regard to material and product responsibilities of design professionals involved in a project-specific application. In simple terms, the geofoam product ('normal' EPS-block) used in lightweight-fill applications is a generic commodity whereas the geofoam products that are optimum to use in compressible-inclusion applications are proprietary. The fact that the lightweight-fill function is a small-strain function whereas the compressible-inclusion function is a large-strain function adds to the difference. As a result, the burden on design professionals in terms of both material specifications and product quality control and assurance is markedly different as well.

8.3 LIGHTWEIGHT-FILL APPLICATIONS

No matter what the specific focus of a given lightweight-fill application (i.e. lateral pressure reduction on an ERS, earthwork on soft ground, slope stabilization, etc.), whenever geofoam is used in such an application it is always a small-strain application. This means that the compressive stresses under service-load conditions must always be kept within the quasi-elastic limit of the geofoam material used so that plastic (non-recoverable) strains do not occur and creep strains are within acceptable limits.

With this broad mandate in mind, EPS-block is and has always been the geofoam product of choice for all lightweight-fill applications based on consideration of both technical performance and product economics. It is essential to know that EPS-block has been a generic, commodity product for decades now even though many EPS molders in the U.S. and elsewhere market their blocks using a tradename and/or give the impression that their blocks are somehow unique or inherently superior to those from other molders.

Furthermore, in areas such as the continental U.S. where there are multiple potential EPS molders who can supply most projects (even those in relatively remote locations), EPS-block is always a thin-profit-margin product as is often the case with a commodity material or product. This is because purchasing decisions by construction contractors (who are usually the buyers on geofoam projects) tend to be made solely on a cost basis. This also means that an EPS molder must and will take every possible step to reduce the cost-per-unit-volume of their final product in order to be competitive. Unfortunately, this sometimes means that if there is a corner to be cut it will be. This places the primary burden of quality specification and vigilance on the design professional.

It is also important to understand that geofoam applications are but one of the many commercial and industrial end-uses of EPS-block, virtually all of which have completely different and much less demanding technical-quality requirement. Thus, while a project requiring EPS-block geofoam may represent a significant production spike for an EPS molder in terms of product volume, this is generally relatively short-lived so when averaged over an extended period of time the geofoam market is typically a minor sales market for an EPS molder. In fact, some smaller EPS molders, in the U.S. at least, intentionally avoid selling to the geofoam market altogether because they do not want the disruption in their normal production schedule for regular customers that a large geofoam order often represents.

The net result of these are other factors is that design professionals involved in geofoam applications of EPS-block bear the total burden of choosing an appropriate material specification for the EPS blocks as well as implementing and enforcing both material quality assurance (MQA) and construction quality assurance (CQA) programs throughout the duration of a project. Failure of design professionals to do so has been the root cause of problems involving the lightweight-fill function (Horvath 1999a, 2010a).

The biggest challenge facing design professionals is the choice of a material specification for EPS-block when used in geofoam applications. Understanding the essential elements of such a specification (Horvath 2001c) requires a detailed understanding of the many ways in which the quality of a finished EPS block can be compromised, beginning with the raw materials used and ending with common errors made in MQA testing in the laboratory (Horvath 2011).

The proposed standard that was developed as part of the Federally-funded research into the lightweight-fill function of EPS-block geofoam (Stark et al. 2004a, 2004b) was crafted with all these issues in mind. This proposed standard used as its cornerstone metrics for grading different types of EPS-block both material stiffness and load-bearing ability as these are the primary technical qualities required in all lightweight-fill applications. Furthermore, this standard was not just some 'pie-in-the-sky' academic exercise. It was used and fine-tuned on several major lightweight-fill projects circa 2000, including the world-famous 'Big Dig' Interstate highway project in Boston, Massachusetts. These projects proved unconditionally that any EPS molder could easily produce EPS-block that met or exceeded this proposed standard on a day-in, day-out basis without resorting to extraordinary measures during production. Consequently, this proposed standard was proven to be reasonable to use in routine practice.

Unfortunately, the standard that evolved under ASTM control in the early years of the 21st century that essentially reflected the viewpoints and input of the EPS industry is a very different document. To begin with, the ASTM standard relies largely on material density as the primary metric for grading different types of EPS-block (Horvath 2012). This was simply a holdover from the decades-old ASTM standard for both EPS-block and XPS used as thermal insulation in non-load-bearing building applications. While material density is the most important metric for non-load-bearing thermal-insulation applications and had been the basis of the ASTM standard for this applications for many decades, it is well-established that material density is an imperfect metric for EPS-block load-bearing and is thus ill-suited as a metric for EPS-block in geofoam applications which are inherently load-bearing in nature (Horvath 2012).

8.4 COMPRESSIBLE-INCLUSION APPLICATIONS

The performance needs of compressible-inclusion applications for reducing lateral pressures on ERSs (both the REP- and ZEP-Wall variants) in terms of geofoam-product compressibility under service loads are diametrically opposite those of the lightweight-fill function. Specifically, in compressible-inclusion applications the most compressible product possible under the lowest stress possible is usually desired as this is inherently a large-strain functional application. Unfortunately, this fundamental fact is something that engineering practitioners and academic researchers alike do not seem to appreciate when selecting a geofoam material and product for compressible-inclusion applications. This appears to be the result of their not being familiar with the stress-strain-time behavior of EPS in both its primary (i.e. block-molded) and derivative (e.g. R-EPS, GPS-PB) forms. As a result, it is common for normal EPS-block to be used as a compressible inclusion which is very inefficient and cost-ineffective (and often technically inefficient and ineffective as well) for the reasons explained and illustrated in detail in Horvath (2010b).

The most cost- and technically effective geofoam material identified to date for compressible-inclusion applications involving ERSs is R-EPS. Note that the manufacturing process that is usually referred to in the EPS industry as 'elasticization' that turns normal EPS-block into R-EPS can be applied to EPS-block of any initial density. Consequently, R-EPS in and of itself is not a single, unique geofoam material (Horvath 1995b). However, it is generally most efficient to use EPS-block of the lowest possible density to make R-EPS which, historically in the U.S. at least, has been material with a density in the range of 11 to 12 kg/m³ (unit weight of 0.70 to 0.75 lb/ft³)[85]. Thus, the only known commercially available compressible-inclusion product in the U.S. to date that consists solely of R-EPS, *TerraFlex*[86], uses EPS-block of this density as its source material.

As an aside, throughout this monograph R-EPS has been referred to as though it were a single material. This is because the only R-EPS known to be manufactured to date specifically for geofoam applications has been manufactured in the U.S. using EPS-block of the above-noted density as the source material. Therefore, to date R-EPS has, in essence, been a unique material. However, if, in the future, R-EPS is manufactured in the U.S. and/or elsewhere using EPS-block source material of a different (presumably lower) density then that R-EPS will have properties different from the R-EPS referenced in this monograph.

Since the 1990s, *TerraFlex*™ has been sold nationally within the U.S. by GeoTech Systems Corporation which also sells the following geofoam-related geosynthetic products:

- *GeoTech Drainage Board*™ which is GeoTech's tradename for GPS-PB;

---

[85] EPS-block with a lower density is not only achievable but is regularly produced in countries other than the U.S. The limiting factor on EPS-block density is achieving sufficient thermal fusion between the prepuff particles during the final-molding stage of manufacture. If there is insufficient *bead fusion* as it referred colloquially in the EPS industry, the final block product will simply not hold together during even normal post-manufacture handling.

[86] This was a Federally-registered trademark, #2381220, between 29 August 2000 and 1 April 2011. At the present time, it is apparently simply a trademark used by the supplier of this product, GeoTech Systems Corporation, without the benefit of legal protection.

- *GeoTech Drainage Panel*™ which is a geocomposite consisting of GPS-PB (i.e. *GeoTech Drainage Board*), a nonwoven geotextile on one face, and an optional geomembrane vapor barrier on the other face; and

- *GeoInclusion*[87] which is a geocomposite consisting of R-EPS (*GeoTech TerraFlex*) combined with the *GeoTech Drainage Panel*.

To date, all four of these products (*GeoTech Drainage Board, GeoTech Drainage Panel, GeoTech GeoInclusion, GeoTech TerraFlex*) have been sold as 4-foot (1219-mm) square panels with total thicknesses as specified by the purchaser (within certain minimum and maximum production limits). The most common thickness of the drainage component (i.e. GPS-PB) is 2 inches (50 mm) whether it is used alone as in the *GeoTech Drainage Board* and *GeoTech Drainage Panel* or as part of the *GeoTech GeoInclusion*.

It is significant to note that the commercial aspects of the *GeoTech Drainage Board, Drainage Panel, GeoInclusion*, and *TerraFlex* products in the U.S. at least are likely to change in the coming years as the two U.S. patents that covered the compressible-inclusion feature of all GeoTech products expired in 2016[88]. This means that all of these products are now generic and able to be manufactured by any EPS block molder. Consequently, the source(s) of these products is likely to change in the future.

In addition, and as a follow-up to what was noted above, as the compressible-inclusion function becomes more widely known and used it is possible that some EPS molder may find it advantageous to make R-EPS using EPS-block with a lower density than the 11-12 kg/m³ (0.70-0.75 lb/ft³) lower-bound that has been an industry standard in the U.S. for many decades. This is because it is likely that design professionals will find that R-EPS made using the lowest feasible density of EPS-block will have the most desirable stiffness properties for both REP- and ZEP-Wall applications.

As a final comment with regard to material and product standards, because the products-of-choice for compressible-inclusion applications that were described above have all been proprietary, it has been incumbent on their manufacturer, GeoTech, to perform the necessary quality-control testing to guarantee their products. Now that these products have recently moved into the generic category, it will be necessary for design professionals to develop their own standards and concomitant project-specific specifications for these products.

---

[87] This was a Federally-registered trademark, #2299122, between 14 December 1999 and 17 July 2010. At the present time, it is apparently simply a trademark used by the supplier of this product, GeoTech Systems Corporation, without the benefit of legal protection.

[88] GeoTech's version of GPS-PB (*GeoTech Drainage Board*) has, by happenstance, approximately the same mechanical (stress-strain-time) properties as GeoTech's R-EPS (*TerraFlex*). In fact, Partos and Kazaniwsky (1987) used *GeoTech Drainage Board* as the compressible inclusion for the Philadelphia, Pennsylvania partial-depth REP-Wall case history that was discussed at length in Chapter 5 as *TerraFlex* did not exist as a commercial product in the 1980s when this project occurred. As a result, GeoTech's GPS-PB-based products, when used as part of a compressible inclusion, were covered by U.S. patents until 2016 even though they were not proprietary intellectual property when used solely for fluid drainage.

Chapter 9

Closing Comments

9.1 SYNOPSIS OF STATE OF KNOWLEDGE

At the risk of oversimplification, the conclusion that can be drawn with respect to using geofoam solely to reduce lateral pressures on ERSs is that whatever the lightweight-fill function can do, the compressible-inclusion can do as well or better and at a lower cost. Thus, when performing a project-specific assessment of the cost-effectiveness of using geofoam with ERSs, a design professional should normally begin by first considering a REP- or ZEP-Wall application.

That having been said, there are a variety of situations where the lightweight-fill function is likely to be preferable. This includes sites underlain by weak and compressible soils where the reduction of vertical stresses on the subgrade adjacent to the ERS is a significant additional benefit that can be exploited in both new construction as well as with existing ERSs. In addition, because an assemblage of EPS blocks is inherently self-stable, it is possible to eliminate the use of an ERS completely in new construction, with the EPS blocks acting as a geofoam wall.

9.2 SUGGESTIONS FOR FUTURE RESEARCH

9.2.1 Overview

At this point in time, the basic concept of using geofoam to reduce lateral pressures on ERSs can be considered to be established so additional proof-of-concept type research is not only no longer required but serves no purpose. However, the writer has come to believe that the state of knowledge with regard to analytical methodologies is not nearly as advanced as it may have seemed even in the relatively recent past. This is best expressed using an anecdote from the writer's personal experience.

An internationally famous academician in geotechnical engineering once stated at a gathering of geotechnical engineers that when he graduated from college he unequivocally knew that he knew everything about geotechnical engineering. It was only as time went on and he gained real-world experience that he realized how little he knew.

The point here with regard to using geofoam to reduce lateral pressures on ERSs is that fundamental research geared toward developing analytical methodologies for use in routine practice is needed more than ever. Specific suggestions are presented in the following sections.

The one broad caution that is raised here concerns the use of results from instrumented, full-scale project applications of any of the technologies discussed in this monograph. On the one hand, actual project applications can be viewed as the ultimate 'ground truth' in terms of how a given technology behaves. However, experience has shown that actual project applications involving ERSs can be impacted significantly by certain soil behaviors, especially the development of apparent cohesion due to matric suction within coarse-grain soil, that are typically neglected in routine analytical methodologies for simplicity and safety.

226

As discussed in Appendix A, serious professional disagreements have developed in recent years over the development and use in practice of a MSEW design methodology that is based solely on the observations of actual MSEWs whose behavior was very likely affected to varying degrees by apparent cohesion in what analytically would typically be considered 'cohesionless' soils. As also discussed in Appendix A, calculated seismic loads are very sensitive to assumptions concerning Mohr-Coulomb cohesion in soil. Consequently, the observation that conventional rigid retaining walls that are adequately designed for gravity loads perform acceptably in small-to-modest earthquakes may derive from the benefit of apparent cohesion within the retained soil as much as any other factor(s).

The writer does not have a simple or easy remedy for dealing with this potential problem when evaluating measurements from actual project applications. As a minimum, it appears to be highly desirable to measure the actual matric-suction profile anytime there is an instrumented full-scale application of any of the geotechnologies presented in this monograph as it will allow for a more accurate assessment of the stress state within the retained soil.

9.2.2 Lightweight-Fill Function

To organize desirable research objectives applicable to this function, it is first useful to summarize the key problem variables for this function:

- full- vs. partial-depth application;

- non-yielding vs. yielding RERS;

- the effect of varying the angle (θ^* in Figure 3.2) defined by the stair-stepped interface between the EPS blocks and retained soil; and

- gravity vs. seismic loading.

In principle, all combinations of the latter three variables involving the basic full-depth application are covered by the analytical methodology developed in Japan beginning in the late 1980s that was discussed at length in Chapter 4. In the past, this methodology was accepted at face value by others for use in practice (e.g. Stark et al. 2004a, 2004b), including by the writer, even though there was no known published supporting research, at least in the English-language literature.

However, both published (Horvath 2000) and unpublished FE analyses performed by the writer in recent years have caused the writer to question whether the basic failure mechanism assumed in the Japanese methodology, i.e. that a classic active earth pressure failure wedge develops within the retained soil adjacent to the assemblage of EPS blocks under all conditions of ERS yielding and loading, is correct or if some other behavioral mechanism actually occurs. Thus, the basic recommendation made here is that fundamental research be conducted into the basic full-depth application of the lightweight-fill function to identify the behavioral mechanism at work within the retained soil under both gravity and seismic loading as it may not be the same (as the Japanese analytical methodology assumes). Only when this has been accomplished can attention be given to exploring the partial-depth application.

9.2.3 Compressible-Inclusion Function

As with the lightweight-fill function, it is first useful to summarize the key problem variables for the compressible-inclusion function:

- REP- vs. ZEP-Wall application;

- full- vs. partial-depth application;

- for partial-depth applications only, the effect of varying the vertical position of the geofoam and geosynthetic tensile reinforcement (if any);

- for the ZEP-Wall application only, non-yielding vs. yielding RERSs;

- for the ZEP-Wall application only, the effect of varying the length of the reinforcements;

- gravity vs. seismic loading; and

- 'normal' vs. expansive soils (primarily for the REP-Wall application but possibly for the ZEP-Wall application as well).

For the basic case of a full-depth, REP-Wall application with a non-yielding RERS subjected to gravity loads only, as discussed in detail in Chapter 5 there is a still-evolving simplified analytical methodology that is available for use. The original version of this methodology was developed by Partos and Kazaniwsky in the mid-1980s and all available information indicates that this methodology in its current form can be applied with confidence in the success of the outcome. However, there is always room for improvement as well as extension to other combinations of problem variables, especially seismic loading, a.k.a. the seismic-buffer concept.

As noted at the outset, this monograph has intentionally been limited to a consideration of 'normal' soils. However, the concept of using a using a compressible inclusion to reduce lateral pressures on ERSs from expansive soils is one with enormous potential given the extensive global distribution of such soils. This was noted more than 20 years ago (Aytekin 1996, 1997) but the concept does not appear to have been developed and progressed to the point where a usable analytical methodology exists even though this subject continues to receive attention from time to time (e.g. Ikizler et al. 2008). Consequently, there is a need for fundamental research along these lines.

This page intentionally left blank.

Chapter 10

Vertical Earth Pressure Reduction

_____

10.1 INTRODUCTION

Although the focus of this monograph is lateral pressures on ERSs, a brief overview is presented in this chapter of the use of compressible inclusions to reduce vertical earth pressures on structural elements bearing on or in the ground. This is both for the sake of completeness as well as in recognition of the fact that the concept of a compressible inclusion to reduce earth loads owes its origins to the vertical-pressure problem. In fact, the concept is so old that it predates modern soil mechanics.

10.2 OVERVIEW

There are two broadly different applications of compressible inclusions to reduce vertical earth pressures on structural elements:

- <u>downward-acting</u> pressures on underground conduits and

- <u>vertical-acting</u> pressures on structural slabs.

The root causal mechanism of the earth loads in each application is very different. In addition, the nature of the structural element in contact with the ground is very different in each application. As a result, both the analytical approach as well as the optimal geofoam product to use differ significantly between the two applications.

10.3 UNDERGROUND CONDUITS

Underground conduit is the generic term used to describe a wide range of structural elements buried in the ground that include relatively small-diameter, nominally circular utility lines and pipes for drainage as well as relatively narrow 'box' culverts and tunnels with a square or rectangular cross-section. The distinguishing features of an underground conduit are that its diameter or width is of the same order of magnitude or less than the height (depth) of the soil cover above its crown and that the conduit is relatively stiff compared to the surrounding ground. As a result, in terms of soil-structure interaction (SSI), the conduit is behaviorally a localized 'hard spot' within the ground, and vertical combined earth and surface-surcharge loads can be greater than those obtained using a simple overburden-stress calculation. This is the result of what is sometimes called _negative arching_ although from a geomechanical perspective this is more akin to downdrag on deep-foundation elements than simply a reverse of the phenomenon of 'true' positive or simple arching that was discussed in Chapter 3.

Before proceeding further, it should be noted that the discussion in this section is limited to a consideration of 'normal' soils, i.e. soils where volumetric behavior related to changes in water content is either non-existent or at least can be neglected.

The SSI behavior of, and concomitant vertical earth forces on, underground conduits was recognized years before modern soil mechanics evolved in the 1920s and 1930s. This prompted academic research into the overall problem of how these forces develop as well as different conduit-placement strategies to minimize these forces.

One of the outcomes of this research was the concept of placing a relatively compressible zone of material...what would nowadays be called a compressible inclusion...above the crown of an underground conduit in order to induce positive arching. In essence, this turns the conduit from being a relative hard spot to a soft spot and reverses the relative displacement between conduit and surrounding ground that would normally exist, with a concomitant significant reduction in vertical earth forces acting on the conduit.

As an aside, note that this is where the diameter or width of an underground conduit relative to the thickness of soil cover above its crown comes into play. The soil cover above an underground conduit must be sufficiently thick relative to the dimensions of the underground conduit so that arching can develop without propagating to, or otherwise interfering or interacting with, the ground surface. In essence, the ground surface must not 'know' or 'feel' that arching has developed some depth below it. This is conceptually identical to Saint-Venant's Principle in solid mechanics as discussed in Chapter 3.

The early compressible inclusions used with underground conduits ranged from simply placing loose, uncompacted soil to bales of hay or straw above the crown. Nowadays, a geofoam compressible inclusion would be the preferred alternative due to the predictable mechanical (stress-strain-time) properties of geofoam materials compared to these early alternatives. In addition, there is no potential for methane generation due to anaerobic decomposition with geofoam materials as there is with organic materials such as hay and straw.

Unfortunately, despite the fact that the use of geofoam compressible inclusions with underground conduits (especially box culverts) has been researched and used for decades, in virtually all of the case histories published to date the geofoam material and product used has been technically inefficient and thus cost-ineffective. Specifically, design professionals have used 'normal' EPS-block, i.e. essentially the same material and product one would use for a lightweight-fill functional application, apparently without realizing or appreciating that the geofoam functions of lightweight fill and compressible inclusion are conceptually as different as night and day. As has been noted repeatedly throughout this monograph, lightweight fill is a small-strain function where product compressibility is to be minimized whereas compressible inclusion is a large-strain function where product compressibility is to be maximized (see Horvath (2010b) as the same issues relative to lateral earth pressures are applicable here with vertical earth forces).

In reality, the most technically efficient and cost-effective geofoam material and product to use as a compressible inclusion above underground conduits is panels of R-EPS such as the *GeoTech TerraFlex* product mentioned in Chapter 8. The reasons for this are explained in Horvath (2010b) and are broadly the same as those presented in Chapter 5 for lateral-pressure reduction on ERSs.

10.4 STRUCTURAL SLABS

The other application considered in this chapter is in virtually all respects a complete opposite of underground conduits. With underground conduits, 'normal' soil reacts displacement-wise to the stiffness of the conduit. Depending on the relative conduit-soil stiffness, the soil above a conduit displaces differently and this has a direct effect on the vertical forces that develop on the conduit. Placing a compressible inclusion above the conduit essentially changes the vertical stiffness of the conduit relative to the soil above it so that the direction of relative soil-conduit displacement reverses. In the process, the vertical earth forces on the conduit are drastically reduced from what would normally develop. This is conceptually akin to shaft resistance in deep foundations. The direction (up or down) in which the shaft resistance acts relative to the deep-foundation element determines whether the shaft resistance acts to support the deep-foundation element or acts as a downdrag load on it.

The problem considered here with structural slabs involves expansive soil and rock, i.e. geomaterials whose volumetric behavior is significantly influenced by changes in water content although vertical confining stress continues to play a role in the overall stress-strain behavior. Consequently, it is the ground that always dominates and 'drives' the problem, not the structural element placed in contact with the ground as with underground conduits. As a result, a slab-on-grade such as the ground floor of a building can be problematic with such soil or rock as the slab simply 'goes for a ride' depending on vertical displacement of the soil or rock subgrade. Thus, the preferred alternative is to use a *structural slab* which, in concept, is structurally designed to span between the primary supporting elements of the overall system (deep foundations founded below the active zone of vertical ground displacement) and not be in contact with the ground.

The problem, of course, is that the finished structural slab cannot be 'wished' into place with a void of some depth between the underside of the slab and the ground surface that can accommodate forecast upward displacement of the ground surface during the design life of the structure. Rather, the slab has to be formed and poured-in-place using the ground as a support for the formwork. However, if the slab is constructed in contact with the ground surface, future ground heave would tend to result in very large compressive stresses acting upward on the underside of the slab. It is impractical from an economic standpoint to structurally design a slab to resist such stresses.

The compromise solution that is used in practice is to construct the slab on a layer of crushable, sacrificial material...which constitutes the compressible inclusion in this application...that is placed between the ground surface and the underside of the slab and acts as part of the formwork during construction. This is a special case of what are called *void formers* in PCC construction.

What makes this application unique is that the compressible inclusion/void former has to be explicitly engineered to serve two conflicting and competing design requirements:

- It must be sufficiently stiff to resist the dead-load stresses of slab construction (human foot traffic followed by the weight of the fluid PCC) with minimal compression.

- It should ideally just disappear after the structural slab cures. Since this is unrealistic, as a pragmatic compromise it should crush and offer minimal resistance after crushing under subsequent ground heave. This is so that the transmitted compressive stresses acting upward on the underside of the cured structural slab are minimized as the slab must be designed to resist these stresses for the life of the structure.

The first compressible-inclusion product developed for this application and one that is still in widespread use to the present uses cardboard as its primary component. However, there can be problems with cardboard-based void formers both during construction (the product can collapse prematurely if wetted prior to PCC placement) as well as after construction (methane generation due to anaerobic decomposition of the cardboard, which can present an explosion hazard, has been found to occur in the U.K. at least as discussed in *Data* (1991)).

The preferred alternative to use nowadays is a geofoam-based product that was developed and engineered specifically for this application. It consists of pieces of EPS-block that when assembled at a project site, it resembles an egg carton. A relatively thin, solid panel made from any number of materials (usually wood or wood fiber) is placed on top of the geofoam to act as the working surface for slab construction. The benefit of this overall product is that it has none of the aforementioned drawbacks of cardboard void formers as it maintains its durability if wetted and will not decompose over time.

The concept of this geofoam product was developed more than two decades ago, more or less contemporaneously, in Canada (*Plasti-Fab GeoVoid®*) and the U.K. (Cordek's *Cellcore* product line). As a result, there is an extensive history of successful project usage both throughout North America and the U.K., including on some very notable projects such as the Channel Tunnel ('Chunnel') between England and France.

Appendix A

Lateral Earth Pressure and Related Analytical Theories

A.1 INTRODUCTION

The intent and function of this appendix are several fold:

- The same relatively small group of fundamental analytical theories and solutions are used for both the lightweight-fill and compressible-inclusion functions as well as all ERS applications considered in this monograph. Although these theories and solutions are classical and thus presented to varying extent and detail in the published literature, it is both useful and efficient to include and discuss them in this monograph so that the reader has a concise reference source within this document.

- There are some subtle implications and aspects to some of the classical theories and solutions that are frequently either not emphasized or even omitted from many texts and are thus often not appreciated by practitioners and academicians alike. It is of interest to note these issues as they can have significant practical importance and implications at times, including for applications considered in this monograph.

- There are some theories and research results of interest and relevance that are not widely known and it is of interest to note these as well. This is especially true of new developments that have specified relevance to applications considered in this monograph.

- To the greatest extent practical, the various and diverse theories and solutions assembled and presented in this appendix are expressed using common notation and terminology[89]. However, in some selected cases the notation used in an original publication is preserved for historical reasons. This is noted where it occurs.

A.2 EARTH LOADS (NON-EXPANSIVE SOILS)

A.2.1 Introduction

First considered are the three classical earth pressure states that develop when 'normal' soils as defined in Chapter 1 comprise the retained soil of an ERS.

---

[89] As quickly becomes apparent to every student of geotechnical engineering, notation is not standardized as it is to a significant extent in other areas of civil engineering specialization, especially structural engineering. Furthermore, within the English language there is no standardization of terminology for most soil properties and analytical parameters. Even within the U.S., there can be regional variations and preferences that sometimes have changed over time. Therefore, the same soil mechanics theory will often have, as a minimum, different notation in one text or reference book compared to another which can be confusing even to experienced geoprofessionals.

A.2.2 Gravity Loading Conditions

A.2.2.1 At-Rest Earth Pressure State

In concept, the development of a lateral earth pressure diagram for the at-rest state is both simple and straightforward as it should, by definition, be both hydrostatic and oriented in the horizontal direction[90]. Assuming a planar and horizontal ground surface[91], at any depth, z, below the ground surface the lateral earth pressure, p_o, is traditionally taken to be simply:

$$K_o \cdot \gamma_{eff} \cdot z \tag{A.1}$$

where K_o is the coefficient of lateral earth pressure at rest.

While the parameter K_o is simple in concept, it is very complex to evaluate in practice. There was an intense flurry of research interest into K_o in the latter decades of the 20th century. As a result, it is now appreciated that K_o is a complex function of both soil strength and loading history. Based on research (Kulhawy and Mayne 1990), K_o in reloading can be estimated using the following empirical equation:

$$K_o = (1 - \sin\phi) \cdot OCR^{\sin\phi} \tag{A.2}$$

where OCR is the current overconsolidation ratio and ϕ is the Mohr-Coulomb friction angle of the soil.

As summarized in Horvath (2004b), there has been some debate in the literature as to whether the value of ϕ used in Equation A.2 should be the peak or constant-volume (critical-state) value. The current thinking favors the use of the constant-volume value, ϕ_{cv}, which is a fundamental property of any soil (P. W. Mayne, personal communication, 2011). This is significantly different from the peak value, ϕ_{peak}, which is the sum of ϕ_{cv} plus a dilatancy component, $\phi_{dilatancy}$. Because this latter component is a function of void ratio and confining stress (Kulhawy and Mayne 1990), it is highly variable for any given soil with the result that ϕ_{peak} is a variable as opposed to a fundamental soil constant.

Regardless of the value of ϕ used in Equation A.2, the problem that arises in practice is that estimating the OCR profile within the a backfill/fill that comprises the retained-soil mass of an ERS that has been created under controlled placement and compaction conditions is relatively complex and always project-specific. Due to mechanical compaction of the soil during construction, the OCR is not only greater than one in some places (typically at shallower depths) but is also highly non-uniform in its depth-wise variation, with this variation depending on the specific compaction 'plant' (equipment) used. This means the resulting lateral earth pressure diagram will also vary nonlinearly with depth.

A rigorous constitutive model and numerical-solution algorithm that captures the overall soil placement and compaction process with great accuracy with respect to K_o has been developed (Duncan and Seed 1986). However, it requires a complex numerical solution using proprietary software. While this in and of itself is not a deterrent to use in

---

[90] This traditional assumption neglects vertical soil-wall shear stresses that are now known to exist even for non-yielding RERSs (Horvath 1990a, Filz and Duncan 1997, Filz 2003). These origin of these shear stresses is discussed in Chapter 2.

[91] The problem of at-rest lateral earth pressures for an inclined ground surface adjacent to an ERS is discussed in Horvath (1990a).

routine practice, this software is not known to have ever been available commercially. In addition, because of the complexity of the constitutive model used, the quantity and quality of the input variables required by this software is significant and well beyond that generated by or otherwise available to routine projects in practice.

Subsequent to development of this model, additional studies were undertaken to develop a simpler, more-practical methodology that could be used in routine practice. The results were published in Duncan et al. (1991) with important corrections in Duncan et al. (1993). While the results are in a chart form and thus amenable for use in routine practice, the quantitative results are dependent on the specific piece of compaction equipment used, information that is generally unknown during the design phase of a typical project[92].

In conclusion, estimating the *OCR* profile to allow direct application of Equation A.2 is not something that is a viable alternative in routine practice at the present time. It may also partially explain why engineers tend to ignore compaction effects in the design of ERSs in routine practice even though the existence and importance of these stresses has been widely known for decades. All of the illustrated applications and example problems in this monograph that require use of the at-rest earth pressure state neglect compaction-induced stresses. However, in any project-specific application these could be included in a straightforward manner as the methodologies presented in this monograph do not inherently preclude their consideration.

A.2.2.2 Active Earth Pressure State

For ERS applications where geofoam is used to reduce lateral pressures, there are only two active earth pressure solutions that are theoretically correct to use although one is actually just an improved version of the other. Both are what are called 'true' retaining-wall solutions in that an ERS is explicitly assumed to exist and, as a result, influence the development of the assumed 'failure wedge' of soil that develops.

These theories are:

- Coulomb's classical 18th-century solution that assumes a planar failure surface within the retained soil. In reality, this surface is always curved[93] to some extent due to friction along the assumed planar soil-wall interface that forms the other of two failure surfaces (planes) that define the failure wedge. The reason for this assumption of planarity was purely pragmatic, to allow development of a closed-form solution using the mathematical tools available to Coulomb at the time. Fortunately, the error resulting from this assumption of planarity, although typically on the unconservative side, is usually so small as to be negligible in any practical situation which is why Coulomb's solution remains viable for use to the present in applications involving the active state.

- One of the so-called 'exact' solutions such as the one developed by Caquot and Kerisel (Kézdi 1975, Clough and Duncan 1991) that explicitly models the curved failure surface within the retained soil. This curved failure surface is typically defined using a relatively simple and mathematically well-defined relationship such as a logarithmic

---

[92] There have been other, earlier attempts to create chart or graphical solutions for estimating compaction-induced stresses that are not discussed or referenced here as they were largely eclipsed and superseded by the work of Duncan et al. (1991).
[93] The sense of the curvature, i.e. concave up or down, depends on the sign of the soil-wall friction angle, δ, which was discussed in detail in Chapter 2. For the more common active earth pressure case of positive δ, the direction is concave upward.

spiral. Note that these more-advanced solutions such as Caquot-Kerisel are really just evolutionary improvements to Coulomb's solution as opposed to being radically different in their formulation and assumptions. They simply took advantage of the approximately 200 years of mathematical and analytical knowledge gained between Coulomb's work in the 18th century and the middle of the 20th century.

Note that Rankine's classical earth pressure solution is intentionally not listed here as an acceptable alternative. The reason is that this is probably one of the most misunderstood, and, as a result, misused, solutions in all of geotechnical engineering. This is because Rankine's solution solves a theory of plasticity half-space problem composed only of rigid-plastic material whose failure is defined by the well-known Mohr-Coulomb strength parameters. Noteworthy is the fact that <u>no</u> ERS is assumed to exist in the formulation of Rankine's problem, unlike with Coulomb's problem and its evolutionary improvements such as Caquot-Kerisel's where an ERS was explicitly assumed to exist and, as a result, control, in part, where failure surfaces developed.

On the other hand, Rankine's problem allows failure surfaces to develop where they want to. Thus, Rankine's solution is, strictly speaking, only appropriate for use when and where so-called 'free-field' conditions exist. This is so that the V-shaped pattern of planar failure surfaces that are an inherent outcome and thus a key part of Rankine's solution can develop without interference from a structural element such as an ERS. The inability of these Rankine failure planes to develop in the vicinity of most types of ERSs thus invalidates the boundary conditions on which Rankine's solution was developed and precludes its use[94].

Consequently, Rankine's solution cannot be used for any of the applications considered in this monograph even though it is acknowledged that it is used routinely (and incorrectly in the vast majority of cases) in both research and practice with ERSs. That Rankine's solution can be used incorrectly and still produce reasonable outcomes is simply a fortuitous coincidence that results from this solution do not differ greatly with those from Coulomb's solution in most applications[95]. This, no doubt, is a major reason why the incorrect application of Rankine's theory has persisted in both research and practice.

In any event, Coulomb's solution, which can be expressed in a single algebraic equation that incorporates all the possible problem variables, is suggested for use with all active earth pressure applications considered in this monograph as it can easily be applied to problems involving complex geometries. On the other hand, the solutions for any of the exact solutions such as that of Caquot-Kerisel are typically found only presented in graphical form as charts or plots which, by virtue of their two dimensionality, can only reflect certain combinations of variables in any one graphic. Experience indicates that this can limit the utility of exact-theory solutions in the types of applications covered by this monograph. It also limits the ability to use such solutions in computer-based calculation

---

[94] The one exception to this is with the RERS category of cantilever retaining walls. The stem geometry of most such walls allows the theoretical Rankine failure planes to form more or less unimpeded so a theoretically sound case can be made that Rankine's solution is conceptually more appropriate to use with this type of ERS that either Coulomb or one of its exact derivatives.

[95] If the soil-wall friction angle δ in Coulomb's solution is zero, the intra-soil failure surface becomes truly planar and the results are identical to those obtained from Rankine's solution. This has led to the incorrect interpretation that Rankine's solution is just a special case of Coulomb's solution with soil-wall friction neglected. In reality, the two solutions have completely different theoretical bases, assumptions, and boundary conditions in their initial formulation. By coincidence, they produce the same calculated results for one specific case that does not occur with actual ERSs as there is simply no such thing as a frictionless ERS in the real world.

tools such as spreadsheets. In addition, the difference in results between Coulomb's solution and any of the exact solutions is insignificant for the active state in any practical context. Therefore, from the perspective of practical significance there is no sacrifice in accuracy from using Coulomb's solution in this situation.

Although the equation for Coulomb's active earth pressure solution appears in numerous textbooks and other reference documents, it is included here both for ease of reference and to present the solution using notation consistent with other equations defined and used throughout this monograph.

To begin with, Figure A.1 defines the notation used for the overall problem geometry[96]. Note that the *geotechnical height*, *H*, of the ERS is always defined as the vertical distance from the point where the surface of the retained soil contacts the interior face of the ERS (Point O_z which is also the origin of the depth variable, *z*, used in subsequent calculations) to the heel of the ERS (Point O'). The geotechnical height of actual ERSs, especially RERSs, is almost always less than the actual or physical height of the ERS, a fact that many textbooks and design manuals do not illustrate or otherwise make clear.

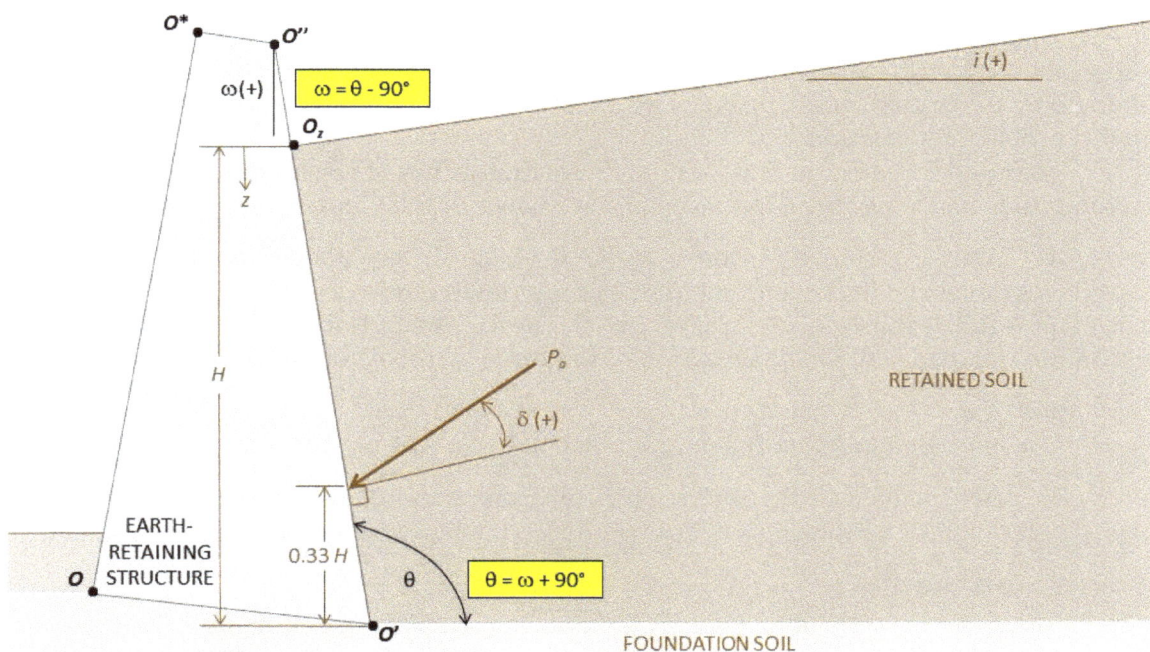

Figure A.1. Definitions Related to Active Earth Pressure State for Gravity and Seismic Loading.

The traditional way of evaluating the resultant active earth force, P_a, is given by the following equation (the assumptions inherent in this expression are explained subsequently):

$$P_a = 0.5 \cdot K_a \cdot \gamma_{eff} \cdot H^2 \tag{A.3}$$

[96] The base of the ERS shown in this figure is depicted as inclined for generality as well as to illustrate that certain variables or dimensions of an ERS relate specifically to its heel (Point O' in the figure).

where the Coulomb coefficient of active earth pressure, K_a, is given by:

$$K_a = \left[\frac{\sin(\theta - \phi) \cdot \csc\theta}{\sqrt{\sin(\theta + \delta)} + \sqrt{\sin(\phi + \delta) \cdot \sin(\phi - i) \cdot \csc(\theta - i)}} \right]^2 \qquad \text{(A.4)}$$

where δ = the soil-wall friction angle, ϕ = the Mohr-Coulomb friction angle of the soil, and other variables have either been defined previously or are defined in Figure A.1[97].

Note that:

- Equation A.4 is valid for $0 \le \theta \le 180°$ and

- the signs of angles i, δ, and ω are important. The more-common positive sense of each angle is shown in Figure A.1. The negative sense of each would be if the angle shown were on the opposite side of the reference line shown with each angle in Figure A.1.

There are several aspects of Coulomb's solution that are not often mentioned in the literature and are even less appreciated in practice that deserve to be noted and discussed. These are explored with reference to Figure A.2 that shows a generic 'failure wedge' (defined by the brown dotted lines) as assumed by Coulomb when formulating and setting up this problem for solution.

First and foremost is that Coulomb's solution is based solely on simple rigid-body vector statics involving three force vectors as shown in this figure: the weight of the soil wedge, $\vec{W}$, and the assumed reactions $\vec{P}$ and $\vec{R}$ along the two planar failure surfaces. For any given problem geometry and set of soil properties, the primary unknown is the angle α that defines the orientation with respect to the horizontal of the soil-on-soil failure plane within the retained soil. The magnitudes of the three force vectors as well as the absolute orientation of $\vec{R}$ are all functions of α.

Coulomb assumed that the unique value of α, α_a, that produced the minimum earth force, $\vec{P}_a$, corresponded to the active earth pressure state. Using an identical model, the failure plane defined by the angle α_p that produced the maximum earth force corresponded to the passive state, $\vec{P}_p$.

Although generally not necessary to know in practical applications, for the purposes of this monograph it is of interest to provide the equation for calculating the angle α_a:

$$\alpha_a = \phi + \tan^{-1}\left(\frac{-\tan(\phi - i) + c_1}{c_2} \right) \qquad \text{(A.5)}$$

Where:

---

[97] It is important to note that the form of Equation A.4 is not unique. Consequently, it is common to find alternative versions of Coulomb's solution in different published references and with different notation used for the variables as well.

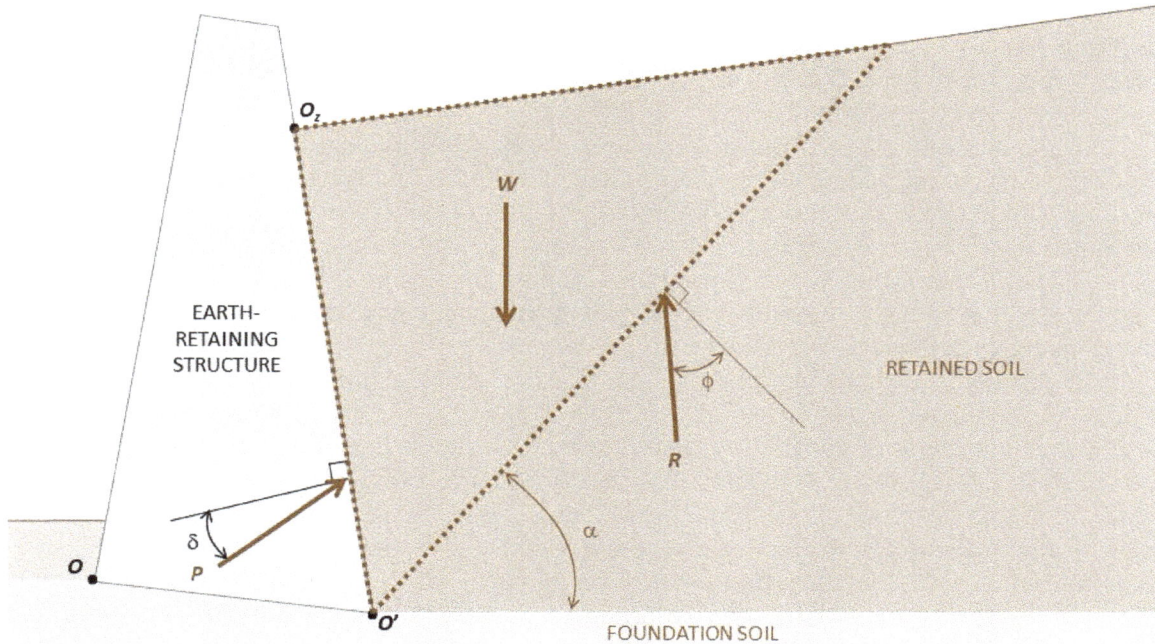

Figure A.2. Generic Coulomb Failure Wedge and Free-Body Diagram of Forces.

$$c_1 = \sqrt{\tan(\phi - i) \cdot [\tan(\phi - i) + \cot(\phi - \omega)] \cdot [1 + (\tan(\delta + \omega) \cdot \cot(\phi - \omega))]} \qquad \text{(A.6)}$$

and

$$c_2 = 1 + \{\tan(\delta + \omega) \cdot [\tan(\phi - i) + \cot(\phi - \omega)]\}. \qquad \text{(A.7)}$$

The reason a planar failure surface was assumed within the retained soil (even though Coulomb himself was apparently aware of the fact that this surface was curved in reality) was to greatly simplify the solution process which invoked force equilibrium only. Had this assumed failure surface been curved, the specific point-of-application of the force vector, $\vec{R}$, along that failure surface would have had to have been included in the solution algorithm, even just to satisfy force equilibrium. This would have been a significant complication. This is also undoubtedly why moment equilibrium was not invoked and satisfied in the solution process as even with planar failure surfaces satisfying moment equilibrium in addition to force equilibrium would have required knowing or assuming the points-of-application of each force vector.

The conclusion of this discussion is that the formulation and concomitant development of Coulomb's solution only produced force vectors with a magnitude and direction but non-unique points-of-application. This means that all that can be rigorously inferred from Coulomb's solution is the <u>magnitude</u> of the active and passive earth <u>forces</u> and their spatial orientation (i.e. angle δ) relative to the planar inside face of an ERS. Where these resultants actually act along the ERS or what earth <u>pressure</u> distribution produces these forces simply cannot be gotten from Coulomb's solution.

Unfortunately, both of these questions (point-of-application and pressure distribution) are necessary information in practice. This is because although only resultant

240

forces are needed to evaluate the geotechnical stability of a yielding RERS, some of these assessments, e.g. safety factor against rigid-body rotation (overturning), invoke moment equilibrium of the overall RERS-soil system so points-of-application of all resultant forces must be known or at least assumed.

In addition, a lateral earth pressure diagram is typically required for structural analysis or design of both yielding and non-yielding ERSs. This is because such a diagram is used as a distributed load per unit width[98] of the ERS in some structural analysis or design methodology. Therefore, in practice it is necessary to both assume a point-of-application of the resultant earth force, P_a, in Figure A.1 as well as develop an assumed pressure diagram that is logical in shape; preferably easy to use in practice; and produces a resultant force of the same magnitude as that given by Coulomb's solution.

It has become standard practice to address these dual and complementary needs of a point-of-application for the resultant earth force and lateral earth pressure distribution by first either assuming or calculating (using some theoretical basis) a pressure distribution and from this determining the location of the resultant force as simply being the geometric centroid of that pressure distribution. It has long been standard practice to assume a depth-wise linear increase in lateral earth pressure that produces the well-known pressure distribution that is triangular (hydrostatic[99]) in shape. This is the reason why Equation A.3, which is essentially the equation of the area of a triangle, is used to calculate the resultant active earth force, P_a.

Interestingly, it has been known from the earliest days of modern soil mechanics that the lateral earth pressure distribution behind a simple RERS as shown in Figure A.1 is actually <u>not</u> triangular in shape, at least for the active state (Terzaghi 1943). Rather, it is parabolic in shape as demonstrated repeatedly over the years by both theoretical considerations based on arching theory (Terzaghi 1943; Handy 1985; Harrop-Williams 1989) as well as innumerable observations of both model and actual structures dating back to Terzaghi's seminal retaining-wall tests in the early 1930s[100].

Nevertheless, experience and research have shown that the customary triangular distribution of active earth pressures, p_a...which is attractive for its simplicity and ease of use and which places the resultant active earth force, P_a, at a height $H/3$ above the heel of the RERS as shown in Figure A.1...to be adequate (i.e. sufficiently accurate) for routine analysis and design purposes, at least geotechnically[101]. Nevertheless, it is essential to at least understand that an assumed triangular lateral earth pressure distribution is solely a matter of convenience and not theoretical correctness.

Using an assumed triangular distribution of lateral earth pressures with depth means that the active earth pressure, p_a, has magnitudes:

$$= 0 \text{ at the ground surface (depth, } z, = 0) \tag{A.8a}$$

$$= \left(K_a \cdot \cos\omega \cdot \gamma_{eff} \cdot H\right) \text{ at the heel of the RERS (depth, } z, = H) \tag{A.8b}$$

---

[98] 'Width' is defined here as the direction perpendicular to Figures A.1 and A.2.
[99] That an assumed linear increase in lateral earth pressure with depth mimics the depth-wise increase in pressure within a fluid is the reason why lateral earth pressure is sometimes referred to as an *equivalent fluid pressure*.
[100] See Terzaghi (1934) in the Supplemental Bibliography section.
[101] Note, however, that under certain circumstances the triangular distribution of active earth pressures might not be adequate for structural assessments of the RERS itself, especially if it is of segmental construction. This is because the triangular distribution can underestimate both lateral pressures and bending moments near the mid-height of the RERS.

with a linear variation between these values. Note that this pressure triangle is distributed over the inside face of the RERS shown in Figure A.1 (a length equal to $H/\cos \omega$ in the general case of an inclined inside face of the RERS) and would thus be inclined from the vertical in cases where the angle ω did not equal zero. Note also that this earth pressure triangle is skewed as well relative to the plane defined by the inside face of the ERS by the soil-wall friction angle δ acting in the same sense (+ or -) as shown for the resultant force P_a.

A.2.2.3 Passive Earth Pressure State

Overall, the passive earth pressure state does not play a significant role in the applications covered in this monograph so is not discussed in the same level of detail as the active state. However, it is useful to note several general observations and conclusions concerning the passive state:

- With the exception of self-yielding RERSs, the passive state provides resistance to one or more geotechnical ULS modes. Consequently, care should always be exercised in both analysis and design not to overestimate passive resistance as this will tend to produce an unconservative and potentially unsafe analytical or design outcome. Overestimation of passive resistance can result from any one or a combination of factors as noted below.

- Consideration should always be given as to whether the soil available to provide passive resistance may be removed in whole or in part by natural forces or human activity during the design life of the structure being analyzed or designed. Thus, it may be prudent to reduce or even completely neglect passive resistance.

- The curvature of the intra-soil failure surface within the retained soil is much more significant for the passive state compared to the active state. Consequently, the deviation from reality of an assumed planar failure surface within the retained soil is much greater for the passive state compared to the active state. The consequence of this is that the error associated with Coulomb's solution for the passive state tends to be significant in most applications. Unfortunately, this error is generally on the unconservative and potentially unsafe side so should not be ignored as with the error involved with the active state. Thus, it would appear that only the solution from an 'exact' solution such as that of Caquot-Kerisel should be used whenever passive earth pressure needs to calculated. However, some publications (e.g. Ebeling and Morrison 1992) suggest that the errors associated with Coulomb's passive-state solution are 'tolerable' if the δ/ϕ ratio is kept at or below certain limits, e.g. 0.5. However, it should be clear that Rankine's solution is <u>never</u> correct to use for passive-resistance calculations for the reasons given previously with respect to the active state.

- Although lateral displacements are not calculated explicitly as part of the routine analysis and design of ERSs, they should always be considered in at least a qualitative sense. In this regard it is important to always keep in mind that mobilization of the passive earth pressure state requires approximately one order of magnitude greater lateral displacement than for the active state. Thus, the lateral displacement required to mobilize the full passive state is generally greater than the actual lateral displacement of most properly designed and well-behaved RERSs under all but extreme-event loading such as from earthquakes.

- As with the active state, the actual distribution of lateral earth pressures for the fully mobilized passive state is not perfectly triangular/hydrostatic as is typically assumed in routine practice. Although there is much less experience from either research or practice as to the exact shape of the pressure distribution, all indications are that it is nominally parabolic in shape as with the active state. Complicating this issue is that many applications where passive earth pressures might be studied in some detail involve FERSs such as anchored bulkheads. Such ERSs deform in addition to having a component of rigid-body lateral displacement so their total, overall lateral displacement pattern is highly non-uniform with depth. Consequently, the development of the passive state is not depth-wise uniform against such ERSs. The passive state will tend to be more developed along those portions of the ERS where total displacements are the largest, e.g. just below the mudline with anchored bulkheads.

A.2.2.4 Arching-Based Concepts

The concept of soil arching in the horizontal direction is now broadly accepted as a much-improved conceptual framework for understanding the development of lateral earth pressures compared to the simplistic rigid-body concepts employed by both Coulomb and Rankine. Unfortunately, the recognition, acceptance, and development of horizontal arching for use with ERSs has largely been limited to academic research as the analytical methodologies used in routine practice for all types of ERSs remain firmly rooted in the traditional, simplistic theories and solutions of the past that were discussed in the preceding section.

The current limitation with regard to horizontal arching is that, with few exceptions, the only application pursued to date has been for the active state with RERSs as this is clearly the ERS application of broadest use and thus greatest interest in practice. Overall, this research has confirmed what has long been known from both model tests and the observations of actual RERSs, that the lateral earth pressure distribution under active displacement of a RERS is depth-wise non-linear. However, the shape of the curved pressure distribution is very dependent on the primary mode of RERS displacement, i.e. translation (sliding), rotation about the bottom, or rotation about the top. Most research to date has focused on the translation mode.

It is important to note that within the overall theoretical framework of horizontal arching there can be variations between and among specific solutions obtained even for a very specific application. For example, there are now several different published solutions just for pure translation of a RERS (Handy 1985, Harrop-Williams 1989, Paik and Salgado 2003, Ertuğrul 2013). This is due to variations in specific assumptions made by each researcher or research team in formulating and solving the problem. Therefore, when using a horizontal arching solution in practice care must be taken to not only use a solution appropriate for the application but to understand the assumptions inherent in that solution.

A.2.2.5 Geosynthetically Reinforced Earth Masses

A.2.2.5.1 Introduction

One of the pressure-reducing alternatives considered in this document is the ZEP-Wall variant of the compressible-inclusion function. As illustrated conceptually and qualitatively in Figure 3.9, fundamental to the ZEP-Wall concept is the use of multiple layers of geosynthetic tensile reinforcement within the retained soil adjacent to the geofoam compressible inclusion.

The current conceptual and mathematical models for modeling the composite behavior of a geofoam + reinforced-soil system for a ZEP-Wall application are all stiffness- and displacement-based and thus require explicit knowledge of the force-displacement-time behavior of the reinforced soil mass in the horizontal direction. This is a significant issue and technical challenge in light of the decades-old use in routine MSE practice of using only strength-based models (albeit with a consideration of time-dependent physical and mechanical properties for the geosynthetic tensile reinforcements) and ULS analysis and design methodologies for all types and applications of reinforced-soil masses whether it be MSEWs, SRWs, reinforced-soil slopes (RSSs), or soil nailing. Thus, there is not a large body of knowledge and experience as to how to explicitly determine the operational stiffness of reinforced-soil masses under service-load conditions well below the ULS. There is even less knowledge and experience of how to do this in a relatively simple manner that is amenable for use in routine practice.

A.2.2.5.2 Overview

Despite the fact that service-load stiffness and displacement are not considered as part of routine MSE practice for analysis and design does not mean that these parameters have not been studied. To date, the most rigorous way in which the stiffness of a reinforced soil mass has been modeled explicitly is as part of relatively complex numerical analysis of a continuum that typically involves a FE solution[102]. Such an approach was used by the writer in early research of the ZEP-Wall concept and its application to lateral earth pressure reduction specifically with non-yielding RERSs (Horvath 1990b, 1991a, 1991b, 1991c).

While the FEM is very effective and can produce theoretically rigorous results if implemented properly, it is much too complex a methodology to use in routine practice. As a result, there have been a number of simpler approaches to modeling the horizontal stiffness behavior of a reinforced soil mass, MSE in particular, that have been developed and proposed over the years. They vary widely in their complexity and theoretical rigor, and, consequently, the expected accuracy of their results.

Because of the impact that modeling horizontal stiffness of a reinforced soil mass has on calculated results for the ZEP-Wall methodology, it is useful to discuss these alternative analytical methods in some detail. In this discussion, it is assumed that the 'gold standard' or best possible computational methodology that provides a baseline against which other analytical methodologies can be compared and evaluated is defined by the above-described FE modeling. Using the FE approach, it is possible to model each layer of tensile reinforcement, as well as the frictional interface with the soil above and below a reinforcement layer, independently and explicitly. It is also possible to use any number of nonlinear models for the mechanical (stress-strain-time) behavior of the soil. Thus, all SLS and ULS modes of behavior of the reinforced-soil system, including reinforcement pullout, are free to develop, and the calculated end result is likely to be as close to actual behavior as can be calculated based on current technology[103].

---

[102] Alternative continuum-solution methodologies based on the FDM as used by the well-known commercial software package *FLAC* could be used as well. However, the writer has no detailed experience with FDM methodologies or software so they will not be discussed further in this monograph.

[103] The caveat to this statement is that the FEM is well-known to produce less-accurate results when portions of the model approach and then undergo the relatively **[continued on following page]**

244

A.2.2.5.3 Simplified FEM Overlay Solution

Based on the writer's personal experience as well as the experience of other researchers as deduced from a review of the literature, the single biggest drawback to using a complete, rigorous numerical analysis of a MSE mass as described in the preceding section in practice is that it is extremely tedious to model each layer of reinforcement, especially for design situations where the details of the reinforcement (number of layers, length of each layer, reinforcement material properties, etc.) are unknown beforehand and will generally vary during the design process. Each time a geometric detail (spacing, length, etc.) related to the reinforcements changes, this typically requires that an entirely new FE mesh be created. While this task has become much simpler and faster in recent years with the advent of more user-friendly interfaces in commercially available software, it still remains tedious. Thus, it is perhaps no surprise that one of the approaches to developing a simpler method for analyzing force-displacement behavior of a reinforced soil mass focused on ways to change how a reinforced soil mass is modeled.

As discussed by Ebeling et al. (1992), the method they used is to overlay a single finite element that only models the aggregate stiffness of all the reinforcement layers on the basic FE mesh that only models the soil. Essentially two distinctly different finite elements occupy the same space and have at least some (but not necessarily all) nodes within that common space in common. The system is then solved using the composite stiffness of both elements acting together. The attraction of this methodology is that changing the reinforcement geometry and properties means changing only those elements...indeed it may be only one element...that contains the reinforcement overlay. Ebeling et al. (1992) demonstrated the overall validity of this approach by using several simple problems as well as one complex actual application.

While this approach is certainly clever and seems to produce reasonable end results, there are still significant problems with its use in practice. First of all, it still requires performance of a FE analysis with all its attendant effort in terms of problem setup and data interpretation. Second, it requires software that has the capability for using overlay elements. Ebeling and his colleagues used a program named *SOILSTRUCT* that was one of the first geotechnical FE codes developed specifically for analyzing various types of ERSs (Horvath 1977). This code was developed originally in the late 1960s in academia and used initially primarily for academic research although in later years it saw use in practice in a modified version by the U.S. Army Corps of Engineers Waterways Experiment Station in Vicksburg, Mississippi. The writer has not investigated whether any of the commercially available FE codes in use at present such as *DIANA*, *PLAXIS*, etc. have the capability to perform this overlay type of analysis.

A.2.2.5.4 Empirical Solution

At the present time, there appear to be two reasonably well-developed alternatives to using a FE approach of any kind for problems involving soil reinforcement. At the opposite end of the spectrum from the baseline rigorous FE solution in terms of theoretical rigor is a purely empirical approach based on measured maximum lateral displacements of the facing of MSEWs. This method has appeared in publications developed by the FHWA for

---

large displacements associated with soil yielding. This is one reason why FDM software has been developed as this approach to modeling continuum behavior produces reportedly superior results when yielding behavior is significant.

many years. It is discussed in some detail in Chapter 5 as it was used for the one of the simplified ZEP-Wall analytical methodologies that have been developed to date. Although this empirical approach is easy to use, the writer's experience is that it has proven to be much too crude and restrictive for the desired purpose of modeling the ZEP-Wall application.

A.2.2.5.5 Theory of Linear Elasticity Solution

The other alternative is the theory of linear elasticity. Rather interestingly, it appears that research into applying elastic theory to the problem of a reinforced soil mass (specifically the MSE application) dates back to the earliest days of developing MSE technology circa 1970. This means that the concept of using a stiffness- and displacement-based SLS analysis and design procedure as is implied by the use of elastic theory was investigated more or less contemporaneously with the traditional strength- and ULS-based methods. As we now know, this latter approach won out so completely and early on in terms of what is used in both research and practice that it is quite possible that many of today's users of MSE technology are not even aware of the fact that a complementary stiffness-based approach using elastic theory was ever developed for MSE more than four decades ago.

There have been at least two independent investigations of the application of elastic theory to the MSE problem. The earlier effort was presented in a paper by Harrison and Gerard (1972). Poulos and Davis (1974) noted that this application of elastic theory to the MSE problem can be considered a special case of the more-general problem of a cross-anisotropic elastic system composed of alternating homogeneous layers. Other cases to this class of problems includes the classic Westergaard problem of a laterally inextensible half-space.

The paper by Harrison and Gerrard (1972) is particularly interesting because they envisaged an elasticity-based stiffness (SLS) approach as complementing the strength-based ULS approach to MSE analysis and design (they were aware of and explicitly referenced the pioneering work of Vidal). In particular, Harrison and Gerrard saw their stiffness-based methodology as providing an insight into MSE behavior under "working" (service) loads for a system that was designed to have some presumably adequate margin-of-safety (safety factors) against strength-based failure (ULS). This is not unlike dealing with, say, a spread-footing foundation subjected to a column load where both settlement under service loads as well as safety factor against a bearing capacity-type ULS failure are routinely calculated. Unfortunately, Harrison and Gerrard's 'vision' for MSE design clearly never evolved into use in research no less routine practice.

Nevertheless, the use of elasticity solutions for MSE applications shows promise as striking a balance between theoretical rigor and ease of use for the ZEP-Wall applications and is explored for this purpose in Chapter 5. Specifically, the solution of Harrison and Gerrard was chosen for this purpose as a reasonably complete presentation of their findings is readily available in their 1972 paper. Their work can be summarized as follows.

To begin with, several parameters are defined:

- The elastic parameters (Young's and shear moduli, Poisson's ratio) of the soil are defined as E_s, G_s and v_s respectively.

- The same three elastic parameters of the reinforcement are defined as E_r, G_r and v_r respectively.

246

- The dimensionless parameter called the *relative reinforcement thickness*, t_r, is defined as the ratio of the total thickness of all reinforcement layers to the total thickness (height) of the retained soil, H.

- The dimensionless parameter called the *relative reinforcement stiffness*, K_r, is defined as:

$$K_r = \frac{t_r \cdot E_r}{E_s}.$$ **(A.9)**

With regard to results, Harrison and Gerrard noted that the composite reinforced soil mass could be visualized analytically as a single pseudo-homogeneous, anisotropic (more specifically, cross-anisotropic) elastic mass whose equivalent homogeneous properties reflected the interaction between the two very distinct components (soil and geosynthetic tensile reinforcement) comprising the mass. This is perhaps the single biggest attraction of their work, the fact that the actual reinforced soil mass that consists of two distinct materials can be modeled in terms of its gross (overall) behavior as if it were an equivalent mass of a single material. Note that this is similar in concept to what Ebeling et al. (1992) accomplished with their FEM-overlay concept discussed previously. The notable difference is that Harrison and Gerard approached this from a closed-form theoretical solution perspective whereas Ebeling et al. used a numerical approach.

With acceptable accuracy, the elastic parameters of this pseudo-homogeneous elastic mass can be simplified and approximated as follows:

- The vertical Young's modulus, E_v, of the overall mass is slightly to somewhat greater than that of the soil alone, E_s. This is consistent with observations of actual MSE structures as reported by Monley and Wu (1993, 1995)[104] where it was found that reinforced soil masses were stiffer under vertical loading compared to otherwise identical soil without reinforcing. However, given that the primary interest in ZEP-Wall applications involving RERSs is in the horizontal stiffness, it is suggested here that $E_v = E_s$ can reasonably, if somewhat conservatively, be assumed for routine use.

- The horizontal Young's modulus, E_h, of the overall mass is defined as:

$$E_h = (K_r + 1) \cdot E_s.$$ **(A.10)**

Not surprisingly, E_h is always greater than that of the soil alone, E_s, which reflects the stiffness, K_r, contributed by the reinforcement.

- The shear modulus, G_v, for shearing occurring in a vertical plane within the overall mass, which is assumed here to be the primary mode of shearing for RERSs, is equal to the shear modulus of the soil, G_s. This is an interesting result as it implies that the reinforcement is wholly ineffective in a purely shearing mode perpendicular to the plane (oriented horizontally here) of the reinforcements. This is perhaps not as unreasonable as it may appear at first as this simply reflects the fact that the layer of soil between two adjacent layers of reinforcement is free to deform in pure shear as if the reinforcement were not there.

---

[104] See also Horvath (1995a).

- The Poisson ratio of the overall mass varies between that of the soil, v_s, and that of the reinforcement, v_r. In any practical problem, there is not likely to be a big difference between the two.

Harrison and Gerard were not the only researchers to pursue use of elastic theory as a model for geosynthetically-reinforced soil masses. The writer is aware of at least one other such effort, the doctoral work of Constantine A. Vokas as summarized in Vokas and Stoll (1987). However, the writer has not studied Vokas' work in detail to see how it might compare with or differ from that of Harrison and Gerard.

There are several criticisms related to using any elastic solution in a geotechnical problem. By far the most significant is that the material-stiffness properties, moduli especially, are typically assumed to be constant throughout the elastic layer and under all load levels. Both issues, but especially the latter, are at odds with the actual behavior of a retained-soil mass of a RERS. Another criticism is that the potential for pullout of the reinforcement is assumed not to occur.

A.2.3 Seismic Loading Conditions

A.2.3.1 Introduction and Overview

Development of analytical methodologies for forecasting seismic-induced lateral earth forces on ERSs actually began very early in the history of modern soil mechanics. The earliest research was conducted during the 1920s in Japan in response to the Great Kantō Earthquake of 1923 ($M_w \cong 7.9 - 8.2$) and focused on traditional gravity retaining walls, the most common type of yielding RERSs.

As a result of two major earthquakes in 1964 (in Niigata, Japan and Anchorage, Alaska), the late 1960s saw the beginning of a more globally extensive surge in interest in geotechnical earthquake engineering that was led by academic researchers located in the U.S. While early research focused on seismic liquefaction which was a signature destructive feature of both 1964 events, attention was given to ERSs as well so it is no surprise that the circa-1920s Japanese work was resurrected from relative obscurity and co-opted for use. The analytical outcome of this early Japanese work was referred to in the late-1960s research as the *Mononobe-Okabe (M-O) Method*, reflecting the names of some, but not all, of the original Japanese researchers in the 1920s.

At this point, it is important to point out that the original, 'true' M-O Method has been altered and reinterpreted in various ways over the approximately 50 years since its emergence into the global mainstream of geotechnical earthquake engineering. As a result, what is currently referred to in U.S.-sourced publications as the *Mononobe-Okabe Method* can actually vary in specific details from author to author and usually deviates from the original in several significant and inconsistent ways that end users may not always be aware of. Consequently, the presentation and discussion in the following sections will first outline the basic elements of the original M-O Method before discussing changes made over the years. In addition, the term 'M-O Method' is only used in this monograph to refer to the original method as defined subsequently.

As it turns out, understanding the basic elements of the M-O Method as well as the alterations made to it over time have become more important than ever. This is because the 21st century has seen greater implementation of seismic design requirements for both yielding and non-yielding RERSs in various building codes and standards that are applied throughout the entire U.S. (Sitar et al. 2012). As a result, all stakeholders involved with

RERSs impacted by these codes and standards need to be aware of the evolution of the various analytical methodologies involved since the 1960s.

Given the great interest in the subject of seismic effects on all types of ERSs, it is not surprising that the body of research and concomitant publication on the subject is already vast and constantly increasing. In addition to further research involving dry coarse-grain soils and yielding RERSs (the focus of the M-O Method), other types of ERSs such as non-yielding RERSs and FERSs[105]; saturated coarse-grain soil; and fine-grain soil have been studied.

Fortunately, there have been a number of publications over the years devoted to synthesizing, summarizing, comparing, and critiquing multiple methodologies that address some focused aspects of the overall problem of ERSs and seismic loading. This makes navigating the volume of published work that spans decades somewhat easier. Ebeling and Morrison (1992) was one such document. Some relatively recent efforts in this regard are:

- Geraili Mikola and Sitar (2013)[106] and Geraili Mikola et al. (2016) who covered nominally-dry coarse-grain soils and RERSs;

- Candia and Sitar (2013)[107] who covered 'cohesive' soils and RERSs; and

- Sitar et al. (2012) and Sitar and Wagner (2015) who covered both yielding and non-yielding RERSs and cohesionless and cohesive soils.

Given the depth and breadth of publications dealing with seismic loads on ERSs, it is well beyond the scope of this monograph to present even a summary of all the different analytical methodologies that have been published. Only those methodologies that are directly relevant to the contents of this monograph are discussed in the following sections.

However, there are two issues to note at this point as they apply to virtually the entire body of published work on this subject:

- With few exceptions, all the analytical methods that have been proposed for use in practice are pseudo-static[108] in nature, i.e. they reduce what is essentially a time-

---

[105] A caution based on the writer's review of the published literature up to the present time is that some authors use the term 'flexible' as it applies to ERSs in a manner that is significantly different from how the writer has defined it for the purposes of this monograph. Specifically, some authors apply the term 'flexible' to RERSs that are free to both translate and rotate due to being supported on a non-rigid base, e.g. on soil as opposed to bedrock. Thus, what these authors term 'flexible RERSs' is actually 'yielding RERSs' in the terminology used in this monograph. As noted in Chapter 2, the writer defines flexible earth-retaining structures (FERSs) as only those ERSs that relatively deformable even under service loads, e.g. sheet-pile walls and bulkheads.

[106] There is some uncertainty as to how to correctly cite work attributed to the first author of this reference. The author's full name as shown on the cover of this reference is "Roozbah Geraili Mikola". Citations of this author's work that have been found online variously use "Geraili Mikola" and "Mikola" as the surname. The writer has chosen to use the former in this monograph.

[107] There is some uncertainty as to how to correctly cite the first author as well. The author's full name as shown on the cover of this reference is "Gabriel Candia Agusti". Citations of this author's work that have been found online all use only "Candia" as the surname. Therefore, the writer has chosen to follow this usage in this monograph.

[108] The writer posits that an argument could be made that the term 'pseudo-dynamic' would be more correct to use in this particular instance. However, the use of 'pseudo-static' appears to be well-established in actual usage so is used here.

dependent, i.e. dynamic, problem to one consisting of a time-independent, i.e. static, system of forces. In the process, what is inherently a three-dimensional (3-D) problem is usually modeled in a 2-D (usually plane-strain) setting. Although computer hardware and software capable of performing true dynamic analyses exist, it is still considered not cost-effective to use in the majority of applications in practice.

- When evaluating and/or comparing analytical methods, it should always be kept in mind that there are two interrelated aspects to the problem, namely, the <u>magnitude</u> of the seismic-induced lateral resultant force and the <u>point-of-application</u> of this force. Alternatively, in terms of lateral earth pressures, there are the <u>magnitudes</u> and <u>distribution</u> of these pressures. The importance of this is that there are always structural ULS considerations for the ERS itself and, for yielding RERSs, multiple geotechnical ULS modes to consider. Not all of these problem components are equally impacted by changes in force/pressure magnitude vs. point-of-application. So, all aspects of a problem need to be evaluated in order to assess the overall impact of an analytical methodology on the problem.

A.2.3.2 Active Earth Pressure State

A.2.3.2.1 Overview of Traditional Perspectives

As with gravity loading, analytical methodologies for seismic loading that are based on the fundamental assumption that an active earth pressure failure wedge develops within the retained soil have historically been assumed to be applicable only to yielding RERSs and, by extension, to FERSs such as anchored bulkheads as well. Consequently, the discussion in this section is limited to yielding RERSs.

This is not a trivial statement. As discussed in subsequent sections dealing with the at-rest earth pressure state and non-yielding RERSs, recent research has blurred the traditional sharp analytical distinction between yielding and non-yielding RERSs that has existed since the evolution of modern geotechnical earthquake engineering began in the late 1960s. Consequently, it is necessary to clearly define the topic being addressed at this point.

A.2.3.2.2 Mononobe-Okabe Method (Original Version)

Overall, the basic concepts and premises incorporated into the M-O Method continue to define the state of practice for use with yielding RERSs and, as indicated above, recent research has actually made a case for extending its use to other ERS categories as well. However, as often occurs with technical concepts that have existed for a relatively long period of time (almost 100 years in this case), some aspects of the original methodology appear to have been forgotten and other aspects have been conflated with later developments by others. In this section, only the original, 'true' version of the M-O Method is discussed. Evolutionary changes and extensions are discussed in following sections.

To begin with, it is important to note that the M-O Method was a product of geotechnical knowledge and research capabilities during the infancy of modern geotechnical engineering in the 1920s. This is not meant to either deprecate or diminish the seminal importance of the outcomes but to establish a level of reality about them and their geomechanical sophistication.

Most relevant is the fact that the key algebraic outcomes were based solely on small-scale, 1-g laboratory testing involving dry coarse-grain soil subjected to pure sinusoidal motion. Furthermore, the simulated RERS effectively had a rigid base and could only 'fail' geotechnically by the mode of rotation about the bottom. Sketches and a detailed discussion of the original testing hardware can be found in Sitar et al (2012), Geraili Mikola and Sitar (2013), and elsewhere. Only in more recent years has it been possible to revisit the problem using modern analytical tools such as centrifuge models and FDM numerical analyses using actual earthquake records as input.

Figure A.3 illustrates the basic variables and free-body system of forces for the geotechnical components of the M-O Method. Intentionally omitted are any surface-surcharge effects on the retained soil as these were not explicitly considered in the development of the M-O Method. Also, not shown are the gravity and inertia forces on the RERS itself.

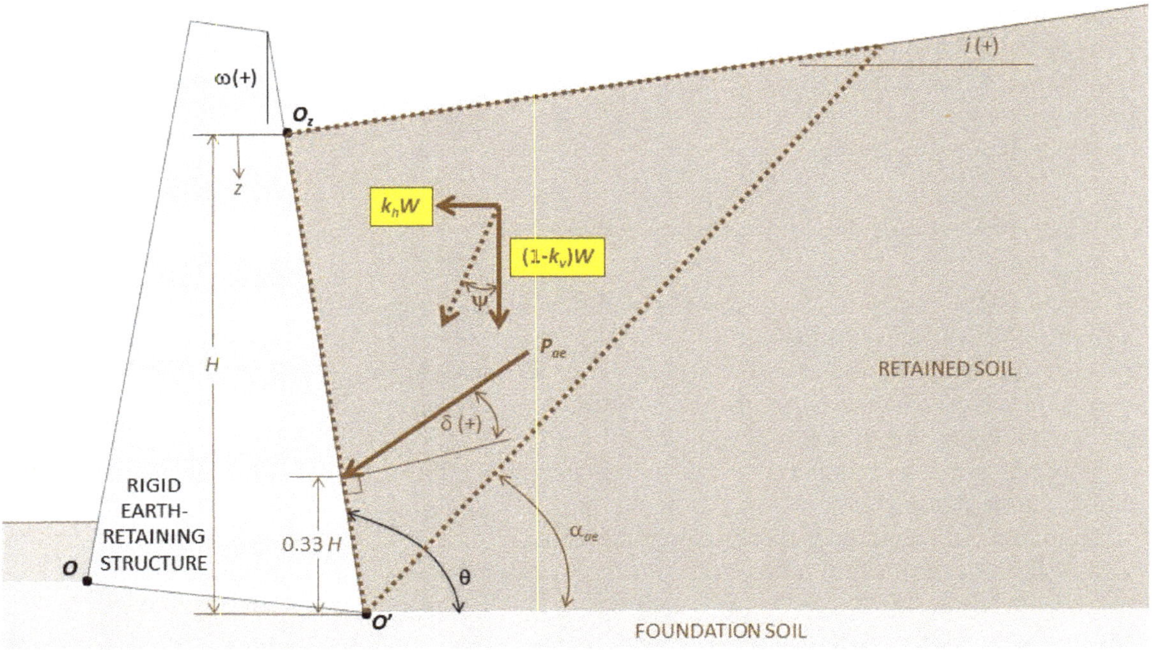

Figure A.3. Mononobe-Okabe Method (Original Version)
Generic Failure Wedge and Free-Body Diagram of Forces.

In the most basic terms, the M-O Method is Coulomb's problem for the active earth pressure state with the addition of seismic-induced inertial effects, both vertical and horizontal, on the assumed failure wedge of soil. Note that it is assumed that the entire failure wedge displaces in-phase and is subjected to the same level of seismic acceleration that is taken to be the free-field value at the surface of the retained soil. Although the inertial effects on the RERS itself are not shown, it is relevant to note the failure wedge of soil and RERS are also assumed to displace in-phase.

The issues of uniform seismic acceleration throughout the entire problem and the in-phase relationship both within the retained soil as well as between the retained soil and RERS are highlighted as subsequent research has found that the geotechnical height of the RERS, H, influences the degree to which these assumptions are reasonable. At some point with increasing H, not only does the more-dominant horizontal ground acceleration vary throughout the vertical column of retained soil but the mass of retained soil does not uniformly displace laterally as a more or less rigid body. However, for the purposes of the present discussion, these assumptions regarding uniformity of both accelerations and displacements are generally considered to be reasonable for typical, traditional 'retaining wall' applications of the M-O Method where the geotechnical height of the yielding RERS is of the order of 6 metres (20 ft) or less.

The earth force acting on the RERS that includes both gravity and seismic effects, P_{ae}, is defined as follows:

$$P_{ae} = 0.5 \cdot K_{ae} \cdot \gamma_{eff}^* \cdot H^2 \tag{A.11}$$

where γ_{eff}^* is the effective soil unit weight adjusted for inertia effects due to the vertical component of seismic acceleration:

$$\gamma_{eff}^* = \gamma_{eff} \cdot (1 - k_v) \tag{A.12}$$

where k_v is the usual dimensionless vertical component of seismic acceleration (i.e. the actual vertical acceleration, a_v, normalized to (i.e. divided by) the gravitational acceleration constant, g) at the ground surface. In the most general case, k_v can be positive, negative, or zero, with the positive sense defined as acting upward.

Note that Equation A.11 implies the classical triangular distribution of lateral earth pressures although, as with Coulomb's original theory, the circa-1920s Japanese laboratory measurements made on the small-scale models used to simulate a yielding gravity retaining wall under seismic loading only measured resultant earth forces.

K_{ae} is the coefficient of active earth pressure under seismic loading and is given by:

$$K_{ae} = \left\{ \frac{\cos^2(\phi - \psi - \omega)}{\cos\psi \cdot \cos^2\omega \cdot \cos(\psi + \omega + \delta) \cdot \left[1 + \sqrt{\frac{\sin(\phi + \delta) \cdot \sin(\phi - \psi - i)}{\cos(\delta + \psi + \omega) \cdot \cos(i - \omega)}} \right]^2} \right\} \tag{A.13}$$

where ψ is termed as the *seismic inertia angle* and defined as:

$$\psi = \tan^{-1}\left(\frac{k_h}{1 - k_v} \right) \tag{A.14}$$

where k_h is the usual dimensionless horizontal component of seismic acceleration at the ground surface (i.e. the actual horizontal acceleration, a_h, normalized to the gravitational acceleration constant, g) and always positive in sign[109]. As can be seen in Figure A.3, the

---

[109] Note that unlike the vertical component of seismic acceleration which can reasonably act either upward (+) or downward (-) in practical applications, it is routinely assumed that the only direction of horizontal seismic acceleration that is relevant to ERS problems is when it acts toward the ERS (to the left in Figure A.3). By convention, this direction is always defined as being positive which is why k_h is only positive in sign.

physical interpretation of ψ is that it defines the angle that the total resultant force of the failure-wedge weight (denoted by the arrow with a dotted line) makes with respect to the vertical.

As will be seen, K_{ae}, either directly or indirectly as defined subsequently, is the metric by which various analytical methodologies for both yielding and non-yielding RERSs are not only compared to each other but to measured results from either 1-g or multi-g (i.e. centrifuge) tests. In this regard, it is important to note that Equation A.13 produces results that not only increase exponentially as a function of seismic intensity (as reflected in increasing values of ψ) but at some point, the solution becomes unstable, i.e. K_{ae} becomes infinite in value. For typical soil properties, this occurs for seismic accelerations, a_h, of approximately 0.7g, i.e. $k_h \cong 0.7$. As will be seen, this was apparently recognized as a serious practical problem almost immediately upon use of the M-O Method in the U.S. in the late 1960s.

Before proceeding to a discussion of later changes to the M-O Method, it is useful to make several additional observations and comments that set the stage for material presented in following sections:

- Largely forgotten over time is that the M-O Method produces only a single resultant earth force, P_{ae}, that includes both gravity and seismic effects lumped together. This is only logical as this reflects how forces were measured in the small-scale laboratory testing that resulted in the method. As will be seen, for various reasons, all subsequent versions of the method decompose the resultant force into a baseline gravity component, P_a, that is assumed to be produced by Coulomb's K_a and a seismic-increment component, ΔP_{ae}, that is assumed to be produced by an incremental lateral earth pressure coefficient, ΔK_{ae}, this is evaluated in various ways depending on the specific methodology being used.

- Arguably the single most important element of the M-O Method that appears to have been forgotten over time, especially by practitioners, is that the total resultant earth force, P_{ae}, is placed at the same $H/3$ point above the heel of the RERS (Point O' in Figure A.3) as P_a in Coulomb's gravity-only problem. Stated another way, both the gravity and seismic components of P_{ae}, P_a and ΔP_{ae} respectively, act at the same $H/3$ point. As will be seen, subsequent versions of the method tend to place the two components at different points-of-application.

- Also forgotten over time is that the results on which the original M-O Method was based in its entirety were obtained using the aforementioned small-scale laboratory models subjected to idealized sinusoidal motion. Among other things, this means that the peak values of k_v and k_h were constant during each test. This has resulted in subsequent uncertainty on how to select values of k_v and k_h in practice as for real earthquakes the peak values of these parameters will typically occur for only one cycle of loading with lower average values characterizing the overall event. This point is also discussed further in the following two sections.

- Simple calculation indicates that, in traditional RERS applications at least, the horizontal earth-force component, which is due to seismic-inertial effects only, has far more influence on (to the point of dominating) overall problem results in terms of geotechnical ULS modes (translation/sliding, rotation/overturning, etc.) than does variation in the vertical earth-force component, with the latter having virtually no effect on calculated outcomes. Stated another way, whether k_v is positive, negative, or zero has very little effect on calculated safety margins against translation, rotation, etc. This is why virtually all analytical methodologies that evolved over the years from the original M-O Method simply ignore k_v completely. That having been said, care needs to be exercised when extending use of the M-O Method or some variant of it to new applications such as those considered in this monograph, especially those involving the lightweight-fill geofoam function. It should never be assumed a priori for new applications that variations in k_v are not significant unless and until proven by calculation or physical testing.

- The angle, α_{ae}, that the soil-on-soil failure plane makes with respect to the horizontal is always flatter than the angle α_a for the same RERS geometry and soil properties under gravity loading only and is defined as follows:

$$\alpha_{ae} = (\phi - \psi) + \tan^{-1}\left(\frac{-\tan(\phi - \psi - i) + c_{1E}}{c_{2E}}\right) \qquad \text{(A.15)}$$

where

$$c_{1E} = \sqrt{\tan(\phi - \psi - i) \cdot [\tan(\phi - \psi - i) + \cot(\phi - \psi - \omega)] \cdot [1 + (\tan(\delta + \psi + \omega) \cdot \cot(\phi - \psi - \omega))]} \text{ (A.16)}$$

and

$$c_{2E} = 1 + \{\tan(\delta + \psi + \omega) \cdot [\tan(\phi - \psi - i) + \cot(\phi - \psi - \omega)]\}. \qquad \text{(A.17)}$$

- Although the laboratory testing that produced the results used to develop the M-O Method was limited to the use of dry coarse-grain soil, theoretical work performed in Japan during the same 1920s timeframe produced a more-general form of the equation (A.13) for K_{ae} that includes the effects of cohesion. This more-general expression for K_{ae}, which can be found in Candia and Sitar (2013) and Sitar and Wagner (2015), is rarely associated with the original M-O Method or even presented and discussed in the published literature although the effect of cohesion on the overall problem has been the subject of numerous publications. In summary, cohesion has a dramatic effect on reducing calculated resultant earth forces as illustrated by Sitar and Wagner (2015). While this has long been known for the gravity component, it applies to the seismic component as well.

A.2.3.2.3 Seed-Whitman Method

Fairly early (1970) in the era of geotechnical earthquake engineering that began in the late 1960s, Seed and Whitman proposed an analytical method (hereinafter referred to as the *Seed-Whitman (S-W) Method*) that was essentially a simplified and modified version of the M-O Method that reflected several considerations by them, the details of which can be found in numerous references such as Geraili Mikola and Sitar (2013), and that can be summarized as follows:

- Observations indicated that actual retaining walls presumably designed only for gravity loads performed 'well' (interpreted to be 'adequately') overall in seismic events where

k_h did not exceed 0.3. Later studies by others (Geraili Mikola at al. 2016) have suggested that this level can be increased to 0.4. An important caveat to this observation is that this does not necessarily apply to situations where the entire RERS may be part of a slope or where the surface of the retained soil defines a significant slope that extends upward well beyond the RERS. In such cases, global slope failure that encompasses or otherwise affects the RERS may govern as opposed to localized behavior of the RERS itself.

- By virtue of the equation (A.13) for K_{ae}, the original M-O Method forecasts exponentially increasing resultant forces and at some point, becomes numerically unstable. This is clearly unreasonable and unrealistic, especially in view of how actual RERSs performed. Therefore, an alternative expression for K_{ae} needed to be developed.

- A reassessment of prior research suggested that the total resultant force acted above the $H/3$ point which means that the seismic increment had to act above this point.

Figure A.4 summarizes the key geotechnical elements and forces of the S-W Method which are explained and discussed below. Note that Seed and Whitman based their methodology on a less-general RERS and retained-soil geometry than that allowed in the M-O Method (Figure A.3). They also assumed that the geometry of the failure wedge under seismic conditions was more or less fixed as a function of the geotechnical height, H, of the RERS although this does not enter into calculations in any explicit way.

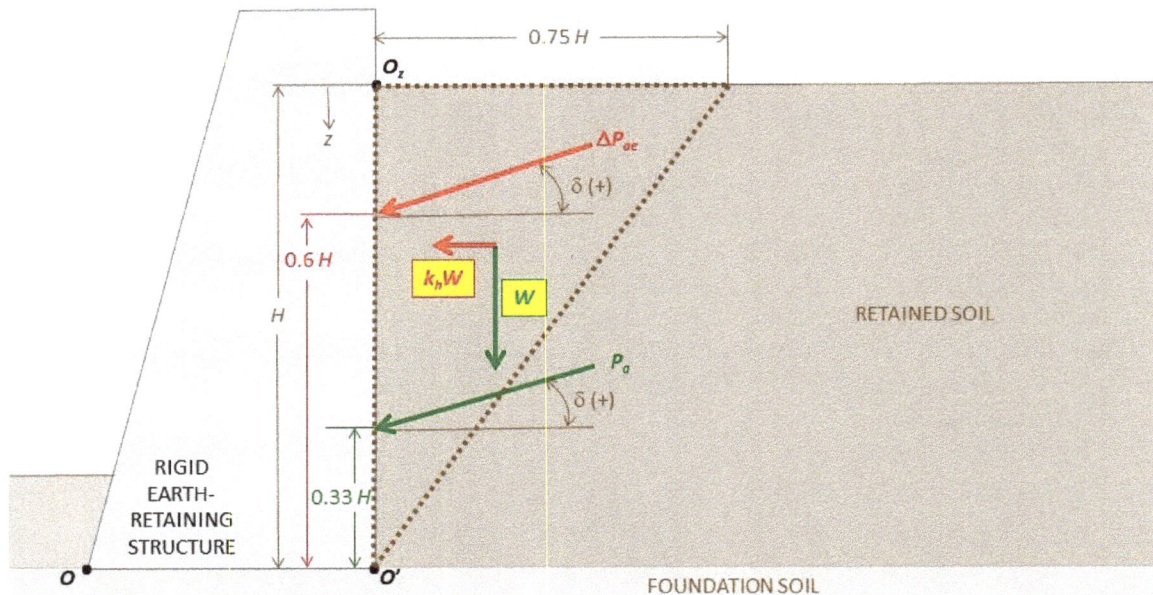

Figure A.4. Seed-Whitman Method - Generic Failure Wedge and Free-Body Diagram of Forces.

The first deviation of note from the M-O Method was that Seed and Whitman explicitly decomposed the single resultant force used in the M-O Method, P_{ae}, into the two artificial components defined previously:

$$P_{ae} = P_a + \Delta P_{ae} . \tag{A.18}$$

Note that when calculating P_a using Coulomb's solution (Equation A.3), the soil unit weight is not corrected for vertical-acceleration effects as the S-W Method ignores the effect of k_v completely. As noted previously, this has negligible effect on calculated results, at least in the conventional RERS applications for which the S-W Method was intended. Note also that the point-of-application of P_a remains at the $H/3$ point that is traditionally inferred for Coulomb's solution.

Seed and Whitman defined ΔP_{ae} as:

$$\Delta P_{ae} = 0.5 \cdot \Delta K_{ae} \cdot \gamma_{eff} \cdot H^2 \tag{A.19}$$

with a point-of-application as shown in Figure A.4. Note that placing the seismic component at a much higher level on the inside face of the RERS than the gravity component is another signature feature of the S-W Method.

ΔK_{ae} in Equation A.19 can be defined rigorously and generically as:

$$\Delta K_{ae} = K_{ae} - K_a \tag{A.20}$$

although Seed and Whitman went a step further and suggested simply taking:

$$\Delta K_{ae} \cong 0.75 \cdot k_h . \tag{A.21}$$

This produces a linear relationship between the seismic-force component and ground acceleration and eliminates the 'runaway' exponential behavior forecast by the M-O Method parameter K_{ae}. As a result, Equation A.19 simplifies to:

$$\Delta P_{ae} = 0.375 \cdot k_h \cdot \gamma_{eff} \cdot H^2 . \tag{A.22}$$

As an aside, it is of interest to note that the 0.6H dimension for the placement of the ΔP_{ae} force shown in Figure A.4 is an approximation of an average location, not an absolute as many references that discuss the S-W Method imply. Wagner and Sitar (2016) noted that Seed and Whitman felt that this force component should be placed somewhere within the range of 0.5H to 0.67H.

Another aspect of Seed and Whitman's work is that they suggested using a value of k_h that is 85% of the peak value, which is referred to as the *peak ground acceleration* (*PGA*), for whatever design event is being assessed (although not stated explicitly, it is implied that 100% of the PGA should be used with the M-O Method). This reflects the point made earlier that the peak values of k_v and k_h generally occur just once in a typical earthquake record and thus do not represent a sustained level of acceleration that would be more representative for design purposes.

As illustrated in Geraili Mikola and Sitar (2013), the net result of these assumptions is that the S-W and M-O methods forecast essentially identical results (using the calculated value of ΔK_{ae} as the metric) up to a PGA of approximately 0.3g. Beyond that, forecasts from these two methods diverge at an ever-increasing rate as forecasts from the S-W Method increase linearly while those from the M-O Method increase exponentially.

In assessing the practical implications of the S-W Method vs. the M-O Method, recall the point made previously that both the magnitude(s) and point(s)-of-application of resultant earth forces are important as both force and moment equilibrium are used when assessing the safety margins of yielding RERSs. In this regard, it is important to note that while the S-W Method forecasts magnitudes of resultant forces that are either about the same (for smaller earthquakes) or less (for larger earthquakes) than forecasts from the M-O Method, the S-W Method places the seismic component of those forces much higher on the RERS than the M-O Method in all cases. Thus, in structural and geotechnical ULS modes where moments are used, the S-W Method will tend to be more demanding than the M-O Method.

A.2.3.2.4 Current Practice (Composite Method)

Overall, it appears that the core concept of the M-O Method, i.e. resultant earth forces based on a Coulomb-like failure wedge modified for inertial effects, has remained the backbone methodology used in routine practice, at least in the U.S. However, over time the M-O Method has been conflated with key elements of the S-W Method. Although most, especially practitioners, still refer to the resulting methodology as the 'Mononobe-Okabe Method' it will be referred to in this monograph as the *Composite Method* to avoid any confusion with the true, original M-O Method that most practitioners at least seem to have forgotten about.

Specifically, the Composite Method uses the artificial separation of gravity and seismic components of the active resultant force from the S-W Method (Figure A.4) but explicitly calculates ΔK_{ae} using Equation A.20, with K_{ae} from the original M-O Method (Equation A.13) and K_a from Coulomb's solution (Equation A.4). ΔP_{ae} is calculated using Equation A.19 with the exception that the soil unit weight used is γ_{eff}^*, i.e. corrected for k_v effects, and not γ_{eff} although there is little practical difference in the calculated outcomes.

Note that because the Composite Method retains the calculation of K_{ae} and, by implication, ΔK_{ae} from the M-O Method (ignoring the Seed-Whitman recommendation for ΔK_{ae} in the process), it is flawed by the same issues as the M-O Method in terms of exponentially increasing values and numerical instability of calculated results for large (> 0.7) values of k_h as noted by Wagner and Sitar (2016) and many others.

One aspect of the Composite Method that is not dealt with consistently is whether the value of k_h used in the various equations is the PGA or some fraction (percentage) of it. That appears to vary with the code that is applied or the professional judgment of the analyst although the current practice appears to favor use of 100% of the PGA which is a further deviation from the S-W Method that recommended use of 85% of the PGA.

Other aspects of the Composite Method are that the more-general problem geometry for the RERS and retained soil shown in Figure A.3 that is part of the M-O Method is allowed. Also, the geometry of the failure wedge is assumed to vary with the angle α_{ae} (Equation A.15) as in the M-O Method.

Because the Composite Method is so widely used in practice, as was the case with the gravity resultant force, P_a, in Coulomb's solution as discussed earlier in this appendix, it is desirable to convert the seismic-increment resultant force, ΔP_{ae}, to a pressure diagram for use in structural analysis and design. Because the assumed point-of-application of this force component is close to the upper-third point of the inside face of the RERS, some (e.g. Geraili Mikola and Sitar 2013) have simply implied that an "inverted triangle" distribution should be used. However, to be rigorous the shape is arguably that of a trapezoid.

Such a trapezoidal diagram has been developed for the purposes of this monograph based on the assumption that the trapezoid should satisfy both force and moment equilibrium with regard to the magnitude of ΔP_{ae} and its assumed point-of-application as shown in Figure A.4. The magnitudes of this lateral earth pressure trapezoid can be shown to be:

$$= 0.8 \cdot \Delta K_{ae} \cdot \cos\omega \cdot \gamma^*_{eff} \cdot H \text{ at the ground surface (Point O}_z \text{ in Figure A.3)} \quad \textbf{(A.23a)}$$

$$= 0.2 \cdot \Delta K_{ae} \cdot \cos\omega \cdot \gamma^*_{eff} \cdot H \text{ at the heel of the RERS (Point O' in Figure A.3)} \quad \textbf{(A.23b)}$$

with a linear variation in pressure between these values. Note that this pressure trapezoid is distributed along the inside face of the RERS shown in Figure A.3 (a length equal to $H/\cos \omega$) and would thus be inclined from the vertical in cases where the angle ω did not equal zero. Note also that this trapezoid is skewed as well by the soil-wall friction angle δ in the same sense as shown for the resultant force ΔP_{ae}.

In assessing the practical implications of the Composite Method vs. the M-O and S-W methods, the net result on both geotechnical and structural ULS modes is outcomes that are more conservative than either the M-O or S-W methods. This is because the magnitudes of the resultant forces are those of the M-O Method while the points-of-application of these forces are those of the S-W Method. The much larger forces and moments of the Composite Method are particularly apparent for larger earthquakes with PGAs exceeding approximately 0.4g.

A.2.3.2.5 Current Trends (Modified Mononobe-Okabe Method)

The last several years have seen a substantial amount of funded research on the subject of earthquake loading on RERSs performed at the University of California, Berkeley (UC-Berkeley) under the overall direction and supervision of Prof. Nicholas Sitar. It appears that the motivation for this work is due, at least in part, to the aforementioned increased attention paid to the subject by building codes and other regulatory documents. These documents have tended to incorporate traditional thinking and analytical methodologies which, for the yielding RERSs currently under discussion, means the Composite Method described in the preceding section. Use of these traditional methodologies has significant cost implications for both proposed/new and existing structures, especially in high-seismic regions such as the West Coast of the U.S., because of the use of the M-O Method equation for evaluating K_{ae}. As noted previously, values for this parameter increase exponentially with increasing seismic acceleration. Consequently, there is motivation to critically review these currently used methodologies to see if their continued use is justified both technically and economically.

The research at UC-Berkeley has addressed both yielding and non-yielding RERSs using physical testing by geotechnical centrifuge and numerical modeling using the FDM (*FLAC*), combined with a critical review of published work going back to the circa-1920s research in Japan. Only the outcomes related to yielding RERSs are discussed here. Relevant reference documents are Sitar et al. (2012), Geraili Mikola and Sitar (2013), Candia and Sitar (2013), Sitar and Wagner (2015), and Geraili Mikola (2016).

The current thinking and trends that the writer deduces from this cited research is that behavior of yielding RERSs cannot be viewed as a single issue. There appear to be at least two significant variables that result in a range of measured and forecast behaviors for such RERSs:

- The geotechnical height of the RERS is important because at some point the traditional assumption reflected in the M-O, S-W, and Composite methods that the entire mass of retained soil moves in-phase and that this motion can be characterized by some percentage of the free-field surface PGA ceases to be reasonable. Research to date suggests that this assumption is reasonable for conventional retaining walls up to a geotechnical height of at least 6 to 7 metres (~20 ft). Consequently, the balance of the comments made in this section of the monograph assume RERSs no greater than this geotechnical height.

- The specific 'flexibility' of the RERS is an important factor as not all yielding RERS are equal in this regard. In this case, flexibility relates to the ability of the overall RERS to both translate and rotate in order to reduce the increased lateral earth pressures due to seismic shaking. To use the specific examples reflected in the UC-Berkeley centrifuge tests, a free-standing cantilever retaining wall supported on soil would be considered much more flexible that a monolithic U-shaped structure (as might be used for a depressed roadway or navigational lock) where the vertical (wall) components could rotate to some degree but translation is restricted[110].

Nevertheless, within this range of observed and forecast behaviors there were some very significant unified outcomes:

- The increase in lateral earth pressures due to the seismic-load component were in all cases triangular in distribution, thus affirming the original M-O Method assumption that the total resultant force can be placed at the $H/3$ point and contradicting the assumption in both the S-W and Composite methods that the seismic-load resultant be placed higher on the inside face of the RERS. Furthermore, this behavior was observed for both coarse- and fine-grain retained soils.

- Using ΔK_{ae} as a metric, the overall trend in increasing seismic-induced load with increasing PGA was linear up to PGAs of more than 0.6g for both coarse- and fine-grain retained soils. This affirms a basic tenet of the S-W Method and confirms the long-held belief that the exponential increase forecast by the M-O Method produces increasingly unreasonable results for PGAs in excess of approximately 0.4g.

- Cohesion within the retained soil appeared to play a modest but noticeable role in reducing the measured and calculated seismic increment of loads.

- An interesting interpretation of the data in Geraili Mikola et al. (2016) showed that regardless of the flexibility, a yielding RERS designed with safety margins considered standard in practice should be able to tolerate seismic loading up to a PGA of approximately 0.3g without requiring any special seismic-related design. This affirms an observation first made by Seed and Whitman decades earlier and since confirmed by

---

[110] The writer would actually consider the latter RERS to be more in the category of a non-yielding RERS due to the restriction against translation as well as the rotational restraint provided by the base slab of the structure. However, given that the walls of the structure in the UC-Berkeley research were designed to allow some rotation and the researchers considered the overall structure to be a yielding RERS, this structure is included in the present discussion. However, it clearly serves as an example of a yielding RERS with minimal yielding and concomitant flexibility in the context of the present discussion.

numerous other researchers in later studies, including by the Principal Investigators at UC-Berkeley who included earthquakes in the early 21st century.

With specific regard to the second item above, although all the RERSs and retained soils studied at UC-Berkeley exhibited a linear relationship between ΔK_{ae} and PGA, the specific slope of that relationship, i.e. what percentage of the PGA should be assumed in calculations, was heavily dependent on the aforementioned relative flexibility of the RERS:

- For less flexible yielding RERSs such as the monolithic U-frame structure, 100% of the PGA (as in the M-O Method) was an upper-bound of the observed and calculated results while 85% of the PGA (as in the S-W Method) reflected the mean of the same data.

- For the more flexible free-standing cantilever retaining wall, 35% of the PGA provided a reasonable fit to the data.

Based on these overall findings, it would appear to be reasonable to propose for use with yielding RERS of conventional construction (i.e. gravity or cantilever design) and 'typical' height used in practice ($\leq$ 7 metres/20 feet) what will be called a *Modified Mononobe-Okabe (MM-O) Method* that incorporates the basic elements of the M-O Method as outlined previously but with ΔK_{ae} based not on the M-O equation (A.13) but the Seed and Whitman recommendation of $\Delta K_{ae} = 0.75 k_h$ (Equation A.21).

It would seem to be prudent that until such time as 'flexibility' in the context under discussion can be quantified in some way that k_h equal to 85% of the PGA as recommended originally by Seed and Whitman is reasonable unless the RERS is judged to be restrained in some way from complete yielding in which case 100% of the PGA would be reasonable. Although the findings of the UC-Berkeley research suggest using design values for k_h as low as 35% of the PGA for RERSs that are completely free to yield, this appears to be too drastic a reduction to implement across the board without further research to support this finding.

Another reason for being cautious at the present time is that the role of the 'foundation' material for the RERS is unclear in the writer's opinion. Specifically, whether the RERs is founded on soil or bedrock may play a role in the percentage or fraction of the PGA best matches observed performance. This may also explain some of the variation in measured results of not only resultant forces but the point of application of those forces as well that has appeared in the literature going back to the seminal 1920s research in Japan. Certainly most, if not all, of the small-scale model tests performed under 1-g conditions replicated 'rigid-base' conditions approximating bedrock whereas all of the recent centrifuge tests and *FLAC* analyses performed at UC-Berkeley modeled soil support.

It would also appear to be reasonable to ignore the effect of k_v on any aspect of the calculations, at least for the geotechnical components. This is well-supported by simple calculations.

One important caveat to these observations is that they were developed primarily for the basic geometric condition where the surfaces of both the retained soil and soil in front of the RERS (in the case of a free-standing retaining wall) are horizontal. Consideration of a sloped retained-soil surface was only a limited part of the UC-Berkeley research (Candia and Sitar 2013). As noted previously in the discussion of the S-W Method, in situations where the RERS is part of an overall slope the global stability of the slope under earthquake loading may dominate the behavior as opposed to localized conditions involving just the RERS.

There is one final issue to address, that of 'cohesion' which, as is well known, is the Mohr-Coulomb strength parameter that has several distinctly different meanings or interpretations in geotechnical engineering. Specifically, there can be:

- 'true' cohesion due to inter-particle cementation, especially for aged soil deposits;

- 'apparent' cohesion due to the matric suction that exists in all unsaturated soils although the significance of this is very dependent on soil-particle size and degree of saturation; and

- 'pseudo-cohesion' that is the theoretical artifact of a total-stress analysis.

In the context of seismic loads on yielding RERSs, cohesion is important as both theoretical and experimental work indicates that it can have an effect that varies from noticeable to profound in reducing not only the seismic component of lateral earth pressures but the gravity component as well. Furthermore, although not explicitly noted in any of the published discussions concerning the performance of actual RERSs in earthquakes, it is quite possible that apparent cohesion due to matric suction has played a role in the observed positive performance of traditional retaining walls in modest earthquakes. This point is raised in Chapter 9 as it relates to interpreting measurements made with instrumented, full-scale applications of the geotechnologies that are the focus of this monograph.

In the writer's opinion, it is this latter issue that should give rise to caution about relying too heavily on observed performance when crafting analytical methodologies. Historically, geotechnical analysis and design methodologies intended for routine use in practice have neglected matric suction when it has a beneficial impact on a problem, primarily for the reason that it is physical phenomena that can rarely be relied on to exist during the design life of a structure. An important and useful lesson in this regard can be drawn from experiences with MSEWs. In recent years, the contentious emergence of a design methodology based solely on the observed behavior of actual MSEWs and that does not satisfy basic equations of static equilibrium has raised some serious issues (Leshchinsky 2009).

A.2.3.3 At-Rest Earth Pressure State

A.2.3.3.1 Overview of Traditional Perspectives

As modern geotechnical earthquake engineering evolved from the 1960s onward, a paradigm quickly developed with regard to non-yielding RERSs where the at-rest earth pressure state is presumed to exist under both gravity and seismic loading. Based on the writer's assessment of the published literature, this paradigm appears to have been broadly accepted by geotechnical engineers, especially practitioners, to the present although recent published work discussed subsequently has seriously challenged key elements of this traditional thinking which is summarized as follows:

- The lateral earth pressures that develop under the seismic loading of non-yielding RERSs are complex in magnitude and distribution, and, ideally, should be evaluated on a site- and project-specific basis using a numerical (FEM or FDM) solution. However, this is impractical for all but very large projects which means that some type of simpler, approximate solution for use in routine practice is not only highly desirable but a pragmatic necessity.

- The M-O Method or any of its derivatives should not be used for the simple reason that an active failure wedge within the retained soil, which is the fundamental assumption of the M-O Method and all derivative methodologies, cannot develop in this case due to the lateral-displacement constraint imposed by the non-yielding RERS.

- The most widely used approximate solution that is deemed appropriate to use with non-yielding RERSs is the one developed by Wood in the early 1970s. Because Wood's solution was not only published relatively early in the evolution of geotechnical earthquake engineering but is relatively easy to use, it quickly became established as the standard for use in routine practice. Summaries of Wood's solution and its results can be found in Ebeling and Morrison (1992), Kramer (1996), Geraili Mikola and Sitar (2013), Wagner and Sitar (2016), and many other references. Wood's solution is based on a linear-elastic continuum model and is not without its limitations as noted by Ebeling/Morrison and Geraili Mikola/Sitar in particular. Nevertheless, it is the current state of practice on routine projects as it is assumed to reflect the best available compromise between theoretical accuracy and ease of use.

- A later, more-general model and solution for seismic earth pressures on RERSs was developed by Veletsos and Younan (1994). It is also based on a linear-elastic continuum model for the ground and allows for yielding that is intermediate between none and full (i.e. that which is necessary to mobilize the active state) so is discussed in some detail later in this appendix in a separate section. However, it is noted here as one limiting case of the Veletsos-Younan problem, that of the non-yielding condition, exactly duplicates the conditions of Wood's solution. The results using the Veletsos-Younan non-yielding limiting case compare favorably with Wood's solution (Veletsos and Younan 1994).

- An analytical methodology developed with non-yielding RERSs (deep basement walls of buildings in particular) in mind and much more recent than Wood's solution is the *Ostadan Method* (Ostadan 2004)[111]. While this method is presented as being design-oriented with practitioners in mind, it should be noted that it requires substantially more effort to use compared to Wood's solution. Specifically, site- and application-specific information including a shear-wave velocity profile and computer analysis of 1-D (vertical) wave transmission through the soil column overlying bedrock with concomitant estimates of seismic accelerations throughout the soil column are required to develop the depth-wise distribution of lateral earth pressures. While none of these requirements are considered extreme by today's standards, the result is that the Ostadan Method takes substantially more effort and cost to use compared to Wood's solution. Given that the Ostadan Method has appeared relatively recently and, based on the one example problem illustrated in Ostadan (2004), does not produce results far removed from Wood's solution, it is unclear to what extent the Ostadan Method has replaced or in the future might replace Wood's solution in routine practice.

---

[111] In some references, Ostadan's work is cited as Ostadan (2005).

A.2.3.3.2 Wood's Solution

As noted in the preceding section, Wood's solution is discussed to varying detail in numerous publications. However, key elements are discussed here for ease of reference.

To begin with, Wood investigated several different cases involving variations in boundary conditions and material properties of the elastic layer undergoing seismic motion. Consequently, Wood's 'solution' actually consists of a range of solutions for different boundary conditions although the results for only one or two cases are typically shown in published works that summarize Wood's research.

The lateral earth pressure distribution for Wood's solution is not a simple shape so the solution is usually expressed in terms of the lateral resultant force[112] per unit width of the RERS, F_{sr}, and its point-of-application on the inside face of the RERS. Note that this is only the seismic component of the total resultant force. For simplicity and conservatism, the limiting upper-bound "wide-backfill" geometric case of Wood's solution is typically given and used:

$$F_{sr} = k_h \cdot \gamma_{eff} \cdot H^2 \qquad \qquad \text{(A.24)}$$

where all terms are as defined previously.

Note that Wood's solution ignores vertical ground accelerations so the effective soil unit weight, γ_{eff}, used in Equation A.24 is always the actual unit weight and not γ_{eff}^* adjusted for vertical acceleration.

The point-of-application of F_{sr} is typically given as a constant equal to a distance of ~0.6H above the heel of the RERS (Geraili Mikola and Sitar 2013, Wagner and Sitar 2016). However, other references (Ebeling and Morrison 1992, Kramer 1996) clearly indicate that this distance it is not constant, even for the wide-backfill limiting case that is usually used for design. That having been said, the range of variation for this limiting case is relatively small but in the writer's judgment appears to be closer to a value of approximately 0.55H although, realistically and given all the uncertainties and approximations involved, is essentially 0.6H when rounded-off.

Although Wood's solution is primarily stated in terms of the lateral resultant force, the depth-wise lateral pressure distribution associated with the solution is shown in numerous publications (Ostadan 2004, Geraili Mikola and Sitar 2013, Wagner and Sitar 2016) so its generic shape is well known. In simple terms, the solution is a curve whose exact shape varies depending on which particular case of Wood's solution is used but, in general, has a peak at or just below the ground surface and decreases to zero or near-zero at the heel of the RERS. It is of interest to note that solutions to the aforementioned Ostadan Method produce results that are broadly similar although they always have a peak at the ground surface and are zero at the heel of the RERS.

A.2.3.3.3 Recent Developments

The increased attention paid to the seismic design of RERSs, especially relatively deep basement walls of buildings, by building codes and other regulatory documents in the

---

[112] Unfortunately, the notation for this resultant force in the published literature is inconsistent, even when the same author is involved. For example, Geraili Mikola and Sitar (2013) call it ΔP_{ae} whereas Wagner and Sitar (2016) call it ΔP_E.

early 21st century provided the impetus to critically evaluate the above-described, decades-long status quo of using Wood's solution for all non-yielding RERSs. The reason is that the seismic demand on a non-yielding RERS using Wood's solution is typically 2 to 2.5 times that using the M-O Method (Wagner and Sitar 2016). Consequently, there are significant economic consequences associated with the analytical methodology used for non-yielding RERSs.

Thus, it is not surprising that the aforementioned recent UC-Berkeley research based on centrifuge testing and numerical modeling using *FLAC* included consideration of non-yielding RERSs although it appears that this aspect of the work may still be ongoing and evolving. The specific references of greatest relevance to non-yielding RERSs are Geraili Mikola and Sitar (2013), Candia and Sitar (2013), and Wagner and Sitar (2016).

The unified outcomes of this research to date can be summarized as follows:

- Geotechnical height of the RERS plays a significant role for non-yielding RERSs, even more so than for yielding RERSs. This is because the former category includes the very common problem of basement walls of buildings (an application mentioned specifically in several of the cited references from UC-Berkeley) that can have geotechnical heights much greater than typical free-standing retaining walls.

- The effect of the greater geotechnical height of many non-yielding RERSs is that the assumption that the entire mass of retained soil:
  - ➤ moves in-phase and
  - ➤ is subjected to the same acceleration (i.e. there is no amplification) defined by the free-field surface motion

  during seismic loading becomes increasingly incorrect. As geotechnical height increases, the column of retained soil increasingly moves out-of-phase and the acceleration varies throughout the soil column due to amplification effects. However, if accelerations are averaged over the geotechnical height, Wagner found that the 13.3 metre (43.6 ft) high RERS with coarse-grain soil he simulated had broadly similar behavior to the 6.5 metre (21.3 ft) RERSs simulated by Geraili Mikola with coarse-grain soil) and Candia with fine-grain soil (Wagner and Sitar 2016), at least for PGAs up to approximately 0.45g.

- Wood's solution grossly overestimated the seismic resultant force for all of the above-described cases studied by Geraili Mikola, Candia, and Wagner (Wagner and Sitar 2016). Although the Ostadan Method was not considered explicitly, because of the overall similarity of its outcomes to those from Wood's solution, it is reasonable to conclude that the Ostadan Method is overly conservative in its outcomes as well.

- In all the above-described cases studied by Geraili Mikola, Candia, and Wagner, both the M-O and S-W methods with 100% PGA provided reasonable agreement with the measured and calculated results in terms of resultant forces (Wagner and Sitar 2016). This is rather interesting given that these methods were developed using assumptions (e.g. development of a failure wedge) that would appear to be applicable only to yielding RERSs.

- Although geotechnical height does not appear to affect the seismic resultant force, it does appear to greatly affect the distribution of the lateral earth pressure attributed to the seismic-load increment. This is likely due to the phase and amplification issues noted previously. While both Geraili Mikola and Candia observed essentially a triangular distribution of the seismic component of lateral earth pressures for the simulated 6.5 metre (21.3 ft) RERSs they studied, Wagner found a lateral earth pressure distribution that was approximately constant with depth for the simulated 13.3 metre (43.6 ft) RERS he studied (Sitar and Wagner 2015, Wagner and Sitar 2016).

The writer's assessment of these findings is that the current state of knowledge with regard to non-yielding RERSs is in a state of flux and, to some degree, confusion. The overall findings of the recent UC-Berkeley research are very much at odds with the traditional thinking for non-yielding RERSs that is based on Wood's solution and, more recently, the Ostadan Method which appears to forecast results broadly similar to Wood's solution, at least in terms of the distribution of the seismic component of lateral earth pressures. There appear to be two major uncertainties at this point in time that require research attention:

- It appears that with increasing geotechnical height the retained soil begins to exhibit depth-wise phase and acceleration variations that significantly affect the lateral earth pressure distribution of the seismic load component in particular. It is likely that this transition is gradual as opposed to abrupt but an attempt should be made to identify at least a range in geotechnical heights over which this is likely to occur.

- It is likely that the nature of the foundation support for the RERS plays some role in its behavior in terms of both the magnitude and distribution of seismic loads. This issue was raised previously for yielding RERSs and may explain some of the differences between the results to date from the UC-Berkeley research where all RERSs were modeled as being founded on soil and other analytical methodologies such as Wood's solution that assumed a rigid base that more closely approximates a RERS founded on bedrock.

Unfortunately, the current state of knowledge presents a confusing picture for the practitioner who is faced with performing designs in the present and cannot wait until some future time when a clearer picture will presumably exist.

A.2.3.4 Intermediate Yielding Conditions

A.2.3.4.1 Introduction

Historically, routine practice for RERS analysis and design under both gravity and seismic loading has been based only on consideration of the limiting restraint conditions of yielding and non-yielding as defined by the classical active and at-rest earth pressure states respectively. In reality, the earth pressure state within a retained soil mass does not instantaneously transition from the at-rest to active state. Rather, a continuous range of nameless earth pressure states intermediate between these two classical limiting states develops as lateral extension of a retained-soil mass occurs and shear strength of the retained soil is mobilized. However, these intermediate states have historically been of little or no practical interest so have received little attention. This is fortuitous as these intermediate states do not lend themselves to quantification using simple analytical models such as those employed by Coulomb (for gravity loading) and Mononobe et alia (for seismic loading) for the active state.

The recognition and use of the compressible-inclusion function with RERSs has caused increased interest in these intermediate earth pressure states because a geofoam compressible inclusion can be designed intentionally to mobilize them in a controlled, predictable fashion, something that has not been practical heretofore. In addition, there can be situations where it is desirable to analyze the pressure-reduction potential of an existing compressible inclusion subjected to different, specific loading conditions. This is especially of interest when seismic loading is involved, e.g. to see how a compressible inclusion sized to reduce gravity loads from the at-rest to active states performs under some level of seismic loading. Therefore, it is of interest to discuss analytical methodologies of potential use for analysis of geofoam compressible inclusions under intermediate yielding conditions.

A.2.3.4.2 Veletsos-Younan Method

A potentially useful theoretical approach and concomitant analytical methodology that provides solutions for a continuous range of yielding conditions starting from the at-rest state and progressing toward the active state was developed and reported by Veletsos and Younan in a series of publications that essentially spanned the decade of the 1990s. It appears that their published work broadly followed the familiar 'publish-or-perish' pattern of funded academic research wherein a relatively modest number of research reports, issued sequentially and each covering a nuanced variation of the same basic problem, were generated. Abstracts of each report were subsequently published in multiple scholarly journals and conference proceedings in order to (in principle) reach the broadest range of multiple target audiences. As is often the case, there is repetition and overlap of content between papers. For reasons that will become apparent in the following section, a synopsis of the different versions (as they will be referred to herein) of their work is presented in this section.

Figure A.5 shows the basic elements of the physical model for what will be referred to in this monograph as the *Veletsos-Younan (V-Y) Method* that is the basis for all of their work presented here[113]. Note that not all boundary conditions shown exist in all versions of the problem.

Overall, it is a 2-D model with its primary component being a viscoelastic layer of:

- finite thickness, H;

- uniform density, ρ_{eff} ($= \gamma_{eff}/g$); and

- stiffness characterized by its shear modulus, G_s. In most problem versions, G_s is depth-wise constant although in one version it increases parabolically with depth starting from zero at the surface.

This viscoelastic layer is subjected to only horizontal base motion. Both idealized harmonic motions and actual earthquake records were considered[114].

---

[113] Veletsos and Younan also solved a related 3-D axisymmetric problem involving a perfectly rigid right-circular cylinder surrounded by a viscoelastic layer (Veletsos and Younan 1995 plus other early 1990s references listed in the Supplemental Bibliography). This problem has no application relative to the contents of this monograph so is not discussed further.

[114] Veletsos and Younan also considered a baseline 'static' case that they caution in their publications is not the same as gravity loading. In their context, 'static' means motion with a circular frequency that is small relative to the natural frequency of the viscoelastic layer. Results at other frequencies are then expressed as multipliers (which can be either greater or less than 1) of the static results.

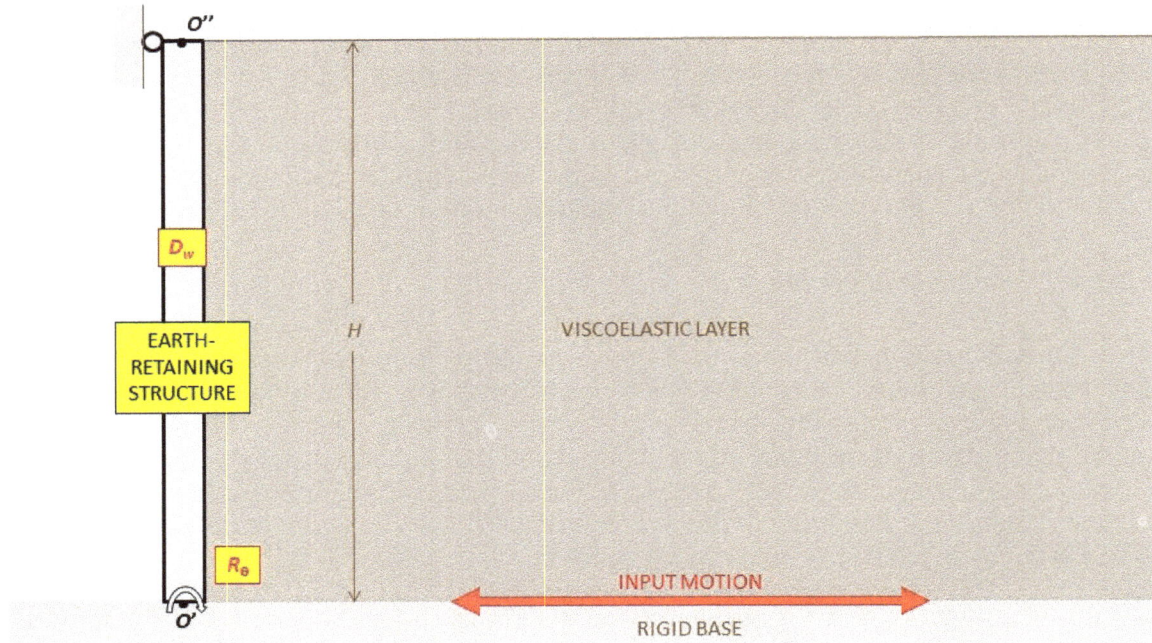

Figure A.5. Basic Elements of the Veletsos-Younan 2-D Model.

One side of the layer is constrained by a generic, <u>massless</u> ERS of uniform thickness, t_w, that is characterized by its flexural stiffness, D_w, that can vary in magnitude from perfectly rigid ($D_w = \infty$) to finite ($0 < D_w < \infty$) depending on the problem version. The boundary conditions on this ERS are defined by its:

- bottom (Point O') fixity against translation (sliding) in all problem versions;

- bottom rotational stiffness, R_θ, that varies in magnitude between perfectly fixed ($R_\theta = \infty$) and finite ($0 < R_\theta < \infty$) depending on the problem version; and

- top (Point O'') unrestrained against either rotation or lateral displacement except for full fixity against lateral displacement only (as depicted by the roller support in Figure A.5) in one problem version only.

Despite the fact that the ERS has a finite flexural stiffness in some problem versions, within the definitions adopted in this monograph the model used in the V-Y Method is considered to apply to RERSs only. This is because the bottom (Point O') of the modeled ERS is always restrained against lateral displacement so does not qualify as a FERS as defined for the purposes of this monograph. Furthermore, as explained by Veletsos and Younan, allowing for $D_w < \infty$ was done simply to approximate the relatively limited flexure of the stem of a traditional cantilever retaining wall (considered a RERS in the context of this monograph), not anything like a true FERS such as a sheet-pile wall.

Table A.1 summarizes the versions of the V-Y Method that the writer has been able to identify from the published literature. Note that the "Version" identification is not something used by Veletsos and Younan in any of their publications reviewed by the writer. This is something created by the writer to better organize the different problem versions solved by Veletsos and Younan. As will be seen in the following section, Versions IIa/b and IIIa (all highlighted in yellow in Table A.1) have potential application with the compressible-inclusion function of geofoams so specific solution details relative to these versions are discussed here.

Table A.1. Versions of the Veletsos-Younan Method.

Version	D_w	R_θ	top	G_s	Reference
I	∞	∞	n/a	constant	Veletsos & Younan (1995)
IIa	∞	varies	free	constant	Veletsos & Younan (1993, 1994)
IIb	∞	varies	free	variable	
IIIa	varies	∞	free	constant	Younan et al. (1997)
IIIb	varies	∞	fixed	constant	

Considering first Version II (which has two sub-versions that differ only in the assumed depth-wise variation of G_s), this models a perfectly rigid RERS that is free to yield in the rotation-about-bottom mode only. As such, this version can be interpreted as approximating the behavior of a classical gravity retaining wall.

The relative stiffness between the lateral deformation of the viscoelastic layer due to horizontal shearing under seismic shaking and the rotational restraint provided at the bottom of the RERS is expressed using a dimensionless parameter, d_θ, that is defined as follows:

$$d_\theta = \frac{G_s H^2}{R_\theta}.$$

(A.25)

Note that for the sub-version of G_s increasing parabolically with depth (Version IIb), the value of G_s used in this equation is the average value which Veletsos and Younan defined to be two-thirds of the maximum value of G_s which is the value at the bottom of the viscoelastic layer.

Note that one limiting case of sub-version IIa is nominally the same as Version I, that of a non-yielding RERS which corresponds to $R_\theta = \infty$ and $d_\theta = 0$. Conceptually, this case is the same as Wood's solution discussed previously and, as noted previously, the Veletsos-Younan results are in excellent agreement with those of Wood.

Results for the other limiting case for both Versions IIa and IIb of $R_\theta = 0$ and $d_\theta = \infty$, which corresponds to a RERS offering no rotational resistance, are not well-defined. As with any solution crafted using the theory of elasticity, material yield ('failure') never occurs so there is no inherent lower-bound value as would be the case with a theory of plasticity solution such as the M-O Method discussed previously. Veletsos and Younan offered no explicit comments concerning what the results from their method might be in the limit as d_θ trends toward infinity. The plotted trends of resultant forces and moments show exponentially decreasing magnitudes of both as the rotational restraint decreases and logic suggests that they would both decrease to zero. They did provide results for different finite values of d_θ up to and including $d_\theta = 5$ so the effect of d_θ on calculated results can be seen for the range considered.

As a final comment on Version II, Veletsos and Younan offered no explicit comment as to how R_θ might be evaluated in a specific application in reality. In principle, R_θ

268

represents the rotational stiffness of the foundation soil or rock underlying the gravity retaining wall.

Considering next Version IIIa, Veletsos and Younan went through a similar process of defining a dimensionless parameter, d_w, for the relative stiffness between the viscoelastic layer and the flexural stiffness of the RERS:

$$d_w = \frac{G_s H^3}{D_w}.$$

(A.26)

As with Version IIa, one limiting case of Version IIIa is nominally the same as Version I, that of a non-yielding RERS which, in this case, corresponds to $D_w = \infty$ and $d_w = 0$. Again, this case is conceptually the same as Wood's solution discussed previously although is does not appear that Veletsos and Younan made an explicit comparison for Version IIIa as they did for Version IIa, at least in the published work reviewed by the writer.

Also, as with Versions IIa/b, the other limiting case of $D_w = 0$ and $d_w = \infty$ is ill-defined. Results were provided for finite values of d_w up to and including $d_w = 40$.

One benefit of Version IIIa compared to Versions IIa/b is that the plotted outcomes presented by Veletsos and Younan in their various publications are easier to quantify for use in an actual application as it is straightforward to define D_w and thus calculate d_w using Equation A.26. Veletsos and Younan used the classical equation for flexure of an elastic plate using 'simple' theory that neglects shear effects:

$$D_w = \frac{E_w \cdot t_w^3}{12 \cdot (1 - v_w^2)}$$

(A.27)

where E_w and v_w are the elastic parameters (Young's modulus and Poisson's ratio respectively of the RERS (wall).

Several examples of calculating d_w for hypothetical-but-realistic RERSs are given in Younan et al. (1997) using values of G_s calculated using its theoretical relationship with shear-wave velocity, V_s:

$$G_s = \rho_{eff} \cdot V_s^2.$$

(A.28)

In summary up to this point and to set the stage for the following section, the primary distinction between Versions IIa/b and III for the purposes of this monograph is the lateral-displacement pattern of the interface between the inside face of the RERS and the viscoelastic layer (= retained soil). Versions IIa/b produce a linear pattern due to assumed rigid-body rotation of the RERS whereas Version IIIa produces a curved pattern similar to that of a classical cantilever beam or column.

As a final comment, for the sake of completeness it is worth noting that the published works of Veletsos and Younan continue to receive attention to the present in research and concomitant publication by others. The results from Version IIIa, which broadly match one of the RERSs simulated in the UC-Berkeley centrifuge testing and *FLAC* analyses, were discussed by Geraili Mikola and Sitar (2013) and Sitar and Wagner (2015). In addition, a number of researchers have compared results from the V-Y Method to those obtained using FEM and FDM software that can perform analyses using non-linear constitutive models in order to define a range of validity of the V-Y Method which uses a linear-elastic constitutive model. One example is Francesco et al. (2010) that studied results using both Versions II and III of the V-Y Method.

A.2.3.4.3 Extended Veletsos-Younan Method: Introduction and Overview

A noteworthy and intriguing aspect of the V-Y Method, and the one of primary relevance to this manuscript, is that it can be extended for use with problems involving geofoam compressible inclusions, both the REP- and ZEP-Wall variants, used in conjunction with a non-yielding RERS such as the basement wall of a building or conventional bridge abutment. This potential does not appear to have been noted and explored to date by anyone other than the writer.

This conceptual extension is based on noting an interesting identity between how the viscoelastic layer in the V-Y Method is allowed to deform in both the original method and its extended application as proposed by the writer. In simple terms, in the V-Y Method, the RERS is allowed to yield either in rotation (Version II) or flexure (Version III) and the viscoelastic layer deforms in such a way that matches the assumed pattern of RERS yielding. In the writer's extension, although the RERS is non-yielding, it is the compressible inclusion between the RERS and viscoelastic layer that is assumed to yield and allow the interface between the compressible inclusion and viscoelastic layer to deform in a pattern that matches either rotation or flexure.

The writer first noticed this behavioral commonality in the mid-1990s, shortly after the publication of Version II of the V-Y Method in Younan and Veletsos (1994). This quickly led to development of the original version of what is referred to in this monograph as the *Extended Veletsos-Younan Method* that was published initially in Horvath (1997b, but not actually printed and distributed until 1998); republished in its original form in Horvath (1998b) in order to reach a broader audience; and republished in a slightly modified/simplified form in Horvath (2008b). This last publication also included some comparisons between the Extended V-Y Method and published work by others in order to illustrate the concept using some numerical examples from both laboratory testing and a case history application.

For ease of reference, details of the original version of the Extended V-Y Method are presented in the following section and summarizes all work published previously in Horvath (1997b, 1998b, 2008b).

As a result of research undertaken in preparation of this monograph, the writer became aware of Version III of the V-Y Method. This offered the potential to develop an alternative version of the Extended V-Y Method. An overview of how this might be done is presented in a subsequent section for the first time in publication. As will be seen, the only difference between the writer's original and alternative versions of the Extended V-Y Method is in the assumed shape of the deformation of the compressible inclusion-viscoelastic layer interface.

A.2.3.4.4 Extended Veletsos-Younan Method: Original Version

As noted in the preceding section, the writer's original version of the Extended V-Y Method is based on Version II of the V-Y Method. In this version, the RERS is yielding in nature but only in the classical rotation-about-bottom mode (Point O' in Figure A.5), with the parameter R_θ defining the rotational stiffness provided by the bottom of the RERS. In turn, this rotation allows the viscoelastic layer to deform in a mode of lateral shearing. This deformation mechanism, which is defined by the rotation of the planar interface[115] between the inside face of the RERS and viscoelastic layer, is shown in Figure A.6a. Note in this figure that the various problem components are not shown to scale but are distorted to emphasize key components and behavior.

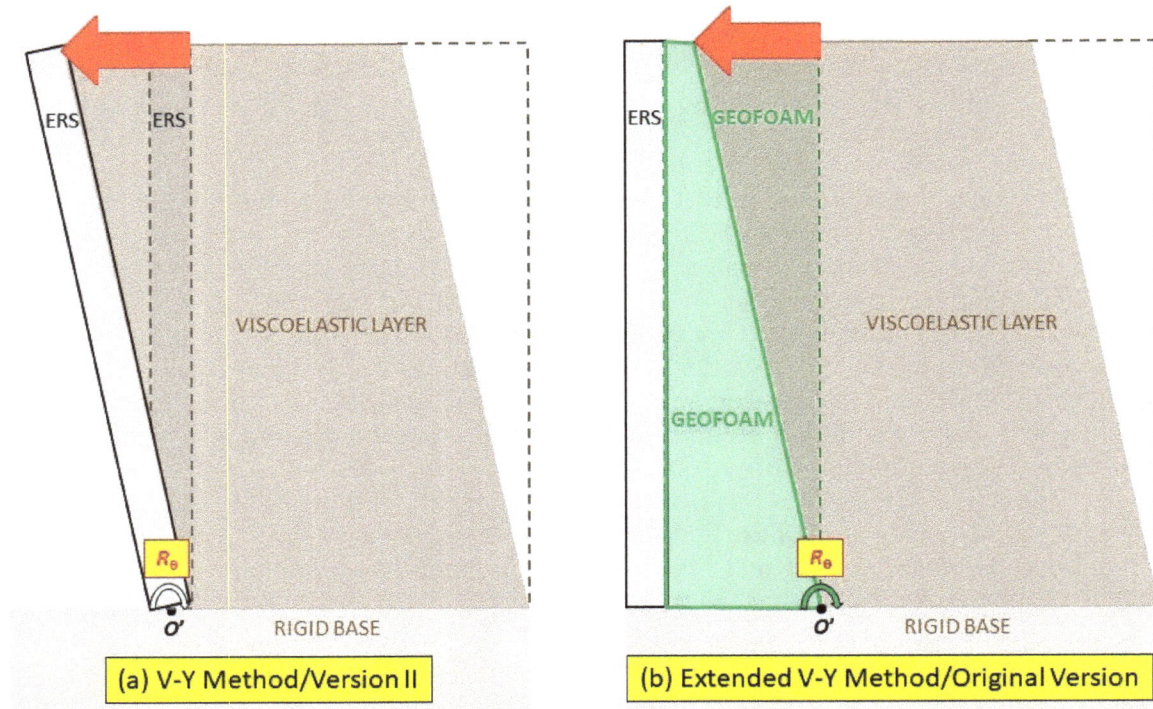

Figure A.6. Behavioral Identity between Veletsos-Younan Method (Version II) and Extended Veletsos-Younan Method (Original Version).

The original version of the Extended V-Y Method developed by the writer in the mid-1990s was intended for use with a non-yielding RERS that, by definition, cannot rotate about its bottom. If, however, a geofoam compressible inclusion is placed between the inside face of the RERS and the viscoelastic layer, either alone in a REP-Wall application (Figure 3.7) or together with geosynthetic tensile reinforcement in a ZEP-Wall application (Figure 3.9), then concomitant deformation of the viscoelastic layer can occur even in the absence of RERS yielding. In essence, the intentionally compressible compressible inclusion provides the yielding potential that the RERS itself is incapable of.

In developing the Extended V-Y Method, it was assumed that the interface between the compressible inclusion and laterally deforming viscoelastic layer would:

- remain planar and

- deform in the mode of rotation-about-bottom

---

[115] A subtle point that is relevant for the discussion that follows is that in the V-Y Method the planarity of this interface during rotation is enforced by the assumed infinite rigidity of the RERS.

as depicted in Figure A.6b for the basic REP-Wall case.

Note that the interface planarity in this case, which is <u>assumed</u>, is subtly different than for the V-Y Method where it is <u>known</u> due to the infinite rigidity of the RERS. Also, it is important to note that the rectangular panel of geofoam is <u>not</u> assumed to rotate as a rigid body as the RERS does in Figure A.6a. Rather, the panel of geofoam remains fixed against rigid-body rotation but is assumed to compress in a pattern that is identical to rigid-body rotation as shown in Figure A.6b.

In summary then, the key assumption made by the writer in developing the original version of the Extended V-Y Method was that the deformation of the viscoelastic layer will be the same whether it is due to a yielding RERS or deformation of a compressible inclusion placed between the viscoelastic layer and a non-yielding RERS. Thus, the conceptual 'heart' of the Extended V-Y Method concept is replicating the deformation pattern, not the actual physical mechanism that produces the deformation pattern.

As already noted, the complete derivation of the original version of the Extended V-Y Method was initially published more than 20 years ago. However, it is repeated here for the sake of completeness as well as to set the stage for the new alternative version in the following section. Note that there are some minor notational changes made in the following derivation compared to what was used in the original publications (Horvath 1997b, 1998b, 2008b) to take advantage of new insights gained in the years since the original derivation was developed.

Figure A.7 defines the key parameters of the original version of the Extended V-Y Method. It is again emphasized that the problem components are shown in distorted scale relative to each other for clarity. In particular, the initial thickness of the compressible inclusion, t_{ci}, is misleading as in practice it would typically be of the order of 150 to 300 millimetres (6 to 12 in) thick, even for a deep basement wall that might be 15 metres (50 ft) high.

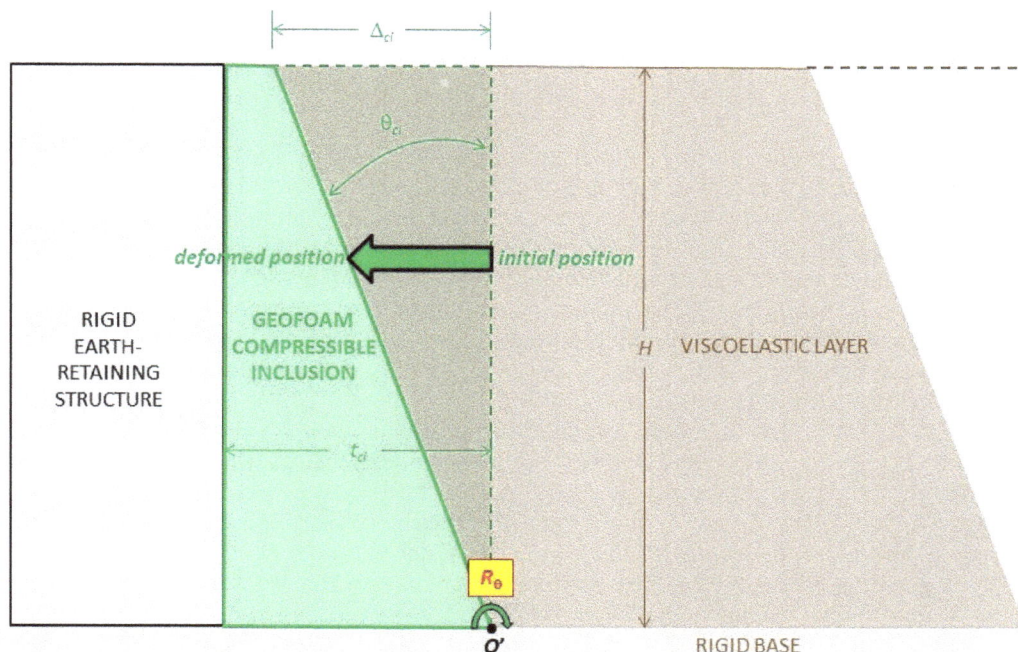

Figure A.7. Key Parameters of the Extended Veletsos-Younan Method (Original Version).

The primary challenge in developing the original version of the Extended V-Y Method was how to express the assumed rotational restraint provided by the compressible inclusion, R_θ, in terms of definable and measurable properties of the compressible-inclusion product. To begin with, the maximum lateral displacement at the surface of the viscoelastic layer due to compression of the compressible inclusion, Δ_{ci}, is given by:

$$\Delta_{ci} = H \cdot \tan\theta_{ci} \tag{A.29}$$

where θ_{ci} is the angular rotation relative to an initial vertical position of the planar interface between the viscoelastic layer and compressible inclusion. Note that Δ_{ci} also represents the maximum compression of the compressible inclusion.

For a unit rotation, i.e. $\theta_{ci} = 1$ radian:

$$\Delta_{ci} = 1.56 \cdot H . \tag{A.30}$$

Assuming linear stress-strain behavior in compression for the geofoam product[116] comprising the compressible inclusion[117]:

$$\sigma_{ci} = E_{ci} \cdot \varepsilon_{ci} \tag{A.31}$$

where σ_{ci} is the horizontal normal stress within the compressible inclusion (assumed constant at a given depth); E_{ci} is the Young's modulus of the compressible inclusion material[118]; and ε_{ci} is the horizontal normal strain within the compressible inclusion (assumed uniformly distributed within the compressible inclusion at a given depth).

But:

$$\varepsilon_{ci} = \frac{\Delta_{ci}}{t_{ci}} \tag{A.32}$$

---

[116] Note that the modulus of a geofoam <u>product</u> is not necessarily the same as the modulus of the geofoam <u>material</u> that comprises the product. To begin with, a product may consist of two or more different materials, usually placed in series in the direction of intended primary compression, each with their own modulus (this is actually the rule rather than the exception with the commercial product that is most efficient for most compressible-inclusion applications). In this case, the product modulus is the relevant composite modulus of two springs in series. Alternatively, the product might consist of one material but that material might have holes through it or be otherwise discontinuous in which case the operative modulus of the overall product is something less than that of the solid material between the holes or discontinuities.

[117] Note that for the simple one-dimensional compression analysis implied here that the constrained modulus, D_{ci}, is arguably more appropriate and correct to use. However, the Poisson ratio for geofoam materials typically used for compressible inclusions is very small and can be taken to be zero in which case it can be shown theoretically that the constrained modulus and Young's modulus are equal. Therefore, Young's modulus is used in the derivation.

[118] For a geofoam product with linear stress-strain behavior, this is the <u>tangent</u> Young's modulus and for a product with non-linear behavior this is the <u>secant</u> Young's modulus, both evaluated at the operative stress level of the particular application, i.e. the stress level to which the compressible inclusion is subjected. Furthermore, because creep is always an important consideration with geofoam materials (which are almost always polymeric), a Young's modulus consistent with the rapid-loading nature of seismic loading should be used.

where t_{ci} is the thickness of the compressible inclusion (assumed uniform with depth in a given application).

Rewriting Equation A.31 using the relationships in equations A.30 and A.32 to replace some of the variables yields:

$$\sigma_{ci} = \frac{1.56 \cdot H \cdot E_{ci}}{t_{ci}} \tag{A.33}$$

which is the same as Equation 12 in Horvath (1997b, 1998b).

At this point, the original 1990s derivation is modified to take advantage of new variables defined and used since the original derivation was made. To begin with, using Equation 5.7 for the dimensionless <u>absolute</u> compressible-inclusion stiffness, K_{ci}:

$$K_{ci} = \frac{E_{ci}}{t_{ci} \cdot \gamma_w}, \tag{A.34}$$

Equation A.33 can be rewritten as:

$$\sigma_{ci} = 1.56 \cdot H \cdot K_{ci} \cdot \gamma_w. \tag{A.35}$$

Next, using the final form of Equation 5.9 for the <u>relative</u> compressible-inclusion stiffness, λ_{ci}:

$$\lambda_{ci} = \frac{K_{ci} \cdot H \cdot \gamma_w}{p_{atm}}, \tag{A.36}$$

Equation A.35 can be rewritten as:

$$\sigma_{ci} = 1.56 \cdot \lambda_{ci} \cdot p_{atm}. \tag{A.37}$$

For the sake of completeness, it is useful to again highlight the difference between the absolute and relative compressible-inclusion stiffnesses, K_{ci} and λ_{ci}. While both are dimensionless parameters and thus universal across different systems of units, they fulfill different roles in practice.

The <u>absolute</u> compressible-inclusion stiffness, K_{ci}, is the preferred way to objectively compare the primary technical characteristic of compressible-inclusion products as the smaller the value of K_{ci} the more compressible the product is in an absolute sense. On the other hand, the <u>relative</u> compressible-inclusion stiffness, λ_{ci}, is the preferred way to objectively compare the performance of different actual or potential compressible inclusions in the same application as the geotechnical height, H, of the application is reflected in the value of λ_{ci}. This is important for application-specific assessments as a compressible inclusion with some specific value of K_{ci} will tend to perform differently for different values of H. This is because the mobilization of various earth pressure states is generally recognized as being a function of H.

Returning now to the derivation process, the lateral resultant force (actually a force per unit width of the RERS due to the 2-D/plane-strain nature of the problem formulation) acting on the compressible inclusion, F_{ci}, is simply the average horizontal normal stress applied to the compressible inclusion times the thickness, H, of the viscoelastic layer in contact with it. Because the rotation of the viscoelastic layer into the compressible inclusion is assumed to be linear as shown in Figure A.7 and the compressible inclusion is assumed to behave in a linear-elastic fashion, the normal stresses in the compressible inclusion will be maximum at the top of the viscoelastic layer where the lateral displacement is maximum and given by Equation A.37. The stresses will be minimum and equal to zero at the bottom of the compressible inclusion which is the point of rotation with no lateral displacement. The displacements and, therefore, the stresses vary linearly in between.

Mathematically, then, F_{ci} can be expressed as:

$$F_{ci} = \left(\frac{\sigma_{ci} + 0}{2} \right) \cdot H \,. \tag{A.38}$$

Substituting Equation A.37 in Equation A.38 yields:

$$F_{ci} = 0.78 \cdot \lambda_{ci} \cdot p_{atm} \cdot H \,. \tag{A.39}$$

The moment per unit width of the RERS, R_{ci}, provided by the compressible inclusion in terms of resisting the rotational deformation of the viscoelastic mass is the resultant lateral force in the compressible inclusion, F_{ci}, times its moment arm about the base of the RERS (Point O' in Figure A.7). Mathematically, this resisting moment is expressed as:

$$R_{ci} = F_{ci} \cdot \left(\frac{2 \cdot H}{3} \right) = 0.52 \cdot \lambda_{ci} \cdot p_{atm} \cdot H^2 \,. \tag{A.40}$$

Because this moment is for one radian of rotation, it is the same as the unit stiffness (i.e. stiffness per unit width of the RERS per radian of rotation) of a fictitious rotational spring that represents the rotational stiffness provided by the compressible inclusion against lateral deformation of the viscoelastic layer. As shown in Figure A.6, this rotational stiffness provided by the compressible inclusion provides the same resistance as the rotational stiffness R_θ assumed in the 2-D model (Figure A.5) that was used as the basis for the original V-Y Method. Therefore:

$$R_\theta = 0.52 \cdot \lambda_{ci} \cdot p_{atm} \cdot H^2 \tag{A.41}$$

which is the desired end result and equivalent to Equation 16 in the original derivation (Horvath 1997b, 1998b) except for the variable substitutions and changes noted previously.

An alternative is to express the final result in terms of the absolute compressible inclusion stiffness by substituting Equation A.36 in Equation A.41 with this outcome:

$$R_\theta = 0.52 \cdot K_{ci} \cdot \gamma_w \cdot H^3 \,. \tag{A.42}$$

Equation A.42 is useful when comparing commercially available compressible-inclusion products for a project-specific application as K_{ci} is a product property that geofoam suppliers can and should provide for their product(s).

To complete this derivation, Veletsos and Younan used the variable d_θ that was defined in Equation A.25 to express the relative stiffness between the deforming viscoelastic layer and the rotational restraint provided by the yielding RERS. For the

Extended V-Y Method, this variable represents the stiffness of the viscoelastic layer relative to the rotational stiffness provided by the compressible inclusion.

Substituting Equation A.41 into Equation A.25 yields:

$$d_\theta = \frac{1.92 \cdot G_s}{\lambda_{ci} \cdot p_{atm}} \qquad (A.43)$$

which will be simplified here[119] to

$$d_\theta = \frac{2 \cdot G_s}{\lambda_{ci} \cdot p_{atm}} \qquad (A.44)$$

with insignificant loss in accuracy given the approximations of estimating the elastic properties of soil in any actual application. How this parameter can be used in analysis and design is discussed in Chapter 5 and was illustrated with some examples in Horvath (2008b).

As an aside, in ZEP-Wall applications where geosynthetic tensile reinforcement is placed within the retained soil (i.e. viscoelastic layer) in addition to using a geofoam compressible inclusion, the presence of the reinforcement is reflected in the value of G_s used in the preceding and subsequent equations. Specifically, G_s would reflect not only the soil properties but the increase due to the reinforcement stiffness.

In any event, if the d_θ result (Equation A.44) is desired in terms of the absolute compressible inclusion stiffness, then Equation A.42 is substituted into Equation A.25 with this result:

$$d_\theta = \frac{1.92 \cdot G_s}{K_{ci} \cdot \gamma_w \cdot H} \cong \frac{2 \cdot G_s}{K_{ci} \cdot \gamma_w \cdot H}. \qquad (A.45)$$

As in the original derivation, it is of some interest to explore the physical interpretation of the d_θ parameter further. To begin with, because we are dealing with a linear-viscoelastic mass, G_s can be eliminated by using the theoretical relationship:

$$G_s = \frac{E_s}{2 \cdot (1 + \nu_s)} \qquad (A.46)$$

where ν_s is Poisson's ratio of the viscoelastic layer.

Equation A.43[120] can then be rewritten as:

$$d_\theta = \frac{0.96 \cdot E_s}{(1 + \nu_s) \cdot \lambda_{ci} \cdot p_{atm}}. \qquad (A.47)$$

[119] The original derivation in Horvath (1997b, 1998b) ended with Equation A.43 which is equivalent to Equation 18 in the original derivation, with some variable changes.
[120] The original 'exact' form of this equation is used here in lieu of Equation A.44 to avoid accumulating round-off.

It is advantageous for the purposes of this derivation to replace λ_{ci} in Equation A.47 with K_{ci} using Equation A.36 which results in:

$$d_\theta = \frac{0.96 \cdot E_s}{\left(1 + \nu_s\right) \cdot K_{ci} \cdot H \cdot \gamma_w}. \qquad \text{(A.48)}$$

This equation can be decomposed into two parts to facilitate interpretation as follows:

$$d_\theta = \left[\frac{0.96 \cdot E_s}{\left(1 + \nu_s\right) \cdot H}\right] \cdot \left[\frac{1}{K_{ci} \cdot \gamma_w}\right] \qquad \text{(A.49)}$$

which, assuming $\nu_s = 0.15$ which is felt to be reasonable for a granular soil, yields

$$d_\theta = \left[\frac{E_s}{1.2 \cdot H}\right] \cdot \left[\frac{1}{K_{ci} \cdot \gamma_w}\right]. \qquad \text{(A.50)}$$

The first term in brackets in Equation A.50 represents the stiffness of the viscoelastic layer in the horizontal direction. As such, it can be interpreted as the Winkler Coefficient of Subgrade Reaction (= E_s/H), i.e. a horizontally oriented axial spring stiffness per unit area, using subgrade modeling concepts based on the simplified elastic-continuum concept investigated in detail by the writer (Horvath 1979, 1983, 1984, 1988, 1989, 2002). Using this concept, it also implies that this term reflects the horizontal stiffness of the material extending a horizontal distance equal to $1.2 \cdot H$[121] back from the RERS that contributes primarily to the stiffness of the viscoelastic layer affecting the problem. On the other hand, the second bracketed term in Equation A.50 is clearly the reciprocal of the horizontally oriented spring stiffness per unit area of the compressible inclusion.

It is clear, then, that the parameter d_θ can be visualized and interpreted as essentially the ratio of the 'soil-spring' stiffness to 'compressible inclusion-spring' stiffness in a horizontal direction. This is conceptually identical and thus consistent with the simplistic models used to analyze compressible-inclusion applications as explored in Chapter 5, i.e. that analysis and design of compressible inclusions can be visualized and modeled as an exercise in finding the unique displacement at which the equivalent spring stiffnesses of the ground and compressible inclusion are matched. As such, this represents a classic statically indeterminate problem in which solution requires not only satisfying static equilibrium of forces but displacements as well based on stiffnesses of problem components.

A.2.3.4.5 Extended Veletsos-Younan Method: Alternative Version

The same basic logic and approach that the writer used more than 20 years ago to create the original version of the Extended V-Y Method based on Version II of the original V-Y Method can, in principle, be used to create an alternative version of the Extended V-Y Method using Version IIIa of the V-Y Method. Once again, the intended application of an

---

[121] The original derivation presented in Horvath (1997b, 1998b) used $1.3 \cdot H$ based on an estimation of Poisson ratio that was felt to be appropriate based on the state of knowledge at that time.

alternative version of the Extended V-Y Method is with non-yielding RERSs with a geofoam compressible inclusion used either alone (REP-Wall) or synergistically with geosynthetic tensile reinforcement (ZEP-Wall).

The fundamental difference between the original and alternative versions of the Extended V-Y Method is the assumed physical mechanism for resistance to lateral deformation of the viscoelastic layer. This, in turn, affects the associated deformation pattern of this interface between the compressible inclusion and the viscoelastic layer that represents the retained soil (and, optionally, embedded reinforcements).

In the original version of the Extended V-Y Method, the resistance was assumed to be pseudo-rotational in nature as defined by the variable R_θ with a concomitant linear interface-deformation pattern. With the alternative version, the resistance is flexural in nature as the compressible inclusion is assumed to have the behavioral characteristics of a cantilever column with a perfectly fixed base and flexural stiffness defined by the variable D_w as shown in Figure A.8. This results in a more-complex (quintic function or fifth-order polynomial) curved deformation pattern of the compressible inclusion-viscoelastic layer interface.

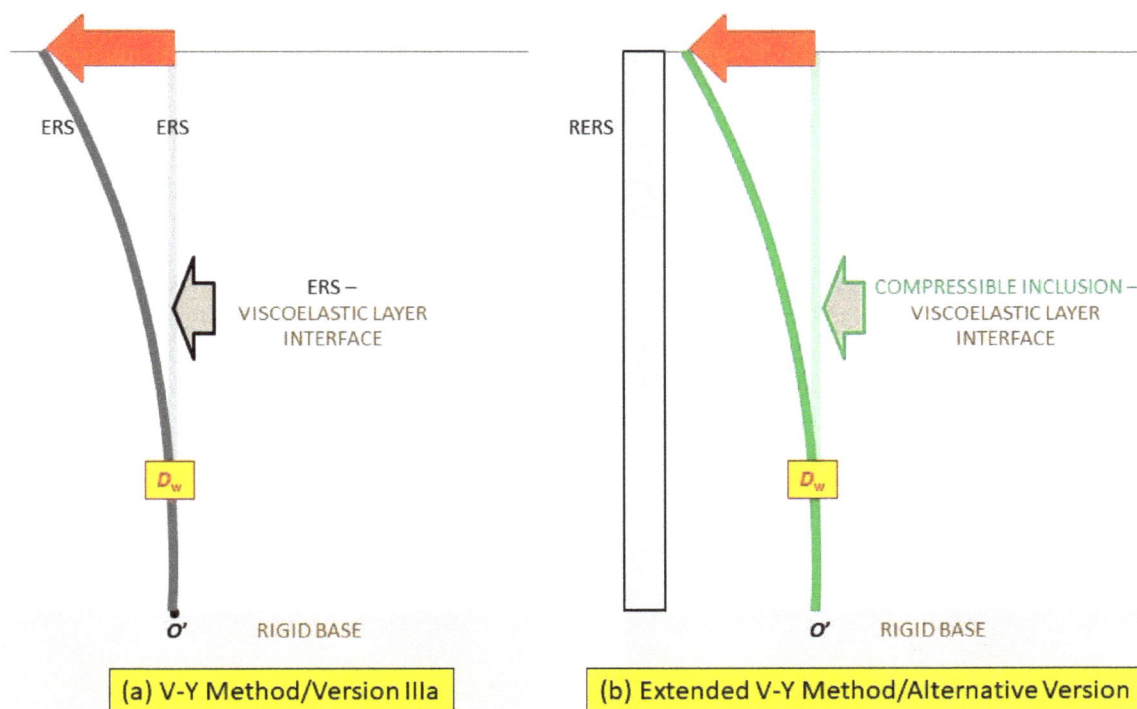

Figure A.8. Behavioral Identity between Veletsos-Younan Method (Version IIIa) and Extended Veletsos-Younan Method (Alternative Version).

Note, however, that the compressible inclusion itself is not assumed to actually bend as a cantilever column. It is simply assumed to have a compression pattern that mimics the bending of a cantilever column. This is conceptually identical to the original version of the Extended V-Y Method where the compressible inclusion was not assumed to actually rotate but simply to develop a compression pattern that mimicked rotation.

It follows that the primary objective of any derivation process for an alternative version of the Extended V-Y Method based on Version IIIa of the V-Y Method would be to develop an algebraic relationship for the equivalent flexural stiffness, D_w, of the compressible inclusion (Equation A.27). Note that while it would be logical to use the actual Young's modulus of the compressible inclusion in this equation, a fictitious thickness, not the actual thickness, of the compressible inclusion would need to be developed algebraically. Again, this is because the compressible inclusion is not actually bending, just mimicking bending.

Consequently, it appears that the primary challenge of developing this alternative version of the Extended V-Y Method would be in developing a rational way to evaluate this fictitious column thickness used to define the equivalent flexural stiffness, D_w, of the fictitious cantilever column that models the behavior of the compressible inclusion. This equivalent flexural stiffness would then be used in Equation A.26 to find the relative stiffness, d_w, between the viscoelastic layer and compressible inclusion as Veletsos and Younan expressed results for Version III of their method in terms of d_w.

It is worth noting that the potential benefit of developing this alternative version of the Extended V-Y Method is that the results presented in Younan et al. (1997) for Version IIIa of the V-Y Method on which this alternative version is based indicate that as the relative stiffness, d_w, increases in magnitude from zero (which is the baseline, limiting, perfectly rigid case), i.e. as the cantilever column becomes more flexible (which means the compressible inclusion becomes more compressible) relative to the viscoelastic layer, the distribution of lateral pressures tends toward triangular in shape and the point of application of the lateral resultant force becomes lower along the inside face of the RERS. This means that the results become closer to those reflected in the M-O Method for a yielding RERS as well as those found in recent research at UC-Berkeley, both as discussed earlier in this appendix.

On the other hand, for the original version of the Extended V-Y Method, the results presented in Veletsos and Younan (1993, 1994) for Version II of the V-Y Method on which this original version is based indicate that as the relative stiffness, d_θ, increases in magnitude from zero (which is again the baseline, limiting, perfectly rigid case), i.e. as the RERS becomes less resistant to rotation (which means the compressible inclusion becomes more compressible relative to the viscoelastic layer), the distribution of lateral pressures tends toward parabolic for Version IIa (G_s uniform with depth) but triangular for Version IIb (G_s increasing with depth). Veletsos and Younan did not show any results of how the depthwise point of application of the lateral resultant force varies in either case.

In any event, further development of an alternative version of the writer's Extended V-Y Method based on Version IIIa of the original V-Y Method is not pursued here.

A.2.3.4.6 Closing Comments

Regardless of whether the V-Y Method in any of its versions is used as originally intended for a yielding RERS without a compressible inclusion or as extended by the writer for use with a non-yielding RERS with a compressible inclusion, it is important to note that there are some practical drawbacks at present to its adoption and use in routine practice. The reason for this pragmatic limitation is based on the fact that, as outlined in Veletsos and Younan (1994) for Version II but applicable for the later Version III as well, calculated results, especially the magnitude and distribution of lateral stresses which are of greatest interest in practical applications, are all heavily dependent on several key variables:

- the mass of the RERS;

- the values of d_θ and d_w;

- the natural frequency of the viscoelastic layer;

- the exact nature (magnitude and frequency) of the base motion to which the viscoelastic layer is subjected which, in theory, will vary from one seismic event to another; and

- the number of modes included in the calculations which involve numerical evaluation of infinite series.

The fact that results are frequency-dependent is of particular interest as the M-O Method discussed previously is frequency-independent.

Consequently, and unfortunately, it is thus not conducive to developing a single, generic lateral pressure diagram for the V-Y Method or, by extension, the Extended V-Y Method as was done previously for the M-O Method. Thus, it appears that solution of either the V-Y Method or Extended V-Y Method needs to be evaluated for each specific application including a specific seismic event or range of events.

While this is not a serious impediment in and of itself, it is an impediment from a pragmatic perspective in terms of use in routine practice as there is no known commercially available software at present for producing results using the V-Y Method and, by extension, the Extended V-Y Method. Not only would such software be highly desirable for project-specific use but it is possible that a research study consisting of a systematic series of parametric analyses may be able to produce a simplified methodology for routine analysis and design purposes. Such a simplified methodology might be of the nature of an earth pressure diagram with a simple shape whose exact shape and magnitudes varied as a function of application-specific input parameters.

A.3 EARTH LOADS (EXPANSIVE SOILS)

A.3.1 Introduction

The subject of 'expansive' soils has received greatly increased attention in recent decades. In the early years of modern soil mechanics, such soils were typically viewed as being regionally localized and relatively anomalous compared to the 'normal' soils that predominated in the relatively wet, temperate climates of Western Europe and North America where modern soil mechanics had its roots and early development. As a result, in technical conferences and the like expansive soils were often referred to as 'special' or 'problem' soils.

It was not until the latter decades of the 20th century, largely through the long-term research and persistent publication of Professor D. G. Fredlund and his colleagues and co-workers that culminated in the publication of a seminal textbook on the subject (Fredlund and Rahardjo 1993), that expansive soils were shown to be a subset of the broader topic of unsaturated soil mechanics. While unsaturated soils are universal, with good approximation in most cases their unique behavioral characteristics in terms of negative pore pressures and matric suction were historically ignored for the most part in routine practice. However, the ascent of geoenvironmental engineering, especially with regard to landfills where unsaturated conditions are the norm within various landfill components, as well as extensive human development in geographic regions where expansive soils predominate during the same late-20th-century timeframe appears to have contributed to the increased awareness of and mainstreaming of unsaturated soil mechanics.

A.3.2 Basic Concepts

Books (e.g. Fredlund and Rahardjo 1993) have literally been written on the subject of unsaturated soils and their behavior so this monograph will not even begin to attempt a comprehensive treatment of the subject. Rather, the focus here is issues solely related to lateral earth pressures.

Traditional soil mechanics was, essentially, formulated originally to be theoretically rigorous for two limiting conditions within a soil mass:

- completely dry and

- completely saturated below the groundwater table with positive water pressures in the voids between soil particles.

While completely dry soils are relatively uncommon in nature, such conditions have traditionally been assumed for soils within the *vadose zone* which is defined as being the depth-wise zone of soil between the ground surface and the groundwater table (defined as the depth at which the water pressure within the soil voids is zero). In many applications to this day, assuming that the soils within the vadose zone are behaviorally dry (i.e. zero porewater pressure) is a satisfactory assumption in many routine geotechnical and foundation applications. This greatly simplifies analyses.

As an aside, note that the vadose zone itself consists of two sub-zones. Soils within the upper sub-zone that begins at the ground surface are partially saturated while soils within the lower sub-zone that extends to the groundwater table are at or close to 100% saturation due to *capillary rise* within the voids of the soil. Consequently, this lower sub-zone is referred to as the *capillary fringe*.

Because the vadose zone is saturated to a varying extent, this partial saturation results in two distinct characteristics:

- What porewater there is within the soil voids is able to sustain tension that will affect the effective stresses within the soil-particle skeleton. The magnitude of this effect is dependent primarily on the size (diameter) of the soil particles (which, in turn, affects the size of the void spaces between particles) and the degree of saturation.

- The water content and, therefore, the degree of saturation within the upper sub-zone in particular is very dynamic, i.e. changing temporally, both with depth and at a given depth due to complex interactions between evapotranspiration loss from the ground surface and the water uptake of any surface vegetation; replenishment from below due to the 'wicking up' of groundwater from the capillary fringe; and intermittent replenishment from above due to surface-water infiltration from precipitation and possibly sources related to human activity such as lawn watering.

In summary, the *stress state* within the vadose zone, or at least some portion of it that is referred to as the *active zone*[122], is temporally dynamic due to temporal fluctuations in the degree of saturation. Both volume change (expansion and contraction) and shear strength are directly related to these changes in stress state. In turn, these changes affect the lateral earth pressures that exist in the ground, independent of the presence or not of an ERS.

As a subset of unsaturated soil mechanics, when the vadose zone consists in whole or in part of certain clay minerals that have a natural tendency to significantly change their properties, especially volume but also strength and permeability, depending on their degree of saturation, this constitutes what is referred to colloquially as *expansive soil*. Note that expansive soil is the combination of a type of soil particle (clay mineral) that inherently has these moisture-sensitive characteristics placed in a setting where moisture changes occur continuously over time that creates the classic expansive-soil scenario. Both ingredients must be present for this to happen.

A.3.3 Relevance to Earth-Retaining Structures

Unsaturated soil mechanics can have two very different impacts on the analysis and design of ERSs. One is more or less benign and routinely ignored in practice although, as will discussed, it can have significant indirect effects on practice. The other completely dominates the problem so cannot be ignored.

The more common and, at least superficially, benign one historically involves the use of non-expansive (typically granular) soils as the backfill or fill material behind an ERS. Such materials often contain a sufficient proportion of smaller particles ('fines') so that the soil exhibits a modest *apparent cohesion* (sometimes referred to as *capillary tension*) due to porewater-tension effects (referred to nowadays as *matric suction*) when partially saturated. However, because granular soils typically do not contain significant amounts of any of the clay minerals that are a necessary component of expansive soils, the only significant artifact of this matric suction is an increase in shear strength due to the cohesion (in the Mohr-Coulomb strength context) added to the fundamental friction between soil particles.

The most common physical manifestation of apparent cohesion is that it allows the soil to stand unsupported on a vertical or near-vertical slope. However, this is not an inherently stable situation as the degree of saturation of the exposed soil can change relatively quickly and with it the magnitude of apparent cohesion. This is the scenario that is a significant causative factor in many utility-trench collapses. Thus, good practice typically ignores this apparent cohesion in any permanent or long-term analysis or design of an ERS as this contribution to soil strength is simply unreliable in any long-term context. As a result, this is the classical situation where it is assumed for analysis or design purposes that the soil is dry for strength-calculation purposes and only has strength based on the Mohr-Coulomb friction angle, ϕ, of the soil.

It is of interest to note that this conservative, but arguably reasonable and justifiable, assumption of neglecting apparent cohesion in this instance has not been universally followed in practice. For example, in recent years the issue of whether or not to include the contribution of apparent cohesion in ERS design, either explicitly or implicitly,

---

[122] 'Active zone' in this context is unrelated to the active earth pressure state and simply means a zone (thickness) of soil beginning at the ground surface that exhibits relatively significant temporal variability in the degree of saturation and concomitant soil stress state and behavior.

282

has been the underlying factor in at times contentious discussions related to a new method for MSEW design called the *K-Stiffness Method* (Leshchinsky 2009, 2010a, 2010b; Leshchinsky et al. 2010a, 2010b). This design method departs from traditional MSEW analytical methodologies as it is based on the observed performance of actual MSEWs as opposed to being based on some hypothesized behavioral model as has been done historically. While basing an analytical methodology on the observed behavior of actual structures is not inherently incorrect, when doing so it is important to understand all the physical behaviors or phenomena that are present in actual structures and to decide whether these behaviors are something that can or should be relied on in the long term or not. This is the basis of the argument made by Leshchinsky and his colleagues.

This issue of apparent cohesion and its impact on ERS performance is relevant to this monograph as it has particular relevance to seismic design as noted earlier in this appendix. As discussed previously, cohesion in the Mohr-Coulomb strength context can significantly reduce forecast seismic loads. Apparent cohesion, which cannot be 'turned off' in nature as it can be ignored for conservatism in design, may well have been a significant factor in the observed acceptable performance of both yielding and non-yielding RERSs in relatively modest seismic events. The point being made here, as Leshchinsky and his colleagues made with respect to MSEW design, is that care should be exercised when extrapolating observations in nature to analytical methodologies intended for use in routine practice that should be inherently conservative when it comes to relying on physical phenomena such as apparent cohesion that can potentially be highly variable during the design life of an ERS that can easily span many decades.

The other impact of unsaturated soil mechanics on ERS analysis and design involves expansive soils. In this case, the matric-suction effects within the active zone can cause significant volume changes of the soil, something that does not occur to any significant extent with granular soils. For example, as the water content decreases the matric suction increases and the soil tends to decrease in volume (shrink). On the other hand, as the water content increases the matric suction decreases and the soil tends to increase in volume (swell). Because the degree of saturation within the active zone changes over time both naturally and as a result of human activity, this means that the volume of expansive soils will also change with time. This is the cause of the well-known phenomena of ground surface shrink and swell (heave) observed in the vertical direction.

The same shrink-swell potential exists in the horizontal direction although this appears to have been less studied to date compared to the vertical direction as most of the issues involving expansive soils are due to volumetric changes in the vertical direction. In addition, the extent to which horizontal shrink-swell can actually occur depends entirely on the physical confinement in that direction.

One limiting case is perfect lateral confinement to the extent that there is no lateral displacement. Under such conditions, when swelling of the overall soil occurs due to an increase in water content, the lateral earth pressures within the ground will be at their maximum and are limited in magnitude only by the passive earth pressure state. This condition typically exists under free-field conditions where the ground is confined by itself so that volume change can only occur in a vertical, one-dimensional fashion due to the stress relief offered by the ground surface. When lateral shrinkage occurs under the same free-field conditions due to a decrease in water content, the lateral stresses will decrease and can, in the limit, reduce to zero. It is at this point that vertical tension cracks that exhibit themselves at the ground surface can develop.

The other limiting case is no lateral confinement where the ground is able to expand laterally without restraint. This is the horizontal equivalent of the so-called *free-swell* condition in the vertical direction. For the most part, this is a theoretical limiting case that is not likely to exist naturally thus would generally require boundary conditions created by human activity for it to occur. Obviously, in this case the lateral earth pressures are zero under all conditions of swell or shrinkage.

When an ERS is placed adjacent to expansive soil, the actual lateral-restraint conditions will be somewhere between these two limiting cases depending on the structural rigidity of the ERS and its geotechnical ability to yield matched to the lateral stress-displacement characteristics of the soil mass in contact with the ERS. This is a classical problem in soil-structure interaction (SSI) where finding the unique solution to a problem requires not only satisfying static force and moment equilibrium but finding the unique pattern of displacements that is compatible for the objects (the ERS and soil mass in this case) that are in contact with each other. Note that this makes ERS problems involving expansive soils inherently more complex than those involving non-expansive soils where some analytical methodology that only considers forces (usually defined by the at-rest, active, and passive states existing either alone or in some combination) and not displacements is adequate for most problem solution in practice.

Complicating solution of a problem involving an ERS and expansive soil is the fact that the lateral earth pressures acting on the ERS will not be constant with time but will vary depending on variations in the water content and concomitant matric suction of the adjacent soil. As noted previously, these variations can occur due to both natural variations as well as human activity which itself might be direct or indirect in nature.

However, the most significant conceptual difference between the use of non-expansive vs. expansive soil adjacent to an ERS is that in the more-common baseline case of non-expansive soil the earth pressure state within the retained soil and the lateral earth pressures the soil exerts on the ERS are completely dependent on the rigidity and yielding conditions of the ERS. In essence, the retained soil develops an earth pressure state that is consistent with the yielding capability of the ERS. The one exception to this is self-yielding RERSs that are discussed in Chapter 6.

On the other hand, with expansive soils it is the soil that is the more dominant system component in that it is the soil that initiates the tendency to displace or not laterally depending on the water-content state and concomitant matric suction within the soil. In this case, it is the ERS that reacts (in terms of lateral displacement and possibly deformation as well) to the soil depending on the combined overall horizontal structural and geotechnical stiffness of the ERS and its supporting system.

In summary, designing an ERS when the retained soil consists of expansive soil presents several formidable challenges compared to the comparable process when non-expansive soils are involved. Perhaps the single biggest difficulty is determining a unique, application-specific 'wet' water-content state of the soil that would provide an upper bound of lateral earth pressures for which an ERS should be designed. This is very difficult as it requires anticipating both natural and artificial (i.e. human) environmental influences for the design life of an ERS which can exceed 100 years in many practical applications.

Experience indicates that this is particularly problematic when non-yielding RERSs such as basement walls of buildings are involved. In such cases, the ERS has no ability to displace or deform to reduce lateral earth pressures from expanding soil so must be designed structurally for the magnitude of pressures from the limiting case of a fully-confined soil. Because these pressures are delimited by the passive state they can be extremely large and, in most cases, are simply not economical to design for. This is consistent with practice where either below-ground walls are simply not constructed in areas with expansive soils or such soils are excavated and replaced with non-expansive soils at some not-insignificant expense.

This also suggests that geofoam compressible inclusions clearly have the potential to be beneficial in this application. The compressible inclusion can accommodate at least some of the lateral swell from an expansive soil and reduce the lateral earth pressures transmitted to the ERS.

A.3.4 Analytical Theories

A.3.4.1 Introduction

The single most important and fundamental aspect of unsaturated soil mechanics is the need to define the stress state in the ground in a way that is theoretically correct for all potential saturation conditions of a soil, both within the vadose zone and below the groundwater table. This is the essential first step for performing assessments of fundamental behaviors such as soil compressibility and shear strength as well as application-oriented behaviors such as lateral earth pressures.

In this regard, the history of modern soil mechanics can be viewed as being developmentally backward in that a relatively simple special case for stress state was defined first in the early decades of the 20th century and only later, mid-century, were there initial attempts to develop a more-general theoretical relationship. This is important to note as to a significant extent these later efforts were driven or at least guided by the fact that the original, traditional relationship for stress state can be expressed in a single equation:

$$\sigma' = \sigma - u \qquad (A.51)$$

where σ' = effective stress, σ = total stress, and u = porewater pressure (always positive in sign). This is the well-known *effective-stress equation*.

For reasons that will become clear, it is desirable to rewrite Equation A.51 with a revised notation for the porewater pressure as follows:

$$\sigma' = \sigma - u_w \qquad (A.52)$$

with $u_w = u$ as defined above. Also, in light of later developments that are discussed subsequently, the quantity $(\sigma - u_w)$ is defined as a *stress-state parameter* (alternatively *stress-state variable*). For this special case, effective stress requires only one stress-state parameter for definition.

Equation A.51 is now recognized as being theoretically correct only for either dry soil or saturated soil below the groundwater level. This equation is also analytically correct for soils within the vadose zone where matric-suction effects can be reasonably ignored and the porewater pressure can reasonably be assumed to equal zero as discussed above, i.e. when granular soils are involved. However, this equation is not correct within the vadose zone in situations where matric suction is significant. Thus, a more-general expression of effective stresses is required.

A.3.4.2 Overview

Efforts to develop a more-general and theoretically rigorous relationship for unsaturated soils began in the 1950s. Fredlund and Rahardjo (1993) trace the evolution of these efforts in some detail. Not surprisingly, almost all of these efforts were based on the premise that a single equation for effective stress of the general form of Equation A.51 can be developed so that Equation A.51 simply becomes a special case of that more-general equation. Only some of the key historical developments as well as current thinking on the subject is summarized here.

A.3.4.3 Bishop's Two-Parameter Equation

The most significant of the early, circa-1950s developments for an all-purpose effective-stress equation is credited to the late Professor A. W. Bishop:

$$\sigma' = (\sigma - u_a) + [\chi \cdot (u_a - u_w)] \tag{A.53}$$

where u_a = pore-air pressure, χ = a dimensionless parameter related to the degree of saturation (= 0 for dry soil and = 1 for saturated soil below the groundwater table), and all other variables are as previously defined. It can be seen that Equation A.53 defaults to the traditional equation (A.52) for the limiting cases of χ = 0 and 1.

As it has turned out, the most enduring aspect of Bishop's equation is that it defined two new stress-state parameters:

- $(\sigma - u_a)$ that is called the *net normal stress* and

- $(u_a - u_w)$ that is called the *matric suction*.

A.3.4.4 Fredlund's Two-Parameter Model

A key conclusion of Fredlund's decades-long research into unsaturated soil mechanics was that it is not possible to develop a single equation for effective stress for the more-general case as it is for the special one-parameter case of dry or saturated soil (Equation A.52). The reason as stated in Fredlund and Rahardjo (1993) and using Bishop's two-parameter equation (A.53) as an example is that it is not possible to come up with a unique value of χ for a given degree of saturation. Extensive research in the decades subsequent to the circa-1950s publication of Bishop's single equation with two stress-state parameters indicated clearly that values of χ depended on soil type and stress path among other variables. As noted in Fredlund and Rahardjo (1993), a theoretically rigorous definition of effective stress should be based on equilibrium conditions alone and not be dependent on a specific soil.

The overall conclusion of Fredlund's work is that the stress-state parameters (it can be shown that there must always be two) necessary to define the behavior of unsaturated soils cannot be combined into a single equation and need to be implemented independently into equations defining soil behavior (shear strength, etc.). Furthermore, Fredlund and Rahardjo (1993) illustrate that there are actually three potential pairs of stress-state parameters that can be used for unsaturated soils depending on whether it is desired to use σ, u_a, or u_w as the reference stress/pressure. They go on to posit that using air pressure, u_a, makes the most sense in which case the two stress-state parameters for net normal stress and matric suction as defined by Bishop are used.

A.3.4.5 Briaud's One-Parameter Equation

In recent presentations (e.g. Briaud 2011)[123], Professor J.-L. Briaud has proposed using the following single equation as a general definition of effective stress under all potential conditions of saturation:

$$\sigma' = \sigma - (\alpha \cdot u_w).$$ **(A.54)**

Note that in this case all effects related to saturation are incorporated into one stress-state parameter (i.e. the right-hand side of Equation A.54) and there is no explicit expression for matric suction.

As defined by Briaud, α is a dimensionless variable that can vary between 0 and 1. Briaud further opined that α is approximately linearly related to the degree of saturation (expressed in dimensionless form and not as a percent). It can readily be seen that the limiting cases of $\alpha = 0$ and $\alpha = 1$ represent dry and saturated (below groundwater level) soils respectively, and that Equation A.54 reverts to the classical effective-stress equation (Equation A.52) for both limiting cases.

This equation can be viewed as a 'back-to-the-future' suggestion by Briaud as, except for a notational change (α for the original β'), it has the same form as what Fredlund and Rahardjo (1993) indicate was the very first suggestion for an all-purpose effective-stress equation applicable to unsaturated soils that was made back in the late 1950s. As a result, there has been 60 years in which researchers have had an opportunity to assess the relationship expressed in Equation A.54 and it has clearly been found to be lacking for the intended purpose.

A.3.4.6 Some Comments Relative to Matric Suction

A potential source of confusion when dealing with unsaturated soils is that plots of matric suction are not rendered in a consistent fashion. Although the parameter of matric suction is generally plotted as the abscissa using a $\log_{10}$ scale, there is inconsistency in the ordinate parameter that is used. In some cases, it is degree of saturation and in other case it is water content, both of which can be depicted either non-dimensionally varying between 0 and 1 or as a percentage varying between 0% and 100%.

However, the greatest potential source of confusion is with regard to water content as it has become common to use the *volumetric water content* which is otherwise rarely used in geotechnical engineering which has historically and near-universally used the *gravimetric water content*. The reason for preferring the volumetric water content in this application is not clear other than, perhaps, the fact that volumetric water content is not open-ended in value whereas gravimetric water content is, at least in theory.

---

[123] The writer attended an online presentation of this cited reference and downloaded a copy of the presentation slides several years ago. Unfortunately, information concerning the original venue and online link for the downloadable version of this cited reference (presentation) is not available. However, two other undated presentations that together appear to contain the material of this cited presentation can be accessed (as of 10 October 2017) at:
- ceprofs.civil.tamu.edu/briaud/Iran%202010.pdf
- ceprofs.civil.tamu.edu/briaud/Stiffened%20Slabs%20on%20Grade%20on%20Shrink%20Swell%20Soils%20Lecture.pdf

A.4 SURFACE LOADS

A.4.1 Introduction

For many of the types of applications covered by this monograph there will be surface or near-surface loading, often referred to as *surface surcharges* in practice, from a variety of static and moving sources such as motor vehicles. These loads should be considered during ERS design although in many practical cases they are relatively small in magnitude and have little effect compared to earth loads.

There are a variety of ways in which surface loads can be dealt with analytically and to some extent the methodology chosen depends on the geometric configuration of the loading. The easiest form of load configuration to deal with by far is the *uniform surcharge* concept where a vertical normal stress of constant magnitude, q_s, is assumed to cover the entire ground surface. This is the only type of surface load considered in this monograph in the interest of simplicity and to focus on the earth loads which are the more important part of the problem in most practical applications. However, all of the concepts presented in this monograph can be used with more-complex renditions of surface loads.

A.4.2 Gravity Loading Conditions

A uniform surcharge under gravity loading as applied to any ERS is simple and straightforward to deal with. For both the active and at-rest earth pressure states, the additional lateral stress caused by the uniform surcharge is assumed to be uniform in magnitude from the ground surface to the base (heel specifically) of the ERS. Thus, the resulting pressure diagram due to the surcharge alone is simply a rectangle. The only thing that differs is the magnitude of the stress as follows:

$$= \left(K_a \cdot \cos\omega \cdot q_s\right) \text{ for the active earth pressure state} \qquad \textbf{(A.55a)}$$

$$= \left(K_o \cdot q_s\right) \text{ for the at-rest earth pressure state.} \qquad \textbf{(A.55b)}$$

Note that these rectangular pressure diagrams are additive to the triangular pressure diagrams for the earth-load components that were discussed previously and that the rectangle for the active state is skewed by the soil-wall friction angle δ in the same sense as shown for the resultant force P_a in Figure A.1.

A.4.3 Seismic Loading Conditions

Different approaches are required to deal with surface surcharges under seismic loading depending on whether active or at-rest conditions are assumed to exist within the retained soil. The active case is more straightforward and is discussed first.

Research indicates that surface-surcharge effects in the active case can be modeled satisfactorily by combining gravity and seismic effects into a single rectangular pressure diagram with a depth-wise constant magnitude of:

$$K_{ae} \cdot \cos\omega \cdot q_s^* \qquad \textbf{(A.56)}$$

where

$$q_s^* = q_s \cdot (1 - k_v) \qquad \text{(A.57)}$$

and is conceptually the surface surcharge modified for vertical acceleration[124].

Note that this rectangular pressure diagram covers surface surcharge only so is in addition to the triangular pressure diagram for the gravity earth load component, P_a, and whichever pressure diagram (M-O triangle, S-W trapezoid, etc.) is used for the seismic earth load increment, ΔP_{ae}.

Dealing with surface-surcharge effects for the at-rest case is less straightforward as it does not appear that any of the traditional methodologies proposed for use with non-yielding RERSs such as Wood's solution have considered this. The writer has thus developed two approximate approaches for use with the at-rest case.

One is to use a concept from the early days of modern soil mechanics where a uniform surcharge can be envisaged as adding an additional fictitious height or thickness to the soil retained by the ERS. The magnitude of this fictitious height increase, ΔH_{eff}, is:

$$\Delta H_{eff} = \frac{q_s}{\gamma_{eff}}. \qquad \text{(A.58)}$$

The total effective height or thickness of soil, H_{eff}, is then:

$$H_{eff} = H + \Delta H_{eff} = H + \frac{q_s}{\gamma_{eff}}. \qquad \text{(A.59)}$$

Note that it is the height H_{eff} that is used in lieu of H in Wood's solution (Equation A.24) or whichever other analytical methodology that is desired.

The alternative approximate approach can be developed by noting that a surface surcharge really represents an additional mass that contributes to the lateral resultant force from seismic shaking such as F_{sr} in Wood's solution. One logical way to put this additional equivalent mass into the overall system is to simply increase the soil unit weight used in Wood's solution (Equation A.24) or other analytical methodology. Specifically, in these equations a unit weight $\gamma_{eff/s}$ should be used in lieu of γ_{eff} where

$$\gamma_{eff/s} = \gamma_{eff} + \frac{q_s}{H}. \qquad \text{(A.60)}$$

Note that this proposed methodology has the effect of distributing a surface surcharge uniformly throughout the thickness H of soil behind the RERS. This is not strictly correct as the surface surcharge really represents a lumped mass at the top of the soil column.

At the present time, there is no information on which the writer can base an opinion as to which of these approximate methods is more accurate or if neither is. Consequently, it would seem logical to calculate the seismic earth force such as F_{sr} for Wood's solution using both methods and to use the one that gives the greater, i.e. more conservative, value of F_{sr}.

---

[124] Note that as with earth loads there are, in principle, three different cases of k_v to consider in each problem: positive or negative in sign or equal to zero. Note also that in routine practice only the $k_v = 0$ case is used as vertical acceleration is found to have negligible effect on calculated outcomes.

Appendix B

Parameters for Load-Reduction Assessment

B.1 INTRODUCTION

An important and arguably necessary issue for meaningful interpretation and assessment of results when using or considering using geofoam to reduce the lateral pressures on an ERS is to have a consistent methodology for both expressing analytical results and as the basis of a metric to compare results to the no-geofoam baseline case. Ideally, this methodology should include as its outcome a single dimensionless parameter that is succinct, insightful, and above all has a sound basis in theory.

This appendix summarizes several methodologies and parameters that have been proposed for this purpose. For historical reasons, methodologies and parameters have been developed separately for gravity and dynamic (seismic) loading so the discussion is divided along those lines.

B.2 GRAVITY LOADING

B.2.1 Background

Although it might seem logical to use lateral pressures acting on the inside face of the ERS as the basis of a methodology for assessing load reduction on ERSs, there is a sound reason for not doing so. The actual distributions of lateral pressure as a function of depth do not, in general, adhere to the classical, simplistic triangular (i.e. hydrostatic or equivalent fluid pressure) distribution that is generally assumed in routine practice to exist for any earth pressure state (active, at-rest, passive). Consequently, there is ambiguity as to how to compare lateral pressure distributions that have different geometric shapes.

A more useful approach is based on using P_h, the lateral resultant (earth) force acting on an ERS. For a given application, this can be evaluated using the following equation:

$$P_h = \int_0^H \overline{\sigma}_h \, dz \qquad \text{(B.1)}$$

where $\overline{\sigma}_h$ is the effective horizontal normal stress (= lateral pressure) acting on the inside face of the ERS as a function of depth, z, below the surface of the ground adjacent to the inside face of the ERS and H is the geotechnical height of the ERS. Essentially, P_h is the area under the geometric shape (triangle, curve, etc.) defining the depth-wise variation in lateral pressure and can be evaluated in any number of ways.

B.2.2 Work of Karpurapu and Bathurst (1992)

While there has been general consensus to date among researchers to use P_h as the basis for developing a parameter for use as a comparative metric, there is a divergence of opinion as to the specifics for doing so. The first known work along these lines was by

Lateral Pressure Reduction on Earth-Retaining Structures Using Geofoam
John S. Horvath, Ph.D., P.E., Life Member.ASCE

290

Karpurapu and Bathurst (1992) who dealt solely with non-yielding RERSs and the REP-Wall version of the compressible-inclusion functional concept to reduce earth loads only. They reported the results of a study that was based on numerical analyses using the FEM.

A primary outcome of their work was to evaluate the efficacy of compressible inclusions in terms of reducing lateral earth pressures from the baseline at-rest state to the active state that was assumed to be the practical achievable minimum. They elected to express their results using a hybrid methodology. Specifically, for each case they analyzed they determined P_h using Equation B.1. They then calculated a parameter K_h that they defined as the coefficient of lateral earth pressure for an equivalent triangular distribution of lateral earth pressure that had the same resultant force as the measured P_h. This can be expressed as:

$$P_h = \int_0^H \overline{\sigma}_h \, dz = \frac{K_h \cdot \gamma_{eff} \cdot H^2}{2} \tag{B.2}$$

where γ_{eff} is the effective unit weight of the retained soil. Rearranging Equation B.2 yields the desired result:

$$K_h = \frac{2}{\gamma_t \cdot H^2} \int_0^H \overline{\sigma}_h \, dz . \tag{B.3}$$

Karpurapu and Bathurst used the results from Equation B.3 to create the desired dimensionless parameter by normalizing each calculated K_h result to the theoretical coefficient of active earth pressure, K_a, that they used as their reference (baseline) condition based on the logic that achieving the active earth pressure state was the desired goal of using a REP-Wall compressible-inclusion with a non-yielding RERS. The results in their paper were then plotted as dimensionless K_h/K_a ratios so for their work this ratio constitutes the desired parameter or metric for assessing the relative effectiveness of compressible inclusions in REP-Wall applications at least.

B.2.3 Work of Horvath (2000)

While the above-described methodology was effective for the limited scope and purposes of the Bathurst and Karpurapu (1992) study, in subsequent work by the writer (Horvath 2000) that dealt with self-yielding RERSs, specifically IABs, an alternative approach was developed that was judged to be more general and thus more useful. This alternative methodology is not limited to IABs or even self-yielding RERSs and is therefore suggested here as a general methodology for use with both the lightweight-fill and compressible-inclusion functions for any type of ERS.

The specific elements of the writer's methodology are:

- The resultant force P_h obtained using Equation B.1 for a specific case (numerical analysis, laboratory measurement, field measurement, etc.) is used directly. The fictitious earth pressure coefficient K_h defined in Equation B.3 is not used at all. This removes the need to rely on using an assumed triangular or any other geometric shape distribution of lateral pressures for any part of the overall methodology and also allows cases with surface surcharges to be included. It also removes any ambiguity or error from dealing with backfills/fills of different composition having different material unit weights.

- The normally consolidated (NC) at-rest, not active, earth pressure state is used as the reference baseline condition for normalization to develop the desired dimensionless parameter for use in assessing results. The writer felt that the NC at-rest state is a more representative and insightful baseline to use because it represents the theoretical no-displacement starting point for a soil mass in any problem dealing with lateral earth pressures. Therefore, the NC at-rest state more clearly illustrates the relative reduction in lateral pressures that can be achieved using some configuration of geofoam, whether for the lightweight-fill or compressible-inclusion (REP- or ZEP-Wall) function. In addition, the NC at-rest state provides a logical basis of comparison for problems such as self-yielding RERSs where lateral earth pressures trending toward the passive state can develop. The resultant force, P_h, for this NC at-rest reference condition is given the notation P_{oNC}.

Based on these assumptions, a new dimensionless parameter, β, called the *relative lateral resultant force* was used to reflect the relative results between analysis subcases by defining:

$$\beta = \frac{P_h}{P_{oNC}}.$$

(B.4)

This means:

- $\beta = 1$ represents the baseline reference condition of the NC at-rest earth-pressure state,

- $\beta < 1$ reflects a reduction in lateral pressure,

- $\beta = 0$ is the lower-bound limiting case of total lateral-pressure reduction, and

- $\beta > 1$ reflects an increase in lateral pressure with the maximum, upper-bound value of β being finite and dependent on the coefficient of passive earth pressure, K_p, in a given application. For example, for the problem considered in Horvath (2000) the maximum value was $\beta = 24$.

Conceptually, this methodology achieves the same overall goal as that used by Karpurapu and Bathurst (1992). It is simply more general and was better suited to the specific elements and requirements of the Horvath (2000) study. It is felt that these same benefits apply to the broader topics considered in this monograph.

B.2.4 Comments

It is important to note that the above-described methodologies of both Karpurapu and Bathurst and the writer focus only on the <u>magnitude</u> of resultant forces applied to an ERS. The point-of-application of the resultant force, which affects bending moments in and global stability of (if it is yielding in nature) the ERS, is not considered. Therefore, care is required when interpreting the results using either of these methodologies as two different problems might happen to produce the same value of either K_h/K_a or β but might not affect the ERS structurally and/or geotechnically in exactly the same way due to different points-of-application of the resultant forces.

B.3 SEISMIC LOADING

There appears to be no logical reason for not using lateral resultant force as the basis for assessing the efficacy of geofoam compressible inclusions for reducing seismic loads on ERSs as is done for gravity loading. Unfortunately, as with gravity loading there is no consensus at the present time as to how this load-reduction metric should be defined.

The writer's preference is to use the same variable, β, and definition (Equation B.4) used for gravity loading. Thus, in this case P_h is the lateral resultant force from seismic loading but P_{oNC} remains the lateral resultant force from gravity loading.

The logic of normalizing seismic forces to gravity forces is several-fold. Not only does it simplify things by having one reference standard but in many practical cases, especially involving remediation or upgrading of existing ERSs to be able to sustain seismic loads, it readily provides a comparison of those seismic loads relative to the gravity loads for which the existing ERS was presumably designed or at least had been/is supporting adequately.

The only alternative proposed to date for seismic loading of which the writer is aware is an outcome of the work of Bathurst and Zarnani over the course of about a decade that is summarized in Bathurst and Zarnani (2013). This work was an outgrowth of the earlier work of Bathurst and Karpurapu (1992) that was restricted to gravity loading. As with Bathurst's work related to gravity loading, the work related to seismic loading was limited to non-yielding RERSs.

Bathurst and Zarnani use the term *isolation efficiency* (no corresponding variable notation was defined), expressed as a percent, to quantify the load-reduction effect or benefit of a geofoam compressible inclusion. It appears that this term derives from the term *seismic isolation* that some (e.g. Athanasopoulos-Zekkos et al. 2011) have used as a synonym for the much more common term *seismic buffer* (defined below). The following definition of isolation efficiency is quoted from Bathurst and Zarnani (2013):

"Isolation efficiency of the seismic buffer is defined as the ratio of change in wall force between rigid and seismic buffer cases divided by peak wall force without the buffer."

Note that *seismic buffer* is a term that has become popular with some for what is simply a REP-Wall compressible-inclusion application specifically for the purpose of reducing seismic loads on a RERS, generally non-yielding. As an aside, it is unclear to the writer who first coined this term although it is clear that Bathurst and Zarnani have certainly popularized it in the 21st century in their several publications on the subject. Note that the term 'seismic buffer' also appears to have other uses in technology that have nothing to do with civil engineering or geofoams so one does have to be clear about contextual usage.

In any event, isolation efficiency as defined by Bathurst and Zarnani differs from the writer's variable β in two significant ways:

- The reference load (denominator in the ratio) for isolation efficiency is the peak lateral (resultant) earth force for the baseline case of a non-yielding RERS without a compressible inclusion but under seismic loading, not the gravity at-rest case as with the writer's variable β. Thus, the reference load for calculating isolation efficiency is

<u>never</u> a constant, even for a given ERS. It varies depending on the particular seismic load history applied to the ERS.

- The numerator in the ratio is the magnitude of the load reduction, i.e. the <u>difference</u> between (presumably peak) lateral forces without and with the compressible inclusion, as opposed to being the magnitude of the reduced force with compressible inclusion.

Note that as a result of the latter item, the qualitative trends of isolation efficiency are the opposite of those with β as summarized earlier in this appendix. Specifically, with isolation efficiency the <u>larger the value</u> the <u>greater the beneficial effect</u> of the compressible inclusion, with the limiting values of 100% being complete load reduction (equivalent to $\beta = 0$) and 0% being no load reduction (broadly similar to $\beta = 1$ but not exactly as isolation efficiency has a dynamic-load basis whereas β always has a gravity-load basis.

It is relevant to note that for a given ERS with a compressible inclusion the isolation efficiency will never be constant. Rather, the isolation efficiency will vary depending on the particular seismic load history applied to the ERS-geofoam-soil system. Thus, it would appear that at best a range of isolation efficiencies can be calculated for a given problem. However, that would still be misleading as the different isolation efficiencies would not be referenced to the same baseline lateral force. As a result, an absolute sense of the efficiency of the geofoam compressible inclusion in reducing seismic loads would not be apparent.

By comparison, by referencing seismic resultant forces to a common reference force as with the writer's method using the β parameter an absolute sense of the efficacy of a compressible inclusion in reducing seismic loads would be readily apparent in all cases as the reference value ($\beta = 1$) never changes.

This page intentionally left blank.

Appendix C

Solutions for Lightweight-Fill Partial-Depth Placement Alternative for Yielding Conditions

---

C.1 GRAVITY LOADS

C.1.1 Theoretical Development

C.1.1.1 Introduction

As discussed in Chapter 4, the analytical methodology for lightweight-fill applications with partial-depth placement proposed in and developed for this monograph is based on the well-known trial-wedge method. The solution presented in this monograph is for the simple, basic problem shown in Figure C.1 (this is the same as Figure 4.7, reproduced here for ease of reference) with θ = 90° for the RERS but could easily be extended to include variations in boundary conditions such as:

- a non-vertical inside face on the RERS (θ ≠ 90°),

- a non-horizontal ground surface for the retained soil ($i \neq 0°$), and

- a surface surcharge on the retained soil ($q_s > 0$).

Note that the relative volume and placement of EPS blocks within the cross-section shown in Figure C.1 is completely arbitrary. However, in order for the solution developed in this appendix to be valid there are some practical constraints that will become clear as the solution methodology is developed.

This problem will be solved in both a 'rigorous' and 'simplified' manner, with the difference being the level of computational effort involved although both methods are easily solvable using generic computational tools that are nowadays universally available. Where and when each solution method is appropriate and the outcome differences between them are discussed and presented subsequently in this appendix.

C.1.1.2 Rigorous Solution

As discussed in Chapter 4, the traditional solution of the trial-wedge method is graphical. While this might still be the optimum solution technique for a specific application that is geometrically irregular and complex, the solution approach chosen here is mathematical in nature. It is referred to as the 'rigorous' solution to the problem at hand for reasons that will become clear subsequently.

The goal of the solution process is to find the angle $\theta^*_{critical} = \theta^*_i$ of the assumed intra-soil failure plane that will produce the maximum value of P^*_i as shown in Figure C.1. This maximum value of P^*_i is taken to be the active earth force, P^*_a.

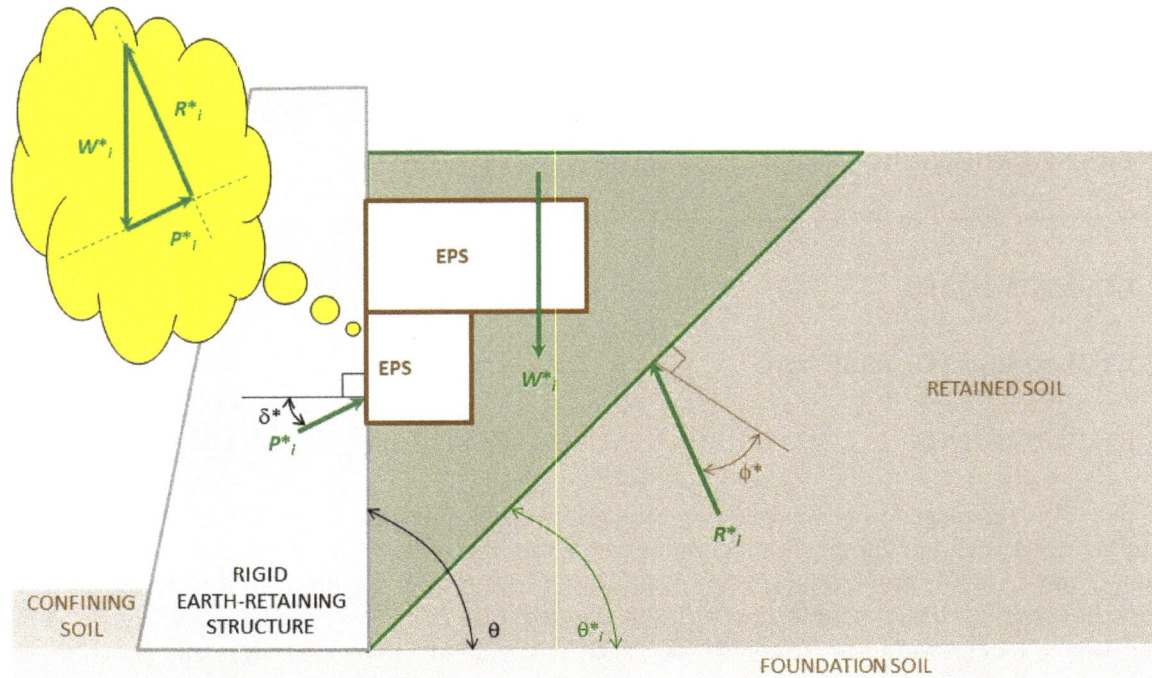

Figure C.1. Lightweight-Fill Function - Partial-Depth Placement Alternative Gravity Loading - Analytical Model Based on Trial-Wedge Method.

This goal is accomplished by first invoking horizontal force equilibrium[125] for the three force vectors (P^*_i, R^*_i, and W^*_i) shown in this figure, with forces acting to the right defined as positive in sign:

$$\Sigma F_H = 0 = [P^*_i \cdot \cos \delta^*] - [R^*_i \cdot \cos (90° - \theta^*_i + \phi^*)] \tag{C.1}$$

from which a solution for R^*_i in terms of P^*_i can be obtained by rearranging the terms of this equation as follows:

$$R^*_i = P^*_i \cdot \frac{\cos \delta^*}{\cos(90° - \theta^*_i + \phi^*)}. \tag{C.2}$$

Next, vertical force equilibrium is invoked with upward forces defined as positive:

$$\Sigma F_V = 0 = -W^*_i + [P^*_i \cdot \sin \delta^*] + [R^*_i \cdot \sin (90° - \theta^*_i + \phi^*)] \tag{C.3}$$

where W^*_i is defined as the weight of the trial wedge which is further defined subsequently.

Substituting Equation C.2 into Equation C.3 and then rearranging, simplifying, and consolidating the terms of Equation C.3 to solve for P^*_i yields the following:

---

[125] Note that all forces in this derivation have dimensions of force per unit length, with the length dimension being perpendicular to the plane of Figure C.1.

$$P*_i = \frac{W*_i}{\sin\delta* + [\cos\delta* \cdot \tan(90° - \theta*_i + \phi*)]}. \tag{C.4}$$

To facilitate solution of Equation C.4, $W*_i$ is artificially divided into two terms:

$$W*_i = W*_{SOILi} - W*_{EPSi} \tag{C.5}$$

where $W*_{SOILi}$ is defined as the weight of trial wedge i assuming it is composed entirely of soil and will thus vary with the particular trial wedge being analyzed according to the following equation:

$$W*_{SOILi} = 0.5 \cdot \gamma_t \cdot (\cot\theta*_i \cdot H) \cdot H \tag{C.6}$$

and $W*_{EPSi}$ is the net weight of the volume of soil replaced by EPS blocks (defined here as V_{EPS} which is known by assumption beforehand) and is assumed to be a constant for a given design alternative under consideration:

$$W*_{EPSi} = (\gamma_t - \gamma_{EPS}) \cdot V_{EPS}. \tag{C.7}$$

Substituting Equations C.6 and C.7 into Equation C.5 and then substituting this expanded version of Equation C.5 into Equation C.4 yields:

$$P*_i = \frac{[0.5 \cdot \gamma_t \cdot (\cot\theta*_i \cdot H) \cdot H] - [(\gamma_t - \gamma_{EPS}) \cdot V_{EPS}]}{\sin\delta* + [\cos\delta* \cdot \tan(90° - \theta*_i + \phi*)]}. \tag{C.8}$$

All variables in Equation C.8 are known except for $P*_i$ and $\theta*_i$. The solution goal is to find the value of $\theta*_i$ that maximizes the value of $P*_i$, defined here as $\theta*_{critical}$, and is subject only to the constraints that:

$$\theta*_{min} \leq \theta*_{critical} \leq \theta*_{max} \tag{C.9}$$

as illustrated qualitatively in Figure C.2 (same as Figure 4.8) where $\theta*_{min} = \phi*$.

There are any number of numerical-solution strategies and tools that might be used for this. For the purposes of the example problems in this monograph, the writer used the *Solver* function in *Excel 2010*.

Note that if the solved-for $\theta*_{critical}$ is equal to either $\theta*_{min}$ or $\theta*_{max}$ this would indicate that the solution is likely invalid as the model used to develop the solution did not apply in that case, presumably because the assumed geometry of EPS blocks interfered with formation of the critical intra-soil failure plan and was thus outside of the range of validity of the model assumptions. In that case, a more-complex numerical analysis such as the FEM would be required.

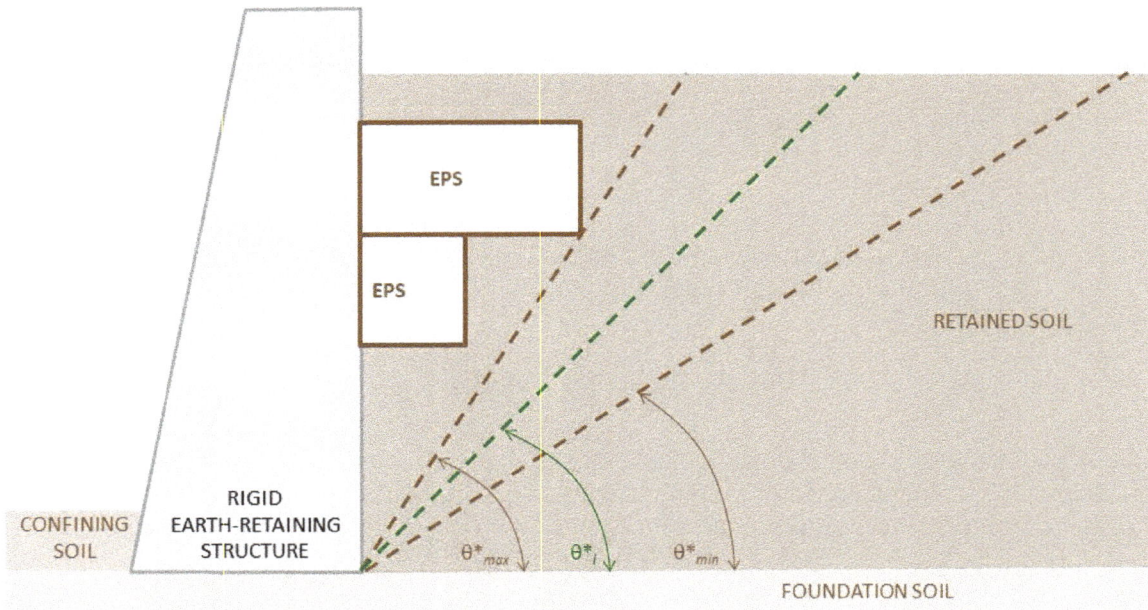

Figure C.2. Lightweight-Fill Function - Partial-Depth Placement Alternative Gravity-Loading - Trial-Wedge Method Solution Limits.

C.1.1.3 Simplified Solution

The rigorous solution presented in the preceding section requires a modest amount of effort to implement using *Excel 2010* or similar analytical tool as it involves solving for two unknowns using one equation through an optimization process. Therefore, the writer also developed an approximate, 'simplified' solution that could be useful for preliminary design when various configurations of EPS blocks are being considered. However, this simplified solution produces results that are always approximate relative to the rigorous solution and can be unconservative (i.e. the resultant earth force is underestimated) in some cases. Consequently, use of this simplified solution should be limited solely to preliminary design with final design always performed using the rigorous solution outlined in the preceding section.

This simplified solution is based on the assumption that $\theta^*_{critical}$ is not an initial unknown to be solved for as in the rigorous solution but is known a priori and corresponds to the theoretical angle based on Coulomb's solution for no interface friction between the retained soil and RERS, i.e. $\delta^* = 0°$,[126] but not to exceed θ^*_{max} as defined in Figure C.2. The calculation of P_a^* is then made using the following steps:

1. Assume a configuration of EPS blocks and calculate their volume, V_{EPS}, noting that this volume will be per unit width of the ERS in the direction perpendicular to Figure C.1. This is the same as the first step in the rigorous method.

2. Calculate $\theta^*_{critical}$ as follows:

---

[126] In this one case the Coulomb solution happens to be the same as Rankine's solution for a perfectly plastic continuum.

$$\theta*_{critical} = 45^0 + \frac{\phi*}{2} \text{ but } \not> \theta*_{max}. \tag{C.10}$$

3. Use the result from Step 2 to calculate the theoretical volume, $V*$, of the failure wedge, noting again that this volume will be per unit width of the ERS in the direction perpendicular to Figure C.1:

$$V* = 0.5 \cdot (\cot \theta*_{critical} \cdot H) \cdot H. \tag{C.11}$$

4. Use the result from Step 3 to calculate the hypothetical all-soil weight of the failure wedge, $W*_{soil}$:

$$W*_{soil} = V* \cdot \gamma_t. \tag{C.12}$$

5. Use the result from Step 1 to calculate the net weight reduction from the EPS blocks, $W*_{EPS}$:

$$W*_{EPS} = (\gamma_t - \gamma_{EPS}) \cdot V_{EPS}. \tag{C.13}$$

6. Use the results from Steps 4 and 5 to calculate the actual weight of the failure wedge, $W*_{actual}$:

$$W*_{actual} = W*_{soil} - W*_{EPS}. \tag{C.14}$$

7. Use the results from Steps 3 and 6 to calculate the reduced equivalent unit weight within the failure wedge, $\gamma*$:

$$\gamma* = \frac{W*_{actual}}{V*}. \tag{C.15}$$

8. Calculate Coulomb's coefficient of active earth pressure, K_a, using Equation A.4 in Appendix A (the notation in the following equation has been changed slightly from that used in Appendix A to be consistent with that used in this derivation as well as the problem geometry assumed in this derivation as shown in Figure C.1):

$$K_a = \left[\frac{\sin(90° - \phi*) \cdot \csc 90°}{\sqrt{\sin(90° + \delta*)} + \sqrt{\sin(\phi* + \delta*) \cdot \sin(\phi* - 0°) \cdot \csc(90° - 0°)}} \right]^2. \tag{C.16}$$

9. Use the results from Steps 7 and 8 together with Equation A.3 in Appendix A to calculate the active resultant earth force, P_a*, acting on the ERS (the notation in the following equation has been changed here to be consistent with that used in this derivation):

$$P_a* = 0.5 \cdot K_a \cdot \gamma* \cdot H^2. \tag{C.17}$$

This completes the calculation sequence.

300

C.1.2 Example Problems

C.1.2.1 Introduction

Application of the rigorous and simplified analytical methodologies for the trial-wedge solutions derived in the preceding sections will now be illustrated using the overall RERS geometry and soil properties from the FE analyses performed for this monograph. Details of this FE model are given in Appendix E and are not repeated here.

As noted previously, the writer used the *Solver* function in *Excel 2010* for all calculations involving the rigorous version of the trial-wedge method presented earlier in this appendix. Solution of the simplified version of the trial-wedge method as well as Coulomb's solution from Appendix A were also incorporated into the same *Excel* spreadsheet for overall calculation efficiency and ease of reference although they could have been performed manually if desired.

Three different examples are illustrated. The first is the all-soil/no-geofoam baseline case that will allow comparison of results from the two versions of the trial-wedge method with exact results from Coulomb's solution. Following this, two cases of partial-depth placement are illustrated to both provide comparison of results from the rigorous and approximate trial-wedge methods as well as to give some general idea of relative load reduction compared to the baseline case.

C.1.2.2 All Soil/No Geofoam

Table C.1 presents the results for this baseline case that was used to verify the programmed rigorous and simplified analytical methods by comparing their results to the known Coulomb solution.

Table C.1. All-Soil/No-Geofoam Baseline Case - Gravity Loading.

Parameter	Results		
	trial wedge		Coulomb
	rigorous	simplified	
P^*_a, kN/m (kips/ft)	77.2 (5.29)	77.2 (5.29)	77.2 (5.29)
$\theta^*_{critical}$, degrees	61.0	64.0	64.0

As can be seen, all analytical methods are essentially in agreement as would be expected although the slight difference in the magnitude of $\theta^*_{critical}$ requires some comment. The reason for this difference is that the value given for Coulomb's solution (and the simplified trial-wedge method as well by virtue of explicit assumption in the methodology development) was calculated for this study, as it is typically done in practice, using Equation C.10. However, strictly speaking this equation is only correct for the case of no interface friction, δ^*, between the inside face of the ERS and retained soil whereas in this example problem $\delta^* = 23°$.[127] Consequently, the error in Table C.1 with regard to $\theta^*_{critical}$ is in the Coulomb and simplified trial-wedge results. The rigorous trial-wedge results are correct.

---

[127] When this example problem was analyzed assuming $\delta^* = 0°$, all three methods shown in Table C.1 produced identical results: $P^*_a = 84.8$ kN/m (5.81 kips/ft) and $\theta^*_{critical} = 64.0°$.

The writer is not aware of a more-general solution of any kind (equation, plot, or table) that has been published explicitly and specifically for calculating $\theta_{critical}$ for Coulomb's solution when $\delta > 0°$, if indeed such a solution exists. However, a solution does exist indirectly using equations developed as part of the Mononobe-Okabe (M-O) Method for seismic loading that is discussed at length in Appendix A. This is because Coulomb's solution can be visualized as a special case of the M-O Method when $k_h = k_v = 0$, i.e. there is no seismic loading. In such a case, the angle of the failure wedge in M-O theory, α^*_{ae}, is the same as Coulomb's $\theta^*_{critical}$.

Consequently, Equation A.15 in Appendix A for calculating α^*_{ae} for M-O theory can be used as a more-general solution for calculating $\theta^*_{critical}$ for Coulomb's solution and thus for the simplified trial-wedge method as well. However, the modest increase in perceived accuracy comes at the expense of the simplicity of using Equation C.10 compared to Equation A.15. Given the relatively small error involved in (incorrectly) using Equation C.10 when $\delta > 0°$ and the fact that the failure surface is actually curved, not planar, anyway it is arguably not worth the effort in most cases in practice to use the much more complex M-O Method equations for α^*_{ae}. Nevertheless, the option to use the M-O Method for a more-accurate assessment of $\theta^*_{critical}$ when $\delta > 0°$ always exists when desired.

C.1.2.3 One-Third Depth of Geofoam

Figure C.3 illustrates the assumed problem geometry used to evaluate the test case of one-third replacement of retained soil with EPS blocks. Also shown in this figure is the upper-bound value, θ^*_{max}, for θ^*_i that is defined by the assumed geometry of the EPS blocks as shown generically in Figure C.2. Although not shown, the lower-bound value, θ^*_{min}, for θ^*_i is 38° in this case (= ϕ^* of the retained soil).

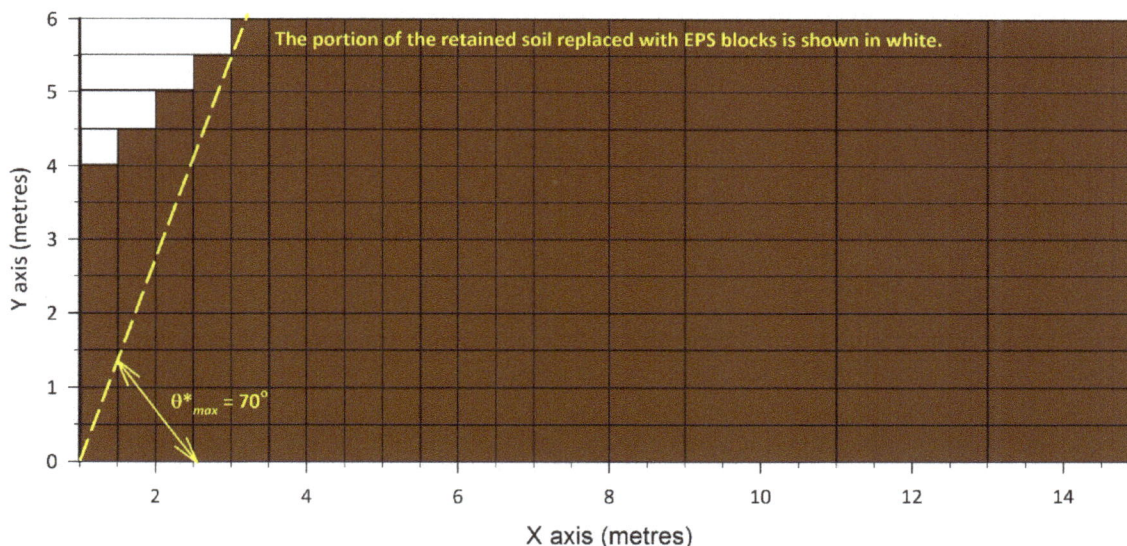

Figure C.3. Lightweight-Fill Function - Partial (1/3)-Depth Placement.

Note that the problem shown in Figure C.3 is actually highly idealized as no soil cover was modeled over the EPS blocks as would almost always be used in practice. This was done intentionally to focus on the primary problem components.

Table C.2 presents the results from the rigorous and simplified analytical methods presented in this appendix. The results for the all-soil/no-geofoam baseline case are also shown for comparison.

Table C.2. One-Third Replacement Case - Gravity Loading.

Parameter	Results		
	trial wedge		Coulomb
	rigorous	simplified	(no-geofoam baseline)
P^*_a, kN/m (kips/ft)	59.5 (4.08)	55.4 (3.79)	77.2 (5.29)
$\theta^*_{critical}$, degrees	57.3	64.0	64.0

Based on the results obtained using the rigorous solution, there is approximately a 23% reduction in the calculated active earth force as a result of the EPS blocks occupying approximately 22% of the volume of the critical failure wedge. The calculated angle for this failure wedge (57.3°) is well within the limits for this case ($\leq 70°$, $\geq 38°$) so the solution is considered valid.

Note that in this case the assumed angle for the simplified analysis is governed by the Coulomb result and not the θ^*_{max} defined by problem geometry (70° in this case). Note also that the simplified analysis is in reasonable agreement with the rigorous analysis and adequate for preliminary assessment purposes. However, consistent with the admonition stated previously, the results for the simplified analysis are unconservative and thus suitable only for preliminary-design purposes.

C.1.2.4 Two-Thirds Depth of Geofoam

Figure C.4 illustrates the assumed problem geometry used to evaluate the test case of two-thirds replacement of retained soil with EPS blocks. Also shown in this figure is the upper-bound value, θ^*_{max}, for θ^*_i that is defined by the assumed geometry of the EPS blocks as shown generically in Figure C.2. Although not shown, the lower-bound value, θ^*_{min}, for θ^*_i is again 38° in this case (= ϕ^* of the retained soil).

Table C.3 presents the results from the rigorous and simplified analytical methods presented in this appendix. The results for the all-soil/no-geofoam baseline case are also shown for comparison. Note that the calculated angle for the failure wedge (49.3°) is just within the limits for this case ($\leq 54°$, $\geq 38°$) so while this solution is considered valid it does indicate that the range of validity of the partial-depth placement solution is being approached, at least for the soil properties and overall problem geometry considered here.

Based on the results obtained using the rigorous solution, there is approximately a 66% reduction in the calculated active earth force as a result of the EPS blocks occupying approximately 58% of the volume of the critical failure wedge.

Note that in this case the assumed angle for the simplified analysis is governed by θ^*_{max} defined by problem geometry (54° in this case) and not by the Coulomb result. Note also that the simplified analysis is again in reasonable agreement with the rigorous analysis and adequate for preliminary assessment purposes.

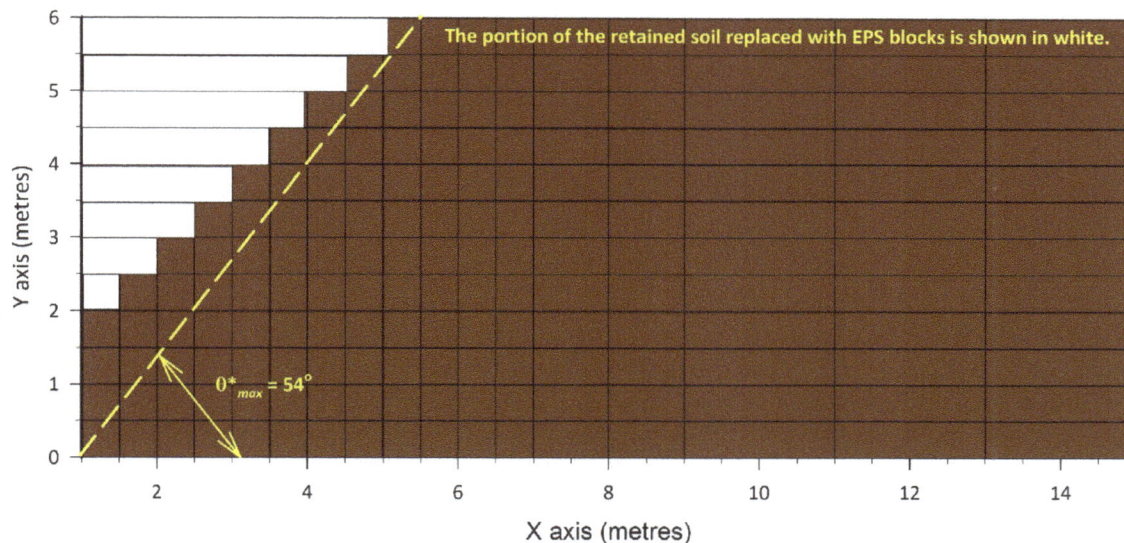

The portion of the retained soil replaced with EPS blocks is shown in white.

$\theta^*_{max} = 54°$

Figure C.4. Lightweight-Fill Function - Partial (2/3)-Depth Placement.

Table C.3. Two-Thirds Replacement Case - Gravity Loading.

Parameter	Results		
	trial wedge		Coulomb (no-geofoam baseline)
	rigorous	simplified	
P_a, kN/m (kips/ft)	26.0 (1.78)	24.6 (1.68)	77.2 (5.29)
$\theta_{critical}$, degrees	49.3	54.0	64.0

C.2 SEISMIC LOADS

C.2.1 Theoretical Development

C.2.1.1 Introduction

Developing a trial-wedge analytical model that includes seismic loads is a conceptually simple and straightforward extension of the preceding discussion for gravity loads in the same sense that the M-O Method is a logical extension of Coulomb's solution.

With reference to Figure C.5 (this is the same as Figure 4.9, reproduced here for ease of reference), the key changes in the resultant forces acting on a typical trial wedge in the seismic case are:

- the addition of a new horizontal force component, $k_h \cdot W^*_i$, that represents the seismic-inertial force due to the horizontal component of shaking of the mass of the soil wedge and

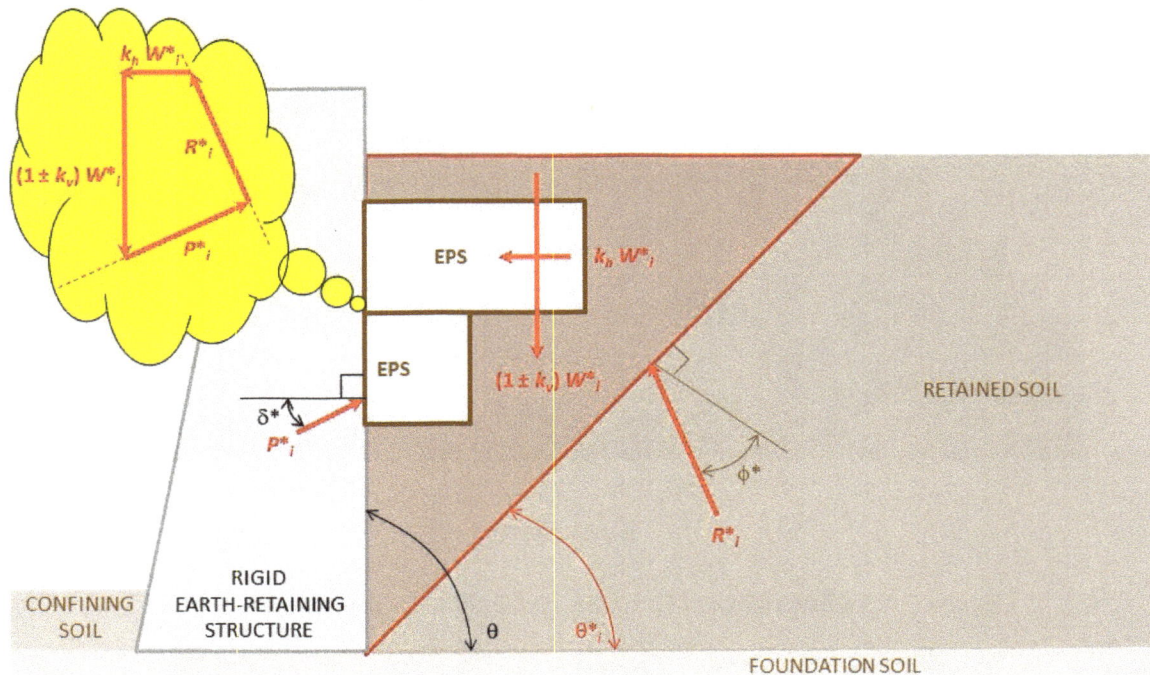

Figure C.5. Lightweight-Fill Function - Partial-Depth Placement Alternative Seismic Loading - Analytical Model Based on Trial-Wedge Model.

- modification of the vertical force component from the gravity-load case that represents the weight of the soil wedge, W^*_i, due to the vertical component of shaking by a factor defined as $(1 - k_v)$. Note that this can have the effect of making the soil + EPS wedge 'weigh' less (if k_v is positive in sign which is defined as upward motion); more (if k_v is negative in sign which is defined as downward motion); or the same (if k_v is zero).

Note that because the trial-wedge model follows the same developmental logic as the M-O Method it suffers from the same flaws, i.e. producing unreasonably high seismic loads for relatively large values of k_h. This subject is discussed at length in Appendix A. It may be possible to develop an alternative version of the trial-wedge method using some of the alternatives to the M-O Method that are discussed in Appendix A. However, this is beyond the scope of this monograph.

The additional horizontal force component due to seismic inertia dominates the overall problem to the extent that most analysis and design methods for RERSs that are intended for use in routine practice ignore the effects of the vertical component of acceleration as noted elsewhere in this monograph. However, for the sake of completeness and due-diligence in investigating this problem, the vertical acceleration effects are included in all solution derivations.

C.2.1.2 Rigorous Solution

Development of the rigorous trial-wedge solution for the seismic-load case follows that presented previously in this appendix for gravity loads with the following modifications that are consistent with the discussion in the preceding section:

- The term $-k_h W^*_i$ is added to Equation C.1 for horizontal force equilibrium. The negative sign results from the assumed sign convention for horizontal forces (positive when acting toward the right in Figure C.5) and the fact that this force is always assumed to act toward the left relative to this figure as this creates the more conservative loading condition. Note that the wedge-weight in this case is never corrected for vertical-acceleration effects.

- The weight term, W^*_i, in Equation C.5 for vertical force equilibrium is multiplied by $(1 - k_v)$.

Details of the subsequent derivation are not presented here as they duplicate those outlined in detail earlier in this appendix for gravity loads. The resulting equation that must be solved to find the angle θ^*_i that yields the maximum value of P^*_i is:

$$P^*_i = \frac{\{[0.5 \cdot \gamma_t \cdot (\cot\theta^*_i \cdot H) \cdot H] - [(\gamma_t - \gamma_{EPS}) \cdot V_{EPS}]\} \cdot [(1 - k_v) \cdot k_h \cdot \tan(90° - \theta^*_i + \phi^*)]}{\sin\delta^* + [\cos\delta^* \cdot \tan(90° - \theta^*_i + \phi^*)]} \qquad \text{(C.18)}$$

Note that this maximum value of P^*_i is defined here as P^*_{ae}, the active resultant earth force that includes both gravity and seismic contributions, and the value of the angle θ^*_i at which it occurs correlates with the angle α_{AE} (called α^*_{AE} here to highlight the fact that it defines a failure wedge that includes both soil and EPS blocks and will thus be different than α_{AE} which is based on an all-soil failure wedge) that is obtained from equations that are part of M-O Method (see Appendix A and Equation A.15). Lastly, as with the gravity-load case, for the purposes of the example problems in this monograph the writer used the *Solver* function in *Excel 2010*.

C.2.1.3 Simplified Solution

For the same reasons given earlier in this appendix for the gravity-load case, the writer also developed a simplified solution for the seismic-load case that could be useful for preliminary design when various configurations of EPS blocks are being considered. As with the gravity-load case, use of this simplified solution yields only approximate results relative to the rigorous trial-wedge method presented in the preceding section so its use should be limited to preliminary design with final design performed using the rigorous solution outlined previously.

The simplified solution for the seismic-load case is based on the assumption that $\theta^*_{critical}$ is known a priori and in this case corresponds to the angle α_{AE} from M-O Method (Equation A.15). The same sequence of calculations as outlined in detail earlier in this appendix for the gravity-load case is performed for the seismic-load case with the following exceptions:

- In Step 2, use $\theta^*_{critical} = \alpha_{AE}$.

- In Step 8, calculate and use K_{ae} from the M-O Method in lieu of K_a from Coulomb's solution.

- In Step 9, substitute the following equation for Equation C.17:

$$P_{ae}* = 0.5 \cdot K_{ae} \cdot (1 - k_v) \cdot \gamma * \cdot H^2 . \tag{C.19}$$

C.2.2 Example Problems

C.2.2.1 Introduction

As with the gravity-load case, application of the rigorous and simplified trial-wedge solutions for the seismic-load case that were presented in the preceding two sections are illustrated using the same overall RERS geometry and soil properties used for the gravity-load examples presented previously in this appendix. Additional assumptions made for seismic loading are k_h = 0.20 and k_v = ±0.10. Note that these are relatively modest accelerations and are within the limits where the M-O Method (and, by extension, the trial-wedge method presented and used here) has been found to produce reasonable results.

Also, as with the gravity-load case, three different examples will be illustrated. The first is the all-soil/no-geofoam baseline case that will allow comparison of results from the trial-wedge methods with exact results from the M-O Method. Following this, two cases of partial-depth placement are illustrated to both provide comparison of results from the rigorous and approximate trial-wedge methods as well as to give some general idea of relative load reduction compared to the baseline case.

C.2.2.2 All Soil/No Geofoam

Table C.4 presents the results for this baseline case that was used to verify the programmed rigorous and simplified analytical methods by comparing their results to the known M-O Method solution. As can be seen, all analytical methods are in complete agreement as would be expected in this case because the M-O solution considers ERS-retained soil friction when evaluating $\theta*_{critical}$ (= α_{AE} in this case).

Table C.4. All-Soil/No-Geofoam Case - Seismic Loading.

Parameter		Results		
		trial wedge		Mononobe-Okabe
		rigorous	simplified	
P_{ae}, kN/m (kips/ft)	+ k_h	116.6 (7.99)	116.6 (7.99)	116.6 (7.99)
	k_h = 0	123.0 (8.42)	123.0 (8.42)	123.0 (8.42)
	- k_h	129.7 (8.88)	129.7 (8.88)	129.7 (8.88)
α_{ae}, degrees	+ k_h	49.6	49.6	49.6
	k_h = 0	50.9	50.9	50.9
	- k_h	51.9	51.9	51.9

C.2.2.3 One-Third Depth of Geofoam

The test case of one-third replacement of retained soil with EPS blocks is the same as that used previously for gravity loading (Figure C.3). Table C.5 presents the results from the rigorous and simplified analytical methods presented in this appendix. The results for the all-soil/no-geofoam baseline case are also shown for comparison.

Table C.5. One-Third Replacement Case - Seismic Loading.

Parameter		trial wedge rigorous	trial wedge simplified	Mononobe-Okabe (no-geofoam baseline)
P_{ae}, kN/m (kips/ft)	$+ k_h$	98.8 (6.77)	97.7 (6.69)	116.6 (7.99)
	$k_h = 0$	103.4 (7.08)	102.2 (7.00)	123.0 (8.42)
	$- k_h$	108.3 (7.42)	106.9 (7.32)	129.7 (8.88)
α^*_{ae}, degrees	$+ k_h$	46.5	49.6	49.6
	$k_h = 0$	47.7	50.9	50.9
	$- k_h$	48.7	51.9	51.9

Based on the results obtained using the rigorous solution, there is approximately a 15-16% reduction in the calculated active earth force as a result of the EPS blocks occupying approximately 15-16% of the volume of the critical failure wedge. The calculated angles for this failure wedge are well within the allowable limits for this case ($\leq 70°$, $\geq 38°$) so the solution is considered valid. As expected, vertical-acceleration effects had only modest influence on the calculated results.

Note that the simplified analysis is again in reasonable agreement with the rigorous analysis but adequate only for preliminary assessment purposes as it always errs on the unsafe side.

C.2.2.4 Two-Thirds Depth of Geofoam

The test case of two-thirds replacement of retained soil with EPS blocks is the same as that used previously for gravity loading (Figure C.4). Table C.6 presents the results from the rigorous and simplified analytical methods presented in this appendix. The results for the all-soil/no-geofoam baseline case are also shown for comparison.

Table C.6. Two-Thirds Replacement Case - Seismic Loading.

Parameter		trial wedge rigorous	trial wedge simplified	Mononobe-Okabe (no-geofoam baseline)
P_{ae}, kN/m (kips/ft)	$+ k_h$	61.5 (4.21)	48.7 (3.34)	116.6 (7.99)
	$k_h = 0$	62.6 (4.29)	48.0 (3.29)	123.0 (8.42)
	$- k_h$	64.1 (4.39)	47.7 (3.27)	129.7 (8.88)
α^*_{ae}, degrees	$+ k_h$	40.0	49.6	49.6
	$k_h = 0$	41.0	50.9	50.9
	$- k_h$	41.9	51.9	51.9

Based on the results obtained using the rigorous solution, there is approximately a 47-51% reduction in the calculated active earth force as a result of the EPS blocks occupying approximately 42-45% of the volume of the critical failure wedge. The calculated angles for this failure wedge are just within the limits for this case ($\leq 54°$, $\geq 38°$) so the solution is considered valid. Again, vertical-acceleration effects had only modest influence on the calculated results.

Note that the simplified analysis is in much poorer agreement with the rigorous analysis in this case due to the much greater difference between the critical angles for the respective failure wedge. However, for preliminary-design purposes the simplified analysis would at least be in the 'ballpark'.

Appendix D

Engineering Versus True Representation of Normal Stress and Strain

D.1 INTRODUCTION

As noted in Chapter 3, the inherent large-strain nature of the compressible-inclusion function raises the question as to whether or not the traditional definitions of stress and strain most often used by both geotechnical and structural engineers in the U.S. at least are appropriate to use when assessing the performance of a compressible inclusion. This is because these definitions are based on the original, undeformed geometry of the object being loaded and are normally considered adequate only for relatively small-strain applications.

The alternative is to use stress and strain formulations that are based on the deformed geometry of the loaded object. This is broadly similar to the analytical decision structural engineers routinely face as to whether to use a *linear analysis* for some structural system that is based on the undeformed geometry of the system or a *nonlinear analysis* that is based on the deformed geometry of the system. The benefit of the latter is that it considers what is commonly known as the *P-Δ effect* and the additional forces and moments associated with it.

The objective of this appendix is to explore some basic concepts of evaluating stress and strain using the traditional civil engineering small-strain approach vs. the less common large-strain approach. For simplicity in the discussion that follows, only normal stress and normal strain are considered as these are the conditions that predominate when loading a compressible inclusion. Also, the problem of a small-scale test specimen of simple geometry (e.g. a right-circular cylinder with a solid cross-section and uniform diameter) subjected to uniaxial loading will be used for simplicity to illustrate the basic concepts presented. However, the discussion is completely general and can easily be extended to any type of object with a complex geometry and loading.

D.2 OVERVIEW

The writer is only aware of the traditional small-strain formulations for stress and strain being used for assessing geofoam behavior in compressible-inclusion applications. However, there are a number of arguments that can be made for at least considering using the alternative large-strain formulations, at least for strain if not for stress as well:

- The simple, basic fact that compressible inclusions are inherently a large-strain problem.

- Polymeric materials in general are often cited as being a broad class of materials for which large-strain formulations are preferable because of their ability, as a group, to undergo large strains prior to rupture.

- Applications in which loads are applied in stages are also cited as being more-rationally analyzed using large-strain formulations. This would certainly apply to many compressible-inclusion applications where there would be immediate, permanent compression of the geofoam under long-term gravity loading. This could be followed by cyclic applications of a surface surcharge and/or perhaps seismic loading at some random time in the future.

It is the writer's experience that civil engineers, at least in the U.S., do not receive much, if any, formal education about large-strain formulations of stress and strain. Therefore, this appendix is intended to be a brief primer and overview of the subject. However, before doing so some cautions are noted:

- Because of the Internet, there is no shortage of websites that can be found that address this subject in varying detail and with varying clarity-of-presentation.

- There is no consistency of notation for either stress or strain among the various websites. Furthermore, the writer has not observed that there is a clear 'best' notation or at least a notation that appears to be used most often. Therefore, the notation used in this appendix is arbitrary but consistent.

- Every such website that the writer has seen shows both the traditional small-strain and alternative large-strain derivations and final results based on traditional structural engineering sign conventions. This means that tensile forces and stresses are positive and the elongation strains they produce are also positive. As is well known, the typical geotechnical engineering sign conventions are the opposite of this. The derivations and results shown subsequently in this appendix reflect geotechnical engineering sign conventions as they are more relevant to this monograph.

- Many websites show a simple equation to convert stresses obtained in one formulation to the other. What very few of these websites make clear is that this simple relationship is based on two key assumptions:

  o the specimen maintains a constant volume as it deforms and

  o the cross-sectional dimension of the specimen is the same from its top to its bottom at any given stage of testing.

The former assumption is nowhere near to being correct for geofoam materials where the gas-filled voids of the cellular structure (which may account for as much as 99% of the total volume of the material initially) are literally crushed when loaded in compression, with the result that the geofoam undergoes significant volume reduction during loading. The latter assumption is not met in many laboratory-testing scenarios of small-scale test specimens as friction between the upper and lower surfaces of the specimen and platens that are part of the testing apparatus tends to cause the middle of a test specimen to deform (either bulge or neck depending on the specific material being tested) differently than ends of the test specimen. Therefore, this simple correlation between small and large strain-based stresses that is typically found online is clearly not applicable to geofoam materials.

D.3 NORMAL STRAIN

As noted previously, civil engineers, at least in the U.S., are educated primarily, if not exclusively, to use what is referred to as *engineering* (a.k.a. *Cauchy*) *strain*. It is based on dimensional changes of a test specimen relative to the original dimensions of the specimen.

Using the geotechnical sign convention of compression positive, *engineering normal strain*, ε, due to uniaxial loading in either compression or tension is defined as:

$$\varepsilon = -\left(\frac{\Delta L_i}{L_o}\right) = \frac{-(L_i - L_o)}{L_o} = \frac{L_o - L_i}{L_o} \tag{D.1}$$

where L_o = original length of the test specimen; L_i = length of the test specimen at an arbitrary point during loading; and $\Delta L_i = (L_i - L_o)$ = change in length of the test specimen at an arbitrary point during loading relative to its original length.

As noted earlier in this appendix, engineering strain, which can be viewed or interpreted as an average or secant strain over some range of what may, in general, be a non-linear stress-strain curve that begins at the origin, is generally considered to be acceptable for use for 'small strains' although what the limiting strain is for acceptability does not seem to be universally agreed upon. However, various websites suggest that strain levels as small as 0.05 (5%) are the limit of 'acceptable' accuracy for engineering strains. Note that this is well below the operational strain range of virtually all compressible-inclusion applications involving geofoam.

The alternative to engineering strain is most commonly called *true strain* although alternative terminology to 'true' exists (e.g. *Hencky, logarithmic, natural*). True strain is defined as an incremental strain at some arbitrary point in the loading sequence <u>relative to the specimen length at that point</u> (as opposed to the specimen length at the start of the test) so can be viewed as a tangent strain at some point on a non-linear stress-strain curve. Thus, in the same way that a non-linear stress-strain curve can be interpreted to have secant vs. tangent moduli (Young's, shear), it can be interpreted to have secant (engineering) vs. tangent (true) strains.

Using the geotechnical sign convention of compression positive, *true normal strain*, *e*, due to uniaxial loading in either compression or tension is defined as:

$$de = -\left(\frac{dL_i}{L_i}\right) \tag{D.2}$$

where d = differentiation operator.

The strain itself is obtained by integrating both sides of Equation D.2:

$$e = \int de = \int_{L_o}^{L_i} -\left(\frac{dL_i}{L_i}\right) = -\ln\left(\frac{L_i}{L_o}\right). \tag{D.3}$$

The correlation between engineering and true normal strain is obtained by expanding the final term in Equation D.3 using relationships defined in Equation D.1:

$$e = -\ln\!\left(\frac{L_i}{L_o}\right) = -\ln\!\left(\frac{(\Delta L_i + L_o)}{L_o}\right) = -\ln\!\left(\left(\frac{\Delta L_i}{L_o}\right) + \left(\frac{L_o}{L_o}\right)\right) = -\ln\bigl((-\varepsilon)+1\bigr) = -\ln(1-\varepsilon) \quad \textbf{(D.4a)}$$

which simplifies to

$$e = -\ln(1-\varepsilon). \tag{D.4b}$$

D.4 NORMAL STRESS

The difference between *engineering stress* and *true stress* is broadly similar to that discussed for strain in that *engineering normal stress* is based on the original cross-sectional area of the test specimen whereas *true normal stress* is based on the cross-sectional area at the specific point in the test for which the stress calculation is made. Note that unless certain assumptions related to volume and dimensional changes that were discussed earlier in this appendix are made and are reasonably valid throughout a test, true-stress calculations require on-specimen displacement measurements that would allow accurate calculation of the test-specimen cross-sectional area to be made. This requirement is certainly applicable to geofoam materials.

It is of interest to note that geotechnical engineers have, for decades, been attempting true-stress calculations for triaxial tests on soil specimens. It is not uncommon for test data to be reduced assuming that the test specimen deforms either as a perfect right-circular cylinder or, more commonly, in some 'barrel' shape, usually defined by some assumed curve such as a parabola, that has its largest diameter at mid-height of the test specimen. However, these true stresses are then plotted against engineering strains that results in a conceptually inconsistent hybrid plot. This, then, raises the question as to whether true strains should be used instead. It would certainly result in plots that were more-rigorously correct and perhaps more insightful as well as discussed in the following section.

D.5 COMBINED EFFECTS

It is beyond the intent and scope of this monograph to perform a comparison between stress-strain tests interpreted using engineering vs. true stress-strain methodologies. However, it is useful to point out some general observations:

- With regard to normal strain, the two methods start out yielding the same results but they diverge with increasing strain, with true normal strains always being smaller than engineering normal strains in both compression and tension.

- With regard to normal stresses, again, the two methods yield the same results initially before diverging with increasing strain. However, the divergence differs from normal strains in that in compression engineering stresses are larger in magnitude than true stresses provided that Poisson's ratio is non-negative. While this might seem to always be the case, it is not. Some closed-cell geofoam materials such as EPS-block have actually been observed to 'neck' at large compressive strains which means that Poisson's ratio is negative at that point which is certainly counter-intuitive. In such cases, the true normal stresses would therefore be larger than the engineering normal stresses. On the other hand, under tensile loading true stresses will always be larger than engineering stresses.

When the same set of stress-strain data are interpreted using both engineering and true definitions, not only will the quantitative results differ but the qualitative results as well. There are countless examples of this that can be found online, usually using the classical tensile test on structural (a.k.a. mild or carbon) steel. The stress-strain curve seen by every civil engineering student, which uses the engineering definitions of stress and strain, shows a relatively complex post-yield behavior of work hardening followed by strain softening to the point of physical rupture. Yet when plotted using the true definitions of stress and strain, the post-yield region is all work-hardening in nature although the slope of the curve is variable up to the point of rupture due to the development of significant necking in the test specimen just before rupture.

Focusing now on how the difference between engineering and true stress-strain affect geofoam materials used in compressible-inclusion applications, it is useful to first review the generic behavior of both 'normal' EPS-block and R-EPS which are the predominant geofoam materials used in this functional application. This is shown in Figure D.1 that was reproduced from Horvath (2010b).

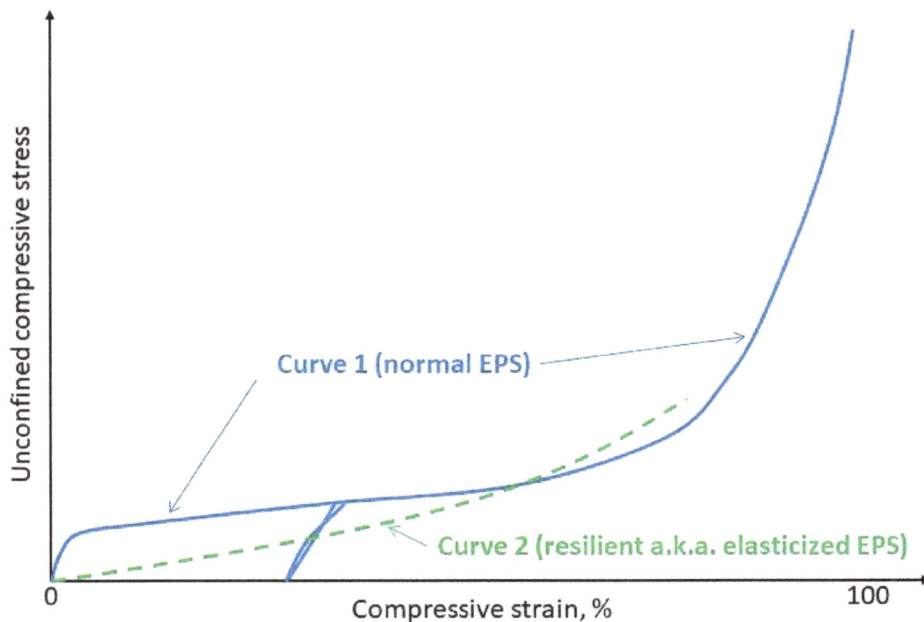

Figure D.1. Generic Engineering Stress-Strain of Compressible-Inclusion Materials.

To begin with, if the strains were recalculated using the true-strain definition, both material curves would shift to the left with a strain of approximately 70%, not 100%, being the asymptotic limit approached by Curve 1 for normal EPS-block.

The effect on stresses is a little less certain. Both geofoam materials shown have relatively small positive values of Poisson's ratio under relatively low stress levels so true stresses would not be expected to be very much different from the engineering stresses depicted in the figure. At higher stress levels, some necking might develop which would tend to make the true stress levels larger than the stresses shown.

Overall, then, the true stress-strain curves would have the same basic shape as the curves shown in Figure D.1 although the slopes of the curves would be steeper which would mean that the calculated moduli would be somewhat larger and the overall product stiffness somewhat greater. This is a potentially significant outcome as all compressible-inclusion functional applications are stiffness-based and thus require an accurate estimate of the stiffness of the geofoam product used.

D.6 CONCLUSIONS

Although the issue of engineering vs. true stress and strain has only been covered in a very introductory and preliminary manner in this appendix, available information suggests that it is a topic worth pursuing further for the compressible-inclusion function. The use of true stress and strain may provide better insight into the behavior of compressible inclusions subjected to relatively complex loading such as:

- gravity loading under staged construction;

- gravity loading followed by application of a surface surcharge, especially if it is cyclic in nature; and

- gravity loading followed by seismic loading.

This is because the behavior of compressible-inclusion applications centers around material stiffness and as noted at the end of the preceding section the operational stiffness of geofoam materials at relatively large strains is different depending on whether it is calculated using engineering vs. true stresses and strains.

Appendix E

Numerical Modeling Using the Finite-Element Method

E.1 INTRODUCTION

Numerical modeling of a continuum using the FEM is used in two basic ways throughout this monograph:

- to provide calculated results that serve as a reference standard to which other analytical methodologies that are also presented in this monograph can be compared. In this case, the FE results are serving as the assumed 'correct answer' in lieu of measured results made on actual ERSs that would, in principle, be the preferred 'gold standard' or 'ground truth' for such comparison; and

- to provide fundamental insight into the basic behavioral mechanisms for certain cases in which geofoam used either alone or synergistically with other geosynthetics can reduce lateral pressures on ERSs. The FEM in geotechnical engineering has a long and rich history of providing fundamental insight into the behavior of complex problems, especially those involving soil-structure interaction (SSI). In such cases, engineering intuition may either be elusive or even incorrect at identifying complex behavior. Time and time again FE analyses subsequently proved to be correct in replicating complex SSI behavior that was either unexpected or even counterintuitive. As far as specific cases relative to the goals of this monograph in which the FEM was used in this way, FE analyses were used to provide basic knowledge about certain applications cases such as partial-depth placement involving the compressible-inclusion function for which no known analytical methodology currently exists. As such, the FE results can be used to provide future direction for the formulation of analytical methodologies for use in routine practice.

Due to the inherent limitations of computer software available to the writer, only gravity-load cases in 2-D were simulated for the FE analyses that were performed specifically for use in this monograph. This appendix contains descriptive background information about the FE software used as well as details of the mesh and material properties for the one- and two-dimensional elements in the mesh. The only specific analytical results presented in this appendix are for the baseline case of an all-soil backfill behind a RERS under both non-yielding and yielding conditions. Other results are presented within the main text of this monograph where relevant.

Before proceeding further, for the sake of completeness it is relevant to note that attempts to use the FEM for both of the above-stated purposes were not wholly successful, specifically with regard to the lightweight-fill function. It was originally intended to use the FEM to critically assess the accuracy of both the decades-old full-depth analytical methodology developed in Japan as well as the writer's newly developed partial-depth analytical methodology that is published for the first time in this monograph.

Unfortunately, the FE results in both cases were disappointing. This appears to be related to the difficulty in accurately modeling the complex nature of the elements of a lightweight-fill application, especially the individual nature of the EPS blocks and the vertical joints between blocks. As a result, the writer's FE analyses did not properly capture the small inter-block gaps and other construction-related features that combine to produce a hard-to-define and even harder-to-model 'slop' in the horizontal direction. This 'slop' results in much more lateral compressibility in the overall geofoam-retained soil system than was able to be modeled numerically. This additional compressibility would be expected to have a significant impact on the calculated outcomes, especially with regard to the lateral earth pressures acting on the ERS.

Because of the disappointing outcomes for the writer's FE analyses that involved the lightweight-fill function, none of these results is presented in this monograph. This led to the writer's emphasis in Chapter 4 that instrumented full-scale installations are a critical necessity for assessing all present and future analytical methodologies related to this function. In the longer term, it is certainly possible that a reliable body-of-knowledge based on field measurements will allow for numerical modeling such as the FEM to be calibrated to provide more-reliable results for the lightweight-fill function. However, at the present time this is a goal that appears to be well in the future. Consequently, for the foreseeable future the results of FE analyses involving ERSs and the lightweight-fill function should always be viewed with some caution.

E.2 SOFTWARE

The program used for all FE analyses presented in this monograph is named *SSTIPNH*™ and was written using the *FORTRAN*[128] programming language. This program is the writer's proprietary microcomputer version of the mainframe program *SSTIPN* that had been developed over a period of many years, originally at the University of California at Berkeley and later at The Virginia Polytechnic Institute and State University at Blacksburg, under the overall direction of Prof. J. Michael Duncan at both institutions. Despite the age of the original software, it contains the same basic analytical 'engine' that is used in current commercially available software.

The basic material model incorporated into *SSTIPNH* for 'solid' materials is Kondner's well-known hyperbolic model that has been a staple of geotechnical FE analyses for many decades. The model parameters can be manipulated to simulate linear-elastic as well as linear-elastoplastic material behavior when desired.

E.3 MODEL DETAILS

E.3.1 Overall Problem

One basic model was used for all FE analyses performed for this monograph. It is broadly similar to the one used in Horvath (2000) that dealt with a self-yielding RERS but without the components specifically related to modeling a bridge with integral abutments. The key features of this model are:

- The overall geometry was intentionally kept very simple in order to focus on the key aspects of overall system behavior for the various problems studied.

---

[128] This word is written in all capital letters in recognition of the fact that it is actually an acronym for the term **For**mula **Tran**slation. It is also in deference to the convention in use at the time the *SSTIPNH* program was originally written. However, it is recognized that later and current versions of this programming language are referred to as *Fortran*.

Lateral Pressure Reduction on Earth-Retaining Structures Using Geofoam
John S. Horvath, Ph.D., P.E., Life Member.ASCE

- The ERS is 6 m (19.7 ft) high (the actual and geotechnical heights are the same) with a planar inside face that is oriented vertically, i.e. θ = 90°. Unlike in Horvath (2000), the ERS itself was not modeled explicitly. Rather, the rigidity and yielding of the ERS were controlled by a combination of nodal boundary conditions along the left side of the mesh that provide overall problem confinement; bar elements (oriented horizontally) that function as linear axial springs that can be used to either allow for ERS yielding or as the geofoam layer in compressible-inclusion applications; and zero-thickness beam (flexure-only) elements (oriented vertically) of varying flexural stiffness to simulate ERSs of varying flexural rigidity. In addition, there was a layer of 1-D interface elements oriented vertically and placed along the inside face of these beam elements that allowed for modeling the interface friction between the simulated ERS and whatever material (soil and/or geofoam) the ERS was retaining. This somewhat-complex combination of elements was used so that one basic mesh could be used to simulate both RERSs and FERSs as well as non-yielding and yielding conditions as desired for different analytical scenarios.

- The retained soil was assumed to be granular in nature; uniform in composition; and with a planar, horizontal ground surface. For all cases analyzed, the soil was placed in 12 layers of equal thickness to broadly simulate filling or backfilling behind an actual ERS. However, no effort was made to simulate compaction of the soil as was done in earlier FE studies by the writer (e.g. Horvath 2000). Rather, each layer of soil was assumed to be in the normally consolidated at-rest state immediately after placement. While simulating soil placement as opposed to using a 'wished-in-place' approach for the finished structure created certain complications such as vertical soil-wall shear stresses even for a non-yielding RERS (as would be expected based on the discussion in Chapter 2), it was felt that using at least a basic level of reality would provide more-realistic results. There was a concern that because a wished-in-place approach is totally devoid of reality, it could produce misleading results that did more harm than good in terms of illustrating overall system behavior.

E.3.2 Mesh

Figure E.1 shows the basic mesh used for all FE analyses performed.

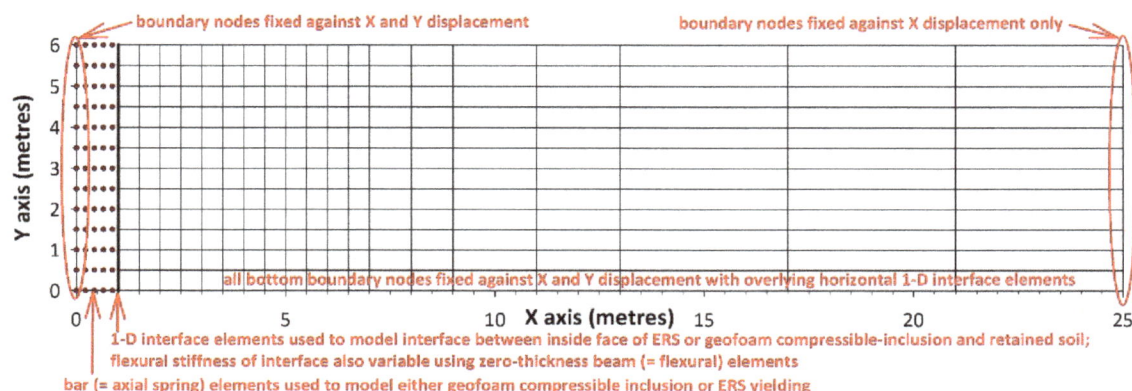

Figure E.1. Finite-Element Mesh.

E.3.3 Material Properties

The material properties for the 2-D solid elements that were used to represent the soil backfill/fill behind the ERSs are summarized in Table E.1.

Table E.1. Hyperbolic Constitutive Model Parameters for 2-D (Solid) Element Materials.

material	parameters									
	γ_t (N/m³)	K	n	R_f	K_b	m	c (kN/m²)	ϕ_1 (°)	$\Delta\phi$ (°)	K_o
soil	19800	450	0.35	0.8	110	0.1	0	38	0	0.38

For analysis cases involving a compressible inclusion, the geofoam panel was not modeled explicitly as a 2-D solid material as in earlier work by the writer (e.g. Horvath 2000) but as one-dimensional (1-D) springs as shown in Figure E.1. This achieved the same basic results as if 2-D elements were used and allowed for a more-general mesh to meet the diverse needs of the FE analyses performed for this monograph.

In all analyses involving a compressible inclusion, R-EPS was assumed as the geofoam material and product (one and the same in this case). Because this material exhibits significant creep under sustained loads, analyses for both rapid-loading/short-term and extended-loading/long-term conditions were performed. As noted in Chapter 5, normally in practice the stiffer rapid-loading/short-term compressible inclusion-product stiffness is used for design to be conservative then the long-term performance is checked (especially when designing for gravity loads) using the softer sustained-loading product stiffness. This latter check is necessary to make sure that the calculated long-term strains are within the operational limits of the geofoam product.

For the analyses performed for this monograph, the R-EPS was assumed to have a rapid-loading/short-term Young's modulus equal to 250 kN/m² (36 lb/in²) and an extended-loading/long-term Young's modulus equal to 125 kN/m² (18 lb/in²). Note that these assumptions are realistic as they are those of a product (*GeoTech TerraFlex* discussed in Chapter 8) that has been commercially available (in the U.S. and Canada at least) since the 1990s.

The material properties for the one-dimensional (1-D) interface elements used to model material interfaces along which sliding could occur are summarized in Table E.2.

Table E.2. Constitutive Model Parameters for 1-D (Interface) Elements.

materials in contact	parameters							
	$K_{s,i}$	$K_{s,u}$	n	R_f	K_n	c_a (kN/m²)	δ_1 (°)	$\Delta\delta$ (°)
soil-soil (also soil-geofoam)	1000	2000	0.5	0.9	100000000	0	38	0
soil-poured PCC	3000	6000	0.5	0.9	100000000	0	34	0
soil-formed PCC	800	1600	0.5	0.9	100000000	0	23	0
formed PCC-geofoam	300	600	1.0	1.0	100000000	0	18	0
geofoam-geofoam	300	600	1.0	1.0	100000000	0	30	0

E.4 RESULTS

E.4.1 Introduction

Most of the results of the FE analyses performed for this monograph are presented and discussed in various sections throughout Chapter 5 where relevant to a particular section of the text. The only results presented in this appendix are for the baseline reference cases of an all-soil backfill with no geosynthetics of any kind present.

The primary purpose for performing and presenting these baseline cases was to check the performance of the mesh and material data as the results from these baseline cases could be compared against hand-calculated solutions of classical earth pressure theories.

E.4.2 Non-Yielding Rigid Earth-Retaining Structures

Figure E.2 shows the results after the simulated retained-soil placement behind a non-yielding RERS. In such a situation, the at-rest earth pressure state would normally be assumed to exist and the classical hydrostatic distribution of lateral earth pressures from this state are shown in this figure. Using the concepts defined in Appendix B and used throughout this monograph for expressing relative lateral resultant forces using the parameter β, this line defines the baseline reference case of β =1. The results from the active state obtained using Coulomb's solution are shown as well using the traditional hydrostatic distribution assumed in practice. In this case, β = 0.52 for the active state which implies that the resultant force is 48% less than that of the theoretical at-rest state.

Two sets of FE results are also shown in Figure E.2. One is for the 1-D interface elements that by definition are along the ERS-retained soil interface. The other is for the first column of 2-D elements within the retained soil adjacent to the ERS-retained soil interface. Because the FEM calculates the stress state at the geometric centroid of a 2-D element, this means that the stresses shown in this case are actually those at a horizontal distance of 250 mm (9.8 in) into the retained soil in this case.

The reason for showing both sets of results is that for the *SSTIPNH* program, the writer's experience from using this program for over 30 years is that 1-D interface elements in any and all applications can sometimes exhibit scatter-like variability in the calculated values along an interface although the overall trend or average is accurate. Therefore, it is the writer's preference to include results from the first column of 2-D elements within the ground away from the ERS as well to acts as a check for the 1-D elements as the stresses calculated in 2-D elements tend to exhibit greater numerical consistency.

In any event, both the 1-D and 2-D element results shown in Figure E.2 are very close in magnitude and agree reasonably well with the theoretical at-rest line, with a calculated β = 1.10 and 1.09 for the 1-D and 2-D element types respectively. The deviation between the FE and theoretical results (β = 1) tends to increase with depth and is likely due to the Poisson effect of placing the retained soil in 12 artificial layers in an idealistic situation of perfectly rigid lateral confinement on both sides of the FE mesh. In addition, there may be some effect of the ERS-retained soil interface friction that develops because of the simulated construction sequence. This interface friction is something that will occur with actual non-yielding RERSs and is thus picked up by the FE analysis but is neglected by classical theory.

320

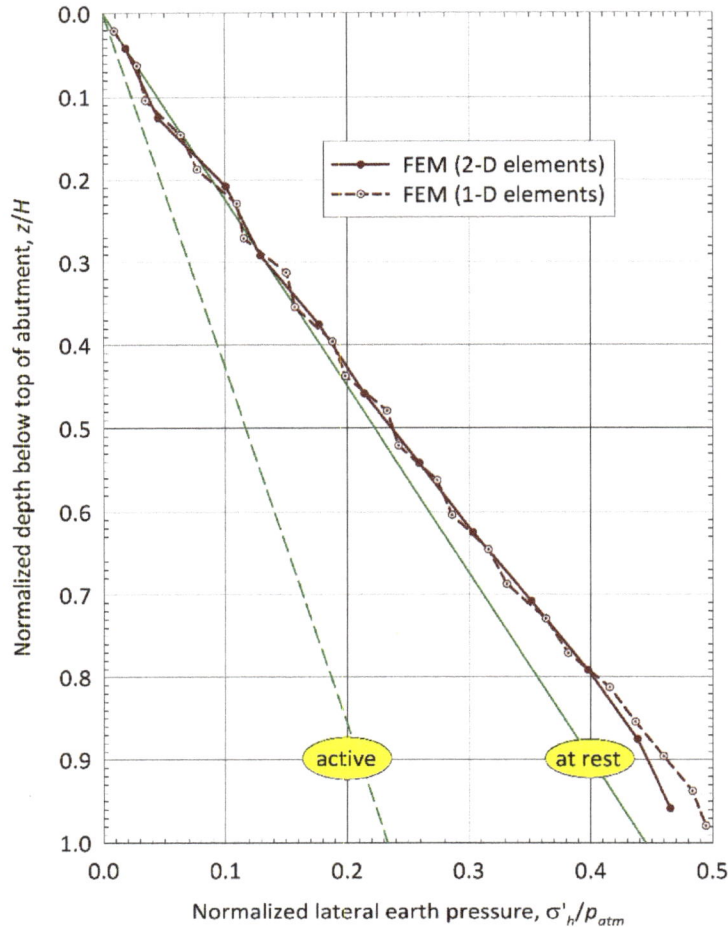

Figure E.2. Calculated Lateral Earth Pressures for Non-Yielding RERS Baseline Case.

While on the subject of ERS-retained soil shear stresses, it is useful to introduce the dimensionless parameter called the *stress level*, *S*, which is but one of many quantities calculated by *SSTIPNH* and similar FE software. Stress level at a point is defined as:

$$S = \frac{\left(\sigma'_1 - \sigma'_3\right)}{\left(\sigma'_1 - \sigma'_3\right)_f} \tag{E.1}$$

where

- $\left(\sigma'_1 - \sigma'_3\right)$ = calculated principal effective stress difference at that point and

- $\left(\sigma'_1 - \sigma'_3\right)_f$ = theoretical principal effective stress difference at failure at that point.

Essentially, stress level is the ratio of the actual (i.e. calculated) shear stress to the shear stress required to cause failure, with the latter being equivalent to the shear strength. For 2-D solid elements in *SSTIPNH*, stress level is calculated based on the average stress state existing at the geometric centroid of an element.

Lateral Pressure Reduction on Earth-Retaining Structures Using Geofoam
John S. Horvath, Ph.D., P.E., Life Member.ASCE

Inspection of Equation E.1 indicates that:

- $S = 0$ defines the isotropic stress state;

- $S = 1$ defines soil 'failure', i.e. full mobilization of shear strength; and

- $0 < S < 1$ defines intermediate conditions.

The most common use of stress levels is to plot and contour them using the FE mesh as the base drawing. Thus, contours of stress levels equal to or approaching one indicate zones where soil failure occurs or is tending to occur. Furthermore, patterns formed by the stress-level contours illustrate actual or potential failure zones (or approximately one-dimensional failure surfaces if the contours are closely spaced) within the soil.

However, care is always required when interpreting contoured stress levels in ERS problems as there is nothing to distinguish a soil failure in the active or passive state. Which is occurring needs to be determined using other results such as displacement patterns and directions.

Figure E.3 shows the contoured stress levels for the non-yielding RERS baseline case (for which the earth pressures acting on the ERS are shown in Figure E.2) using conventional discrete-solid-line contouring with a contour interval, ΔS, equal to 0.1[129]. Note that in the absence of ERS-soil friction induced by the backfill/fill process, i.e. if the RERS in this case were 'wished-in-place', the uniform, horizontal contours observed in the right-hand portion of the mesh away from the ERS-retained soil interface (which reflect pure K_o conditions) would continue right up to the ERS-retained soil interface. However, as can be seen, the existence of shear stresses along this interface causes localized distortion of the stress-level pattern.

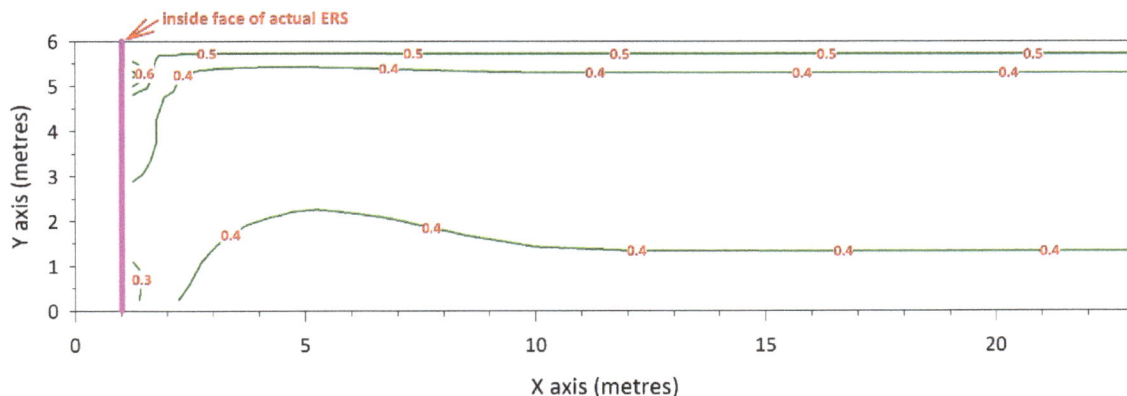

Figure E.3. Line-Contoured Stress Levels, *S*, for Non-Yielding RERS Baseline Case.

---

[129] As will be seen later in this appendix, the alternative of continuous-color contouring is useful in situations where stress levels are complex and rapidly varying.

322

E.4.3 Yielding Rigid Earth-Retaining Structures

E.4.3.1 Background

The other baseline case considered in this appendix is that of a yielding RERS. Unlike the non-yielding case, there is no unique result for the yielding case as:

- There many potential yielding modes when one considers the three basic modes defined in Chapter 2 either acting individually or in various combinations.

- For any of these modes or modal combinations, there is a theoretically infinite range of yielding magnitudes, Δ, that can occur when starting from the at-rest condition ($\Delta = 0$) discussed in the preceding section.

Therefore, the FE analyses performed for yielding RERSs had to be inherently limited in scope.

The issue of yielding ERSs becomes even more complex if the assumption of ERS rigidity is relaxed and FERSs are considered. As noted in Chapter 2, the yielding pattern associated with FERSs is significantly more complex than that associated with RERSs. For FERSs, the depth-wise variation in Δ is always non-linear (with RERSs it is at most linear with the two rotation modes) and the pattern of this depth-wise variation tends to vary with the particular FERS type, application, and soil conditions. Thus, the yielding patterns of FERSs cannot be neatly categorized by mode as with RERSs.

More importantly, empirical relationships between Δ and ERS performance that are at least approximately valid for RERSs may not be so for FERSs. This fact is explored in detail in Chapter 5 dealing with the compressible inclusion function. In that chapter, it is illustrated that the yielding pattern associated with the compressible-inclusion function mimics that of a FERS, not RERS.

That having been said, there is extensive historical use going back to the 1980s of empirical correlations involving the relationship between Δ and lateral earth pressure state for RERSs in simplified analytical methodologies for geofoam compressible-inclusion applications even though the latter display the behavior of FERSs. Thus, it is relevant to investigate the basic behavior of yielding RERSs as an initial, first-order step into exploring the behavior of geofoam compressible inclusions.

For the purposes of this monograph, it was felt that FE analyses involving only the yielding mode of pure translation would provide the greatest insight into applications involving the compressible-inclusion function. First of all, of the three modes of yielding discussed in Chapter 2, only translation produces a nominally parabolic distribution of lateral earth pressures that is qualitatively identical to that observed for the basic full-depth REP-Wall case of a compressible inclusion.

An additional consideration is the fact that the simplified analytical methodology dating back to the 1980s that is used to calculate the required minimum thickness of the compressible inclusion necessary to mobilize the active earth pressure state, $(t_{ci})_a$, in the basic full-depth REP-Wall case is based on the dimensionless ratio of minimum wall yield to geotechnical wall height, $(\Delta/H)_a$, necessary to mobilize the active state. This ratio, which is expressed as either a decimal number or percent, appears in various text and reference books such as Clough and Duncan (1991) which was the reference used for this purpose by the writer both in past as well as the present studies.

Before presenting and discussing the FE analyses performed, it is important to understand how Δ is defined when using the $(\Delta/H)_a$ ratio. In the specific case of the values tabulated in Clough and Duncan (1991, Table 6.6), it is defined as either the uniform value for the entire RERS yielding in the translation mode or the value at the top of the RERS for the rotation-about-bottom mode. For the purposes of this monograph, recognizing this definition is especially important for two reasons:

- As noted above and explored in detail in Chapter 5, the concept that $(\Delta/H)_a$ exists is used as an essential component of the simplified analytical method for sizing a compressible inclusion that dates back to work by Partos and Kazaniwsky in the 1980s and that the writer has modified over the years and continued to use to the present.

- As explored in detail in Chapter 5, for the basic full-depth REP-Wall case that uses a compressible inclusion of uniform thickness, the resulting depth-wise variation of Δ is neither uniform (as would be consistent with pure translation of a RERS) nor linearly varying (as would be consistent with rotation-about-bottom of a RERS). Rather, Δ varies in a relatively smooth, non-linear manner as a function of depth, broadly in the shape of a parabola that mimics the distribution of lateral earth pressure. Therefore, it is unclear what single value of Δ from the actual compression of a compressible inclusion should be used in the $(\Delta/H)_a$ ratio that is used to develop the simplified analytical methodology presented in Chapter 5. This issue is explored in that chapter.

E.4.3.2 Results of Analyses

For the purposes of the FE analyses performed for this monograph and presented in this appendix, the yielding case was enabled in the FE model by selecting a uniform stiffness magnitude for the horizontal bar (axial-spring) elements in the mesh (see Figure E.1) combined with creating a flexurally rigid interface between these bar elements and the retained soil. The result was that pure translational yielding of any desired magnitude Δ could be simulated.

For the initial analysis using this model, the bar-element stiffness was chosen to produce $\Delta \cong 25$ mm (I in) which is a Δ/H ratio $\cong 0.004 \cong 0.4\%$. This is the criterion for the active earth pressure state for "loose sand" per Table 6.6 in Clough and Duncan (1991).

The calculated lateral earth pressures for this analysis are shown in Figure E.4 along with the theoretical at-rest and Coulomb-active pressures shown previously in Figure E.2. In this case, the FE 1-D element results are more-scattered compared to the non-yielding case (Figure E.2) although the overall trend agrees with the 2-D element results. However, the more significant outcome is that the FE results exhibit the classical parabolic distribution of horizontal arching that was discussed in Chapter 3 as opposed to the hydrostatic distribution assumed in routine practice for the active earth pressure state. This outcome was not unexpected as it is well-known (even if not widely used) that arching theory provides a much more realistic distribution of lateral earth pressures that develop from RERS translation than the traditional triangular (hydrostatic) assumption.

Given the relevance of horizontal arching to this discussion, Figure E.5 compares the FE results (2-D elements only) to the closed-form solutions for translation-mode arching behind RERSs presented by Handy (1985) and Harrop-Williams (1989) that were previously noted in Chapter 3. The agreement between and among results is considered reasonable.

324

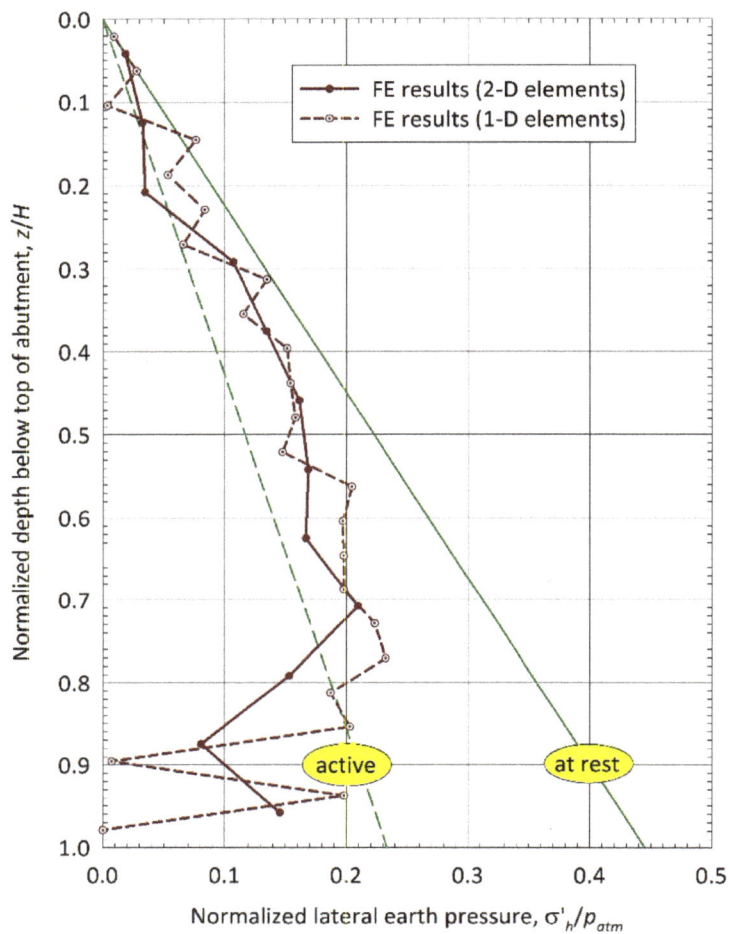

Figure E.4. Calculated Lateral Earth Pressures for Yielding RERS Baseline Case (Δ/H = 0.004).

Lateral Pressure Reduction on Earth-Retaining Structures Using Geofoam
John S. Horvath, Ph.D., P.E., Life Member.ASCE

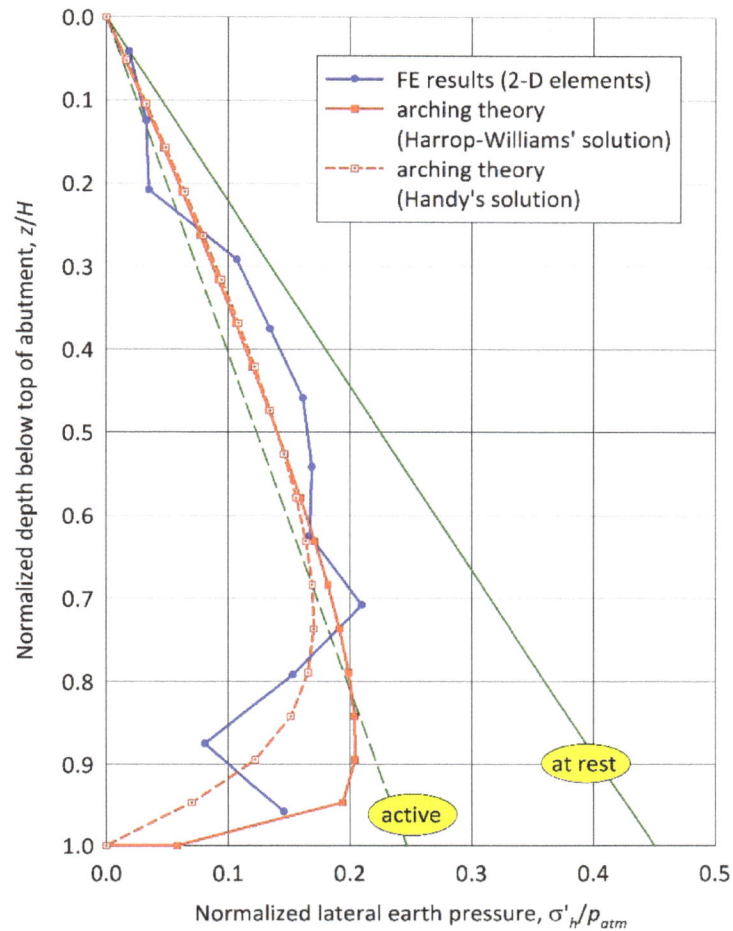

Figure E.5. Calculated Lateral Earth Pressures for Horizontal Arching with a Yielding RERS.

Although both the FE and horizontal-arching results have a depth-wise variation in lateral pressures that differ substantially from the traditional triangular/hydrostatic distribution of active earth pressures normally assumed with Coulomb's solution, the lateral resultant forces do not differ greatly between any of the methods shown in Figures E.4 and E.5. This is illustrated in Table E.3 where the β values for all five results shown collectively in these two figures are compared. This may explain, at least in part, why the hydrostatic distribution of lateral earth pressure, which has long been the backbone of routine practice, has produced RERS designs that have generally performed satisfactorily even though the distribution itself is incorrect.

Table E.3. Comparison of β Values for Yielding RERS Case.

Coulomb/active	FE results		arching theories (translation mode)	
	2-D elements	1-D elements	Harrop-Williams	Handy
0.52	0.53	0.57	0.56	0.48

Figure E.6 shows the line-contoured ($\Delta S = 0.1$ as before in Figure E.3) stress levels, S. The red dashed line shows the trend of the highest stress levels, i.e. equal to or approaching 1, and suggests the development of a zone of failure or near-failure broadly defining a failure wedge. This is more apparent in Figure E.7 that shows the same results using continuous-color contouring. In this case, red indicates $S = 1$ and a white dashed line is used to indicate the trend of a developing failure zone within the retained soil.

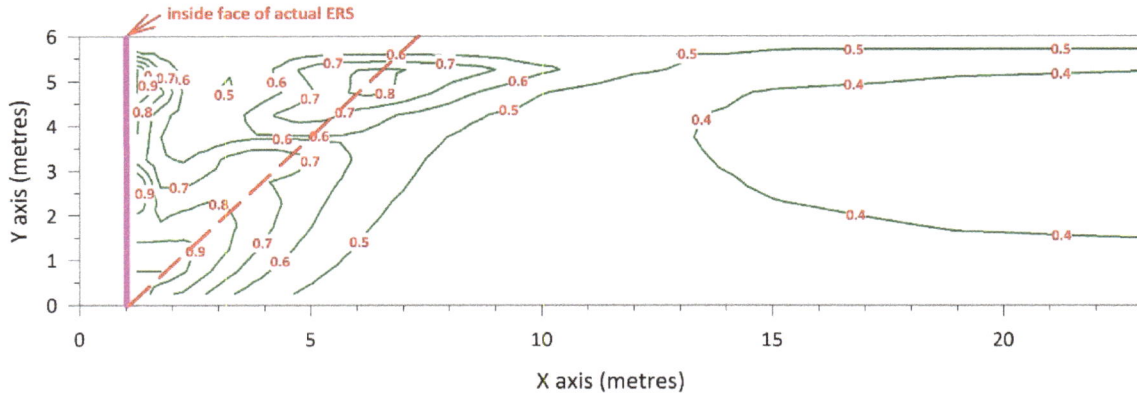

Figure E.6. Line-Contoured Stress Levels, S, for Yielding RERS Baseline Case.

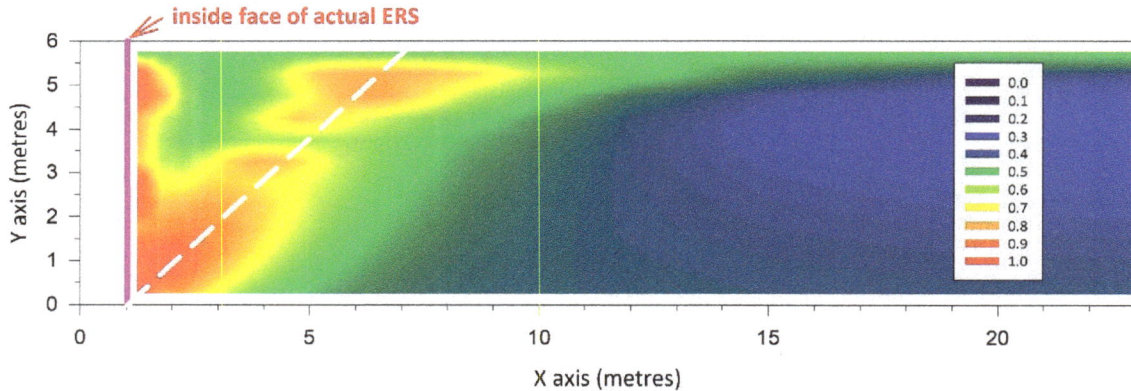

Figure E.7. Color-Contoured Stress Levels, S, for Yielding RERS Baseline Case.

Two additional analyses were performed to explore the sensitivity of the FE results to translation magnitude. Specifically, the stiffnesses of the bar (axial-spring) elements that control the translation magnitude were varied to produce uniform lateral displacements, Δ, of approximately 12.5 mm (0.5 in) and 6 mm (0.25 in) that correspond to Δ/H ratios $\cong$ 0.002 (0.2%) and $= 0.001$ (0.1%), respectively. These are the $(\Delta/H)_a$ ratios given in Clough and Duncan (1991) for "medium-dense sand" and "dense sand" respectively.

Figure E.8 shows the calculated lateral earth pressures acting along the inside face of the RERS for the three Δ/H ratios analyzed using the results from the 2-D elements only. Also included in this figure are the calculated results from the non-yielding baseline case (shown previously in Figure E.2) which is the $\Delta/H = 0$ limiting case. The resulting β values for each Δ/H ratio are indicated in the figure legend and clearly indicate that the active earth pressure state ($\beta \cong 0.5$ per Table E.3) is achieved only at the largest ratio in this case (i.e. the results shown previously in Figure E.4 and E.5).

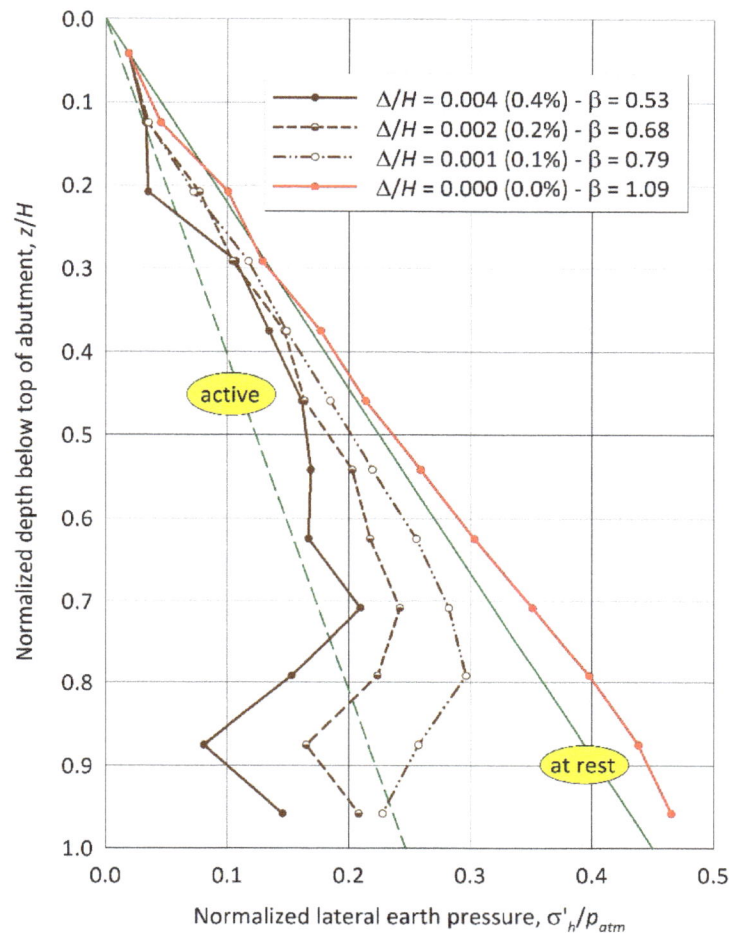

Figure E.8. Influence of Yielding Magnitude, Δ, for Yielding RERS Baseline Case.

The relationship between Δ/H and β is better seen in Figure E.9 where the non-linear transition from at-rest to active state is qualitatively identical to that which has been observed historically, with the only difference being that traditionally the ordinate (vertical axis) is scaled in lateral earth pressure as opposed to β as in this case but the meaning is identical[130].

---

[130] One behavioral element missing from the FE results shown in Figure E.9 that is observed with actual RERSs is the localized dip in the force-displacement curve that occurs just before trending toward a constant value of force with increasing displacement. This is because in actual soils the shear strength transitions from peak to constant-volume (critical-state) conditions, something that was not replicated in the soil model used in the writer's FE analyses.

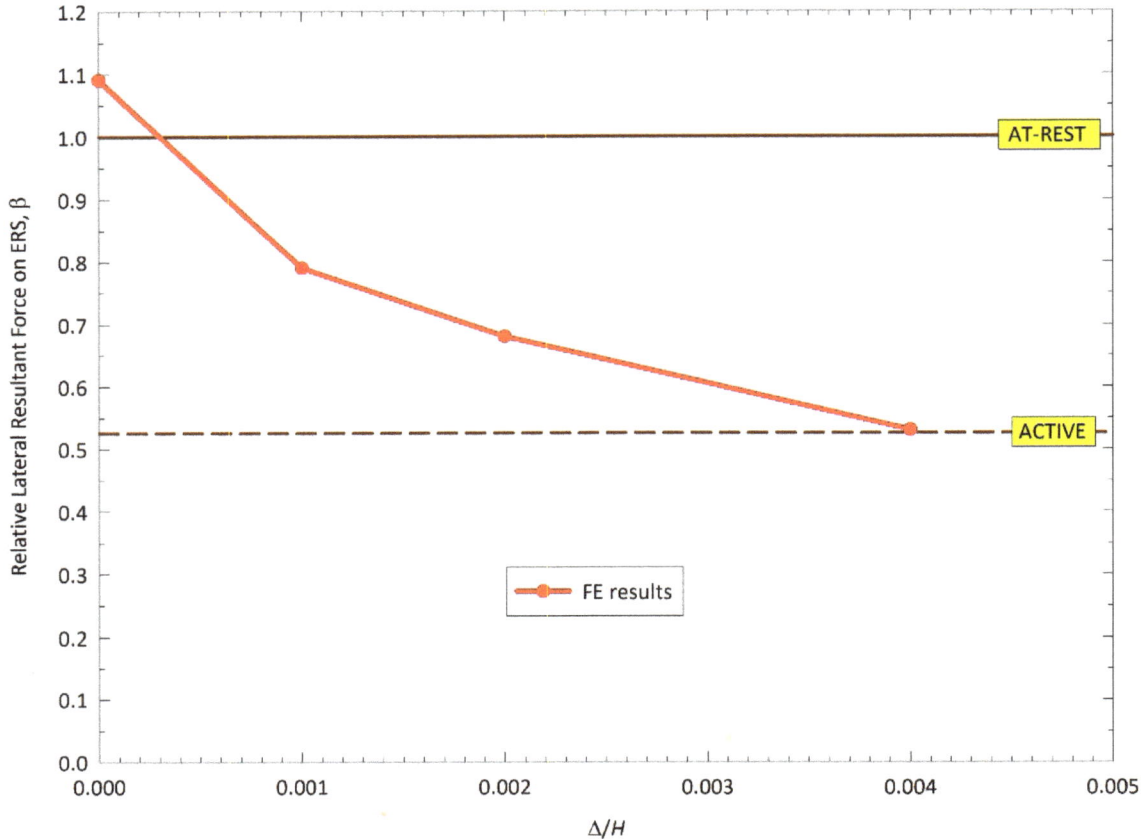

Figure E.9. Relationship between Δ/H and β for FE Model Used.

To conclude this discussion, the properties of the coarse-grain soil assumed for the FE analyses performed for this monograph were chosen to reflect what the writer felt would be a medium-dense consistency. The empirical $(\Delta/H)_a$ correlations in Clough and Duncan (1991) indicate that a value of 0.002 (0.2%) would be the minimum sufficient to mobilize the active earth-pressure state for such soils. However, the FE analyses clearly indicate that a larger ratio of 0.004 (0.4%), which correlates with "loose sand" per Clough and Duncan (1991), is required in this case.

It is emphasized that the results presented in this appendix should not be interpreted as invalidating the values presented in Clough and Duncan (1991) as the FE results are obviously sensitive to the specific mesh, constitutive model of the soil, and soil properties used. Nevertheless, it does suggest that the $(\Delta/H)_a$ ratios found in the published literature be used with some caution and understanding that they are approximate.

329

BIBLIOGRAPHY OF CITED REFERENCES

Note

Where available, a link to access a copy of a cited publication online is provided and highlighted in **bold** font. Only links where a copy can be accessed at no cost are provided. The most recent date on which the writer accessed such links is also given as information.

Ahmad, F. (1991). Supplemental comments on "Influence of Lateral Boundary Movements on Earth Pressure" by K. Z. Andrawes, A. McGown, and F. Ahmad. *Performance of Reinforced Soil Structures*, Thomas Telford Ltd., p. 381.

Andrawes, K. Z. (1991). Supplemental comments on "Application of Boundary Yielding Concept to Full Scale Reinforced and Unreinforced Soil Walls" by K. Z. Andrawes, K. H. Loke, K. C. Yeo, and R. T. Murray. *Performance of Reinforced Soil Structures*, Thomas Telford Ltd., p. 93.

Andrawes, K. Z., Loke, K. H. and Murray, R. T. (1992). "The Behaviour of Reinforced Soil Walls Constructed by Different Techniques". *Grouting, Soil Improvement and Geosynthetics*, American Society of Civil Engineers, pp. 1237-1248.

Andrawes, K. Z., Loke, K. H., Yeo, K. C. and Murray, R. T. (1991a). "Application of Boundary Yielding Concept to Full Scale Reinforced and Unreinforced Soil Walls". *Performance of Reinforced Soil Structures*, Thomas Telford Ltd., pp. 79-83.

Andrawes, K. Z., McGown, A. and Ahmad, F. (1991b). "Influence of Lateral Boundary Movements on Earth Pressure". *Performance of Reinforced Soil Structures*, Thomas Telford Ltd., pp. 359-364.

Andrawes, K. Z., Yeo, K. C., and Loke, K. H. (1993). "Behaviour of Geogrid Reinforced Soil Walls Subjected to Lateral Boundary Yielding". *Retaining Structures*, Thomas Telford Ltd., pp. 549-558.

Arellano, D., Stark, T. D., Horvath, J. S., and Leshchinsky, D. (2011a). *Guidelines for Geofoam Applications in Slope Stability Projects.* Preliminary draft final report (internal), National Cooperative Highway Research Program Project 24-11(02), Transportation Research Board, Washington, DC.

Arellano, D., Stark, T. D., Horvath, J. S., and Leshchinsky, D. (2011b). *Guidelines for Geofoam Applications in Slope Stability Projects.* Final report (internal), National Cooperative Highway Research Program Project 24-11(02), Transportation Research Board, Washington, DC.

Arellano, D., Stark, T. D., Horvath, J. S., and Leshchinsky, D. (2013a). *Guidelines for Geofoam Applications in Slope Stability Projects.* National Cooperative Highway Research Program Research Results Digest 380, Transportation Research Board, Washington, DC.

Arellano, D., Stark, T. D., Horvath, J. S., and Leshchinsky, D. (2013b). *Guidelines for Geofoam Applications in Slope Stability Projects.* Final report (public), National Cooperative Highway Research Project 24-11(02), Transportation Research Board, Washington, DC.

Athanasopoulos-Zekkos, A., Lamote, K., and Athanasopoulos, G. (2011). "Seismic Isolation of Earth Retaining Walls Using EPS Compressible Inclusions - Results from Centrifuge Testing". PowerPoint presentation at the 4th International Conference on Geofoam Blocks in Construction Applications, Lillestrom, Norway.

Aytekin, M. (1996). "Use of Geofoam with Expansive Soil". Proceedings of the Second International Conference in Civil Engineering on Computer Applications, Research and Practice (ICCE-96), University of Bahrain, Bahrain, Vol. 2, pp. 541-546.

Aytekin, M. (1997). "Numerical Modeling of EPS Geofoam Used with Swelling Soil". *Geotextiles and Geomembranes*, Vol. 15, Nos. 1-3, pp. 133-146.

Bathurst, R. J. and Alfaro, M. C. (1996). "Review of Seismic Design, Analysis and Performance of Geosynthetic Reinforced Walls, Slopes and Embankments". Proceedings of the Third International Symposium on Earth Reinforcement (IS-Kyushu '96), Fukuoka, Kyushu, Japan.

Bathurst, R. J. and Zarnani, S. (2013). "Earthquake Load Attenuation Using EPS Geofoam Buffers in Rigid Wall Applications". *Indian Geotechnical Journal*, Vol. 43, No. 4, pp. 283–291.

Bathurst, R. J., Keshavarz, A., Zarnani, S., and Take, A. (2007a). "A Simple Displacement Model for Response Analysis of EPS Geofoam Seismic Buffers". *Soil Dynamics and Earthquake Engineering*, Vol. 27, No. 4, pp. 344-353.

Bathurst, R. J., Nernheim, A., Walters, D. L., Allen, T. M., Burgess, P., and Saunders, D. D. (2009). "Influence of Reinforcement Stiffness and Compaction on the Performance of Four Geosynthetic-Reinforced Soil Walls". *Geosynthetics International*, Vol. 16, No. 1, pp. 43-59.

Bathurst, R. J., Zarnani, S., and Gaskin, A. (2007b). "Shaking Table Testing of Geofoam Seismic Buffers". *Soil Dynamics and Earthquake Engineering*, Vol. 27, No. 4, pp. 324-332.

Briaud, J.-L. (2011). "Unsaturated Soil Behavior for the Practicing Engineer". PowerPoint presentation made online, 12 May 2011.

Broms, B. B. (1971). "Lateral Earth Pressures Due to Compaction of Cohesionless Soils". Proceedings of the 4th Conference on Soil Mechanics and Foundations, pp. 373-384.

Candia Agusti, G. and Sitar, N. (2013). *Seismic Earth Pressures on Retaining Structures in Cohesive Soils.* Report UCB GT 13-02, University of California - Berkeley, Department of Civil and Environmental Engineering, Geotechnical Engineering, Berkeley, CA. **www.ce.berkeley.edu/sites/default/files/assets/users/sitar/GT%2013-02%20Candia%20and%20Sitar%20Report.pdf** Accessed 8 September 2017.

Card, G. B. and Carder, D. R. (1994). *A Literature Review of the Geotechnical Aspects of the Design of Integral Bridge Abutments.* Project Report 52, Transport Research Laboratory, Crowthorne, Berkshire, U.K.

Chen, T.-J. and Fang, Y.-S. (2008). "Earth Pressures Due to Vibratory Compaction". *Journal of Geotechnical and Geoenvironmental Engineering*, Vol. 134, No. 4, pp. 437-444.

Clough, G. W. and Duncan, J. M. (1991). "Earth Pressures". Chapter 6 in *Foundation Engineering Handbook*, 2nd ed., H.-Y. Fang (ed.), Van Nostrand Reinhold.

Costa, Y. D., Zornberg, J. G., Bueno, B. S. and Costa, C. L. (2009). "Failure Mechanisms in Sand over a Deep Active Trapdoor". *Journal of Geotechnical and Geoenvironmental Engineering*, Vol. 135, No. 11, pp. 1741-1753.

"Data Collection Will Clarify *Clayboard* Doubts" (1991). *Ground Engineering*, April.

Duncan, J. M. and Seed, R. B. (1986). "Compaction-Induced Earth Pressures under K_o-Conditions". *Journal of Geotechnical Engineering*, Vol. 112, No. 1.

Duncan, J. M., Williams, G. W., Sehn, A. L., and Seed, R. B. (1991). "Estimation Earth Pressures Due to Compaction". *Journal of Geotechnical Engineering*, Vol. 117. No. 12, pp. 1833-1847.

Duncan, J. M., Williams, G. W., Sehn, A. L., and Seed, R. B. (1993). Closure to "Estimation Earth Pressures Due to Compaction". *Journal of Geotechnical Engineering*, Vol. 119, No. 7, pp. 1172-1177.

Ebeling, R. M. and E. E. Morrison, Jr. (1992). *The Seismic Design of Waterfront Retaining Structures.* Technical Report ITL-92-11/NCEL TR-939, U.S. Department of Defense.

Ebeling, R. M., Patev, R. C., and Mosher, R. L. (1996). *Case Histories of Earth Pressure-Induced Cracking of Locks.* Technical Report ITL-96-9, U.S. Army Corps of Engineers, Waterways Experiment Station, Vicksburg, MS.

Ebeling, R. M., Peters, J. F., and Mosher, R. L. (1992). "Finite Element Analysis of Slopes with Layer Reinforcement". Proceeding of Stability and Performance of Slopes and Embankments - II, American Society of Civil Engineers, pp. 1427-1443.

Ebeling, R. M., Peters, J. F., and Mosher, R. L. (1997). "The Role of Non-Linear Deformation in the Design of a Reinforced Soil Berm at Red River U-Frame Lock No. 1". *International Journal for Numerical and Analytical Methods in Geomechanics*, Vol. 21, pp. 756-787.

Edgar, T. V., Puckett, J. A., and D'Spain, R. B. (1989). "Effects of Geotextiles on Lateral Pressure and Deformation in Highway Embankments." *Geotextiles and Geomembranes*, Vol. 8, No. 4, pp. 275-292.

England, G. L. (1994). "The Performance and Behaviour of Biological Filter Walls as Affected by Cyclic Temperature Changes". *Serviceability of Earth-Retaining Structures*, American Society of Civil Engineers, pp. 57-76.

England, G. L. and Dunstan, T. (1994). "Shakedown Solutions for Soil Containing Structures as Influenced by Cyclic Temperatures - Integral Bridge and Biological Filter". Proceedings of the Third International Conference on Structural Engineering, Singapore.

England, G. L., Dunstan, T., Tsang, C. M., Mihajlovic, N., and Bazaz, J. B. (1995). "Ratcheting Flow of Granular Materials". *Static and Dynamic Properties of Gravelly Soils*, American Society of Civil Engineers, pp. 64-76.

Ertuğrul, N. A. (2013). *Effect of Soil Arching on Lateral Soil Pressures Acting upon Rigid Retaining Walls*. Master Thesis submitted to the Graduate School of Natural and Applied Sciences, Middle East Technical University, Ankara, Turkey.
etd.lib.metu.edu.tr/upload/12615994/index.pdf
Accessed 5 September 2017.

Fahey, M. and Carter, J. P. (1993). "A Finite Element Study of the Pressuremeter Test in Sand Using a Nonlinear Elastic Plastic Model". *Canadian Geotechnical Journal*, Vol. 30, No. 2, pp. 348-362.

Fang, Y.-S. and Ishibashi, I. (1985). "Static Wall Pressures with Various Wall Movements". *Journal of the Geotechnical Engineering Division*, Vol. 112, No. 3, pp. 317-333.

Filz, G. M. (2003). "Compaction-Induced Lateral Earth Pressures and Vertical Shear Forces Acting on Non-Moving Retaining Walls." Proceedings of Earth Retention Systems 2003: A Joint Conference sponsored by American Society of Civil Engineers Metropolitan Section Geotechnical Group, Deep Foundations Institute, and ADSC, pp. 81-97.

Filz, G. M. and Duncan, J. M. (1997). "Vertical Shear Loads on Nonmoving Walls; I: Theory". *Journal of Geotechnical and Geoenvironmental Engineering*, Vol. 123, No. 9, pp. 856-862.

Francesco, L., Foti, S., Lancellota, R., and Mylonakis, G. (2010). "Dynamic Response of Cantilever Retaining Walls Considering Soil Non-Linearity". Proceedings of the Fifth International Conference on Recent Advances in Geotechnical Earthquake Engineering and Soil Dynamics, St. Louis, MO.
scholarsmine.mst.edu/cgi/viewcontent.cgi?article=3029&context=icrageesd
Accessed 25 September 2017.

Fredlund, D. G. and Rahardjo, H. (1993). *Soil Mechanics for Unsaturated Soils*. Wiley-Interscience.

Geraili Mikola, R. G. and Sitar, N. (2013). *Seismic Earth Pressures on Retaining Structures in Cohesionless Soils*. Report UCB GT 13-01, University of California - Berkeley, Department of Civil and Environmental Engineering, Geotechnical Engineering, Berkeley, CA.
www.ce.berkeley.edu/sites/default/files/assets/users/sitar/GT%2013-01%20-%20Geraili%20and%20Sitar%20%28corrected%29.pdf
Accessed 8 September 2017.

Geraili Mikola, R. G., Candia, G. and Sitar, N. (2016). "Seismic Earth Pressures on Retaining Structures and Basement Walls in Cohesionless Soils". *Journal of Geotechnical and Geoenvironmental Engineering*, Vol. 142, No. 10.
www.researchgate.net/publication/303504074_Seismic_Earth_Pressures_on_Retaining_Structures_and_Basement_Walls_in_Cohesionless_Soils
Accessed 18 September 2017.

Handy, R. L. (1985). "The Arch in Soil Arching". *Journal of the Geotechnical Engineering Division,* Vol. 111, No. 3, pp. 302-318.

Harrison, W. J. and C. M. Gerrard (1972). "Elastic Theory Applied to Reinforced Earth". *Journal of the Soil Mechanics and Foundations Division*, Vol. 98, No. SM12, pp. 1325-1345.

Harrop-Williams, K. O. (1989). "Geostatic Wall Pressures". *Journal of the Geotechnical Engineering Division*, Vol. 115, No. 9, pp. 1321-1325.

Hoppe, E. J. (2005a). *Field Study of Integral Backwall with Elastic Inclusion.* Report FHWA/VTRC 05-R28, Virginia Transportation Research Council, Charlottesville, VA. **vtrc.virginiadot.org/PubDetails.aspx?id=297029** Accessed 6 January 2018.

Hoppe, E. J. (2005b). "Field Study of Integral Backwall with Elastic Inclusion". Proceedings of IAJB 2005: The 2005 FHWA Conference on Integral Abutment and Jointless Bridges.

Hoppe, E. J. (2006). *Field Measurements on Skewed Semi-Integral Bridge With Elastic Inclusion: Instrumentation Report.* Report FHWA/VTRC 06-R35, Virginia Transportation Research Council, Charlottesville, VA. **vtrc.virginiadot.org/PubDetails.aspx?id=297249** Accessed 6 January 2018.

Hoppe, E. J. and Bagnall, T. M. (2008). *Performance of a Skewed Semi-Integral Bridge: Volume I: Field Monitoring.* Report FHWA/VTRC 08-R20, Virginia Transportation Research Council, Charlottesville, VA. **vtrc.virginiadot.org/PubDetails.aspx?id=296679** Accessed 6 January 2018.

Hoppe, E. J. and Eichenthal, S. L. (2012). *Thermal Response of a Highly Skewed Integral Bridge.* Report FHWA/VCTIR 12-R10, Virginia Center for Transportation Innovation and Research Charlottesville, VA. **vtrc.virginiadot.org/PubDetails.aspx?id=298171** Accessed 6 January 2018.

Hoppe, E. J. and Gomez, J. P. (1996). *Field Study of an Integral Backwall Bridge.* Report VTRC 97-R7, Virginia Transportation Research Council, Charlottesville, VA. **vtrc.virginiadot.org/PubDetails.aspx?id=296792** Accessed 6 January 2018.

Horvath, J. S. (1977). "Finite Element Analysis of a Cut and Cover Tunnel Constructed Using Slurry Trench Walls". Proceedings of the Sixth Soil Mechanics Seminar, Columbia University, Department of Civil Engineering and Engineering Mechanics, New York, NY.

Horvath, J. S. (1979). *A Study of Analytical Methods for Determining the Response of Mat Foundations to Static Loads.* Ph.D. dissertation, Polytechnic Institute of New York, Department of Civil Engineering, Brooklyn, NY.

Horvath, J. S. (1983). "Modulus of Subgrade Reaction: New Perspective". *Journal of Geotechnical Engineering*, Vol. 109, No. 12, pp. 1591-1596.

Horvath, J. S. (1984). Errata in "Modulus of Subgrade Reaction: New Perspective". *Journal of Geotechnical Engineering*, Vol. 110, No. 8, p. 1171.

Horvath, J. S. (1988). *Historical Review and Critique of Mathematical Models for Plate- and Beam-Type Foundation Element Subgrades*. Research Report CE/GE-88-4, Manhattan College, School of Engineering, Civil Engineering Department, Geotechnical Engineering Program, Bronx, NY.

Horvath, J. S. (1989). "Subgrade Models for Soil-Structure Interaction". Proceedings of the Foundation Engineering Congress, American Society of Civil Engineers, pp. 599-612.

Horvath, J. S. (1990a). *A Study of Some Miscellaneous Wall Problems*. Research Report CE/GE-90-1, Manhattan College, School of Engineering, Civil Engineering Department, Geotechnical Engineering Program, Bronx, NY.

Horvath, J. S. (1990b). *The Use of Geosynthetics to Reduce Lateral Earth Pressures on Rigid Walls; Phase I: Concept Evaluation*. Research Report CE/GE-90-2, Manhattan College, School of Engineering, Civil Engineering Department, Geotechnical Engineering Program, Bronx, NY.

Horvath, J. S. (1991a). "Using Geosynthetics to Reduce Surcharge-Induced Stresses on Rigid Earth-Retaining Structures". Preprint Paper 91-0096, Transportation Research Board 70th Annual Meeting, Washington, DC.

Horvath, J. S. (1991b). "Using Geosynthetics to Reduce Earth Loads on Rigid Retaining Structures". Proceedings of Geosynthetics '91, Industrial Fabrics Association International, pp. 409-424.

Horvath, J. S. (1991c). "Using Geosynthetics to Reduce Surcharge-Induced Stresses on Rigid Earth-Retaining Structures". *Transportation Research Record 1330*, Transportation Research Board, Washington, DC, pp. 47-53.

Horvath, J. S. (1991d). *Developments in Thick-Geosynthetics Technology: 1991 Update*. Research Report CE/GE-91-1, Manhattan College, School of Engineering, Civil Engineering Department, Geotechnical Engineering Program, Bronx, NY.

Horvath, J. S. (1995a). Discussion of "Tensile Reinforcement Effects on Bridge-Approach Settlement" by G. J. Monley and J. T. H. Wu, *Journal of Geotechnical Engineering*, Vol. 121, No. 1, pp. 93-94.

Horvath, J. S. (1995b). *Geofoam Geosynthetic*. Horvath Engineering, P.C., Scarsdale, NY.

Horvath, J. S. (1995c). "Geoinclusion". *Fabrics & Architecture*, Vol. 7, No. 5, pp. 38-39.

Horvath, J. S. (1996a). "The Compressible Inclusion Function of EPS Geofoam: A State-of-Art Review". Notes prepared for distribution at a presentation to Construction Project Consultants, Inc., Tokyo, Japan.

Horvath, J. S. (1996b). "The Compressible Inclusion Function of EPS Geofoam: An Overview". Proceedings of the International Symposium on EPS Construction Method (EPS Tokyo '96), EPS Construction Method Development Organization, Tokyo, Japan, pp. 67-75 [in Japanese].

Horvath, J. S. (1996c). "The Compressible Inclusion Function of EPS Geofoam: An Overview." Proceedings of the International Symposium on EPS Construction Method (EPS Tokyo '96), EPS Construction Method Development Organization, Tokyo, Japan, pp. 71-81.

Horvath, J. S. (1997a). Discussion of "Numerical Study of Parameters Influencing the Response of Flexible Retaining Walls" by H. H. Vaziri, *Canadian Geotechnical Journal*, Vol. 34, No. 1, p. 166.

Horvath, J. S. (1997b). "The Compressible Inclusion Function of EPS Geofoam". *Geotextiles and Geomembranes*, Vol. 15, Nos. 1-3, pp. 77-120.

Horvath, J. S. (1997c). Discussion of "Analyses of Active Earth Pressure Against Rigid Retaining Wall Subjected to Different Modes of Movement" by H. Matsuzawa and H. Hazarika, *Soils and Foundations*, Vol. 37, No. 4, p. 133.

Horvath, J. S. (1998a). *The Compressible Inclusion Function of EPS Geofoam: An Overview of Concepts, Applications, and Products*. Research Report CE/GE-98-1, Manhattan College, School of Engineering, Civil Engineering Department, Geotechnical Engineering Program, Bronx, NY.

Horvath, J. S. (1998b). *The Compressible-Inclusion Function of EPS Geofoam: Analysis and Design Methodologies*. Research Report CE/GE-98-2, Manhattan College, School of Engineering, Civil Engineering Department, Geotechnical Engineering Program, Bronx, NY.

Horvath, J. S. (1998c). *Mathematical Modeling of the Stress-Strain-Time Behavior of Geosynthetics Using the Findley Equation: General Theory and Application to EPS-Block Geofoam*. Research Report CE/GE-98-3, Manhattan College, School of Engineering, Civil Engineering Department, Geotechnical Engineering Program, Bronx, NY.

Horvath, J. S. (1999a). *Lessons Learned from Failures Involving Geofoam in Roads and Embankments*. Research Report CE/GE-99-1 (revised), Manhattan College, School of Engineering, Civil Engineering Department, Geotechnical Engineering Program, Bronx, NY, 1999.

Horvath, J. S. (1999b). "Geofoam and Geocomb: Lessons from the Second Millennium A.D. as Insight for the Future". Preprint paper for the 13th Geosynthetics Research Institute Conference, Philadelphia, PA.

Horvath, J. S. (1999c). "Geofoam and Geocomb: Lessons from the Second Millennium A.D. as Insight for the Future". Proceedings of the 13th Geosynthetics Research Institute Conference, Philadelphia, PA.

Horvath, J. S. (1999d). *Geofoam and Geocomb: Lessons from the Second Millennium A.D. as Insight for the Future*. Research Report CE/GE-99-2, Manhattan College, School of Engineering, Civil Engineering Department, Geotechnical Engineering Program, Bronx, NY.

Horvath, J. S. (2000). *Integral-Abutment Bridges: Problems and Innovative Solutions Using EPS Geofoam and Other Geosynthetics.* Research Report CE/GE-00-2, Manhattan College, School of Engineering, Civil Engineering Department, Geotechnical Engineering Program, Bronx, NY.

Horvath, J. S. (2001a). *Geomaterials Research Project - Geofoam and Geocomb Geosynthetics: A Bibliography through the Second Millennium A.D.* Research Report CGT-2001-1, Manhattan College, School of Engineering, Center for Geotechnology, Bronx, NY.

Horvath, J. S. (2001b). *Geomaterials Research Project; Non-Earth Subgrade Materials and Their Thermal Effects on Pavements: An Overview.* Research Report CGT-2001-2, Manhattan College, School of Engineering, Center for Geotechnology, Bronx, NY.

Horvath, J. S. (2001c). *Geomaterials Research Project; Concepts for Cellular Geosynthetics Standards with an Example for EPS-Block Geofoam as Lightweight Fill for Roads.* Research Report CGT-2001-4, Manhattan College, School of Engineering, Center for Geotechnology, Bronx. NY.

Horvath, J. S. (2001d). *Geomaterials Research Project; Cellular Geosynthetics 2001: Geofoam Lightweight Fills and Beyond.* Research Report CGT-2001-5, Manhattan College, School of Engineering, Center for Geotechnology, Bronx, NY.

Horvath, J. S. (2002). *Soil-Structure Interaction Research Project; Basic SSI Concepts and Applications Overview.* Research Report CGT-2002-2, Manhattan College, School of Engineering, Center for Geotechnology, Bronx, NY.

Horvath, J. S. (2003a). *Innovative Aspects of the Use of Expanded Polystyrene (EPS) on Boston's 'Big Dig'.* Research Report CGT-2003-1, Manhattan College, School of Engineering, Center for Geotechnology, Bronx, NY.

Horvath, J. S. (2003b). "An Overview of the Functions and Applications of Cellular Geosynthetics". Proceedings of IeC GEO3 - International e-Conference on Modern Trends in Geotechnical Engineering: Geotechnical Challenges and Solutions, Indian Institute of Technology at Madras, India.

Horvath, J. S. (2003c). "Controlled Yielding Using Geofoam Compressible Inclusions: The New Frontier in Earth-Retaining Structures". Proceedings of IeC GEO3 - International e-Conference on Modern Trends in Geotechnical Engineering: Geotechnical Challenges and Solutions, Indian Institute of Technology at Madras, India.

Horvath, J. S. (2004a). *Geomaterials Research Project - A Technical Note re Calculating the Fundamental Period of an EPS-Block-Geofoam Embankment.* Research Report CGT-2004-1, Manhattan College, School of Engineering, Center for Geotechnology, Bronx, NY.

Horvath, J. S. (2004b). *Integrated Site Characterization and Foundation Analysis Research Project - A Technical Note re Effect of K_{onc} Assumption on Site-Characterization Algorithm for Coarse-Grain Soil.* Research Report CGT-2004-2, Manhattan College, School of Engineering, Center for Geotechnology, Bronx, NY.

Horvath, J. S. (2004c). "Integral-Abutment Bridges: A Complex Soil-Structure Interaction Challenge". Proceedings of Geo-Trans 2004, American Society of Civil Engineers.

Horvath, J. S. (2004d). "Cellular Geosynthetics in Transportation Applications". Proceedings of Geo-Trans 2004, American Society of Civil Engineers.

Horvath, J. S. (2004e). "Controlled Yielding Using Geofoam Compressible Inclusions: New Frontier in Earth-Retaining Structures". Proceedings of Geo-Trans 2004, American Society of Civil Engineers.

Horvath, J. S. (2005). "Integral-Abutment Bridges: Geotechnical Problems and Solutions Using Geosynthetics and Ground Improvement". Proceedings of IAJB 2005: The 2005 FHWA Conference on Integral Abutment and Jointless Bridges.

Horvath, J. S. (2008a). "Seismic Lateral Earth Pressure Reduction on Earth-Retaining Structures Using Geofoams". Proceedings of Geotechnical Earthquake Engineering and Soil Dynamics IV, American Society of Civil Engineers.

Horvath, J. S. (2008b). "Extended Veletsos-Younan Model for Geofoam Compressible Inclusions Behind Rigid, Non-Yielding Earth-Retaining Structures". Proceedings of Geotechnical Earthquake Engineering and Soil Dynamics IV, American Society of Civil Engineers.

Horvath, J. S. (2010a). "Emerging Trends in Failures Involving EPS-Geofoam Fills." *Journal of Performance of Constructed Facilities*, Vol. 24, No. 4, pp. 365-372.

Horvath, J. S. (2010b). "Lateral Pressure Reduction on Earth-Retaining Structures Using Geofoams: Correcting Some Misunderstandings." Proceedings of ER2010: Earth Retention Conference 3, American Society of Civil Engineers.

Horvath, J. S. (2011). *Geomaterials Research Project; Manufacturing Quality Issues for Block-Molded Expanded Polystyrene Geofoam.* Research Report CEEN/GE-2011-2, Manhattan College, School of Engineering, Civil and Environmental Engineering Department, Geotechnical Engineering Program, Bronx, NY.

Horvath, J. S. (2012). *Geomaterials Research Project; The Evolution of Generic Material Standards for Block-Molded Expanded Polystyrene (EPS) Used for Small-Strain Geofoam Applications in the U.S.A.* Research Report CEEN/GE-2012-1, Manhattan College, School of Engineering, Civil and Environmental Engineering Department, Geotechnical Engineering Program, Bronx, NY.

Horvath, J. S. (2014). *The Effect of Vertical Forces on Anchored-Bulkhead Behavior: A White Paper.* John S. Horvath Consulting Engineer, Scarsdale, NY.

Huntington, W. C. (1957). *Earth Pressures and Retaining Walls.* Wiley.

IAJB 2005 (2005). *Proceedings; The 2005 FHWA Conference; Integral Abutment and Jointless bridges (IAJB 2005).* Baltimore, MD.

Ikizler, S. B., Aytekin, M., and Nas, E. (2008). "Laboratory Study of Expanded Polystyrene (EPS) Geofoam Used with Expansive Soils". *Geotextiles and Geomembranes*, Vol. 26, No. 2, pp. 189-195.

Inglis, D., Macleod, G., Naesgaard, E., and Zergoun, M. (1996). "Basement Wall with Seismic Earth Pressures and Novel Expanded Polystyrene Foam Buffer Layer". Proceedings of the 10th Annual Symposium of the Vancouver Geotechnical Society, The Canadian Geotechnical Society, Richmond, BC, Canada.

Ingold, T. S. (1979). "The Effects of Compaction on Retaining Walls". *Géotechnique*, Vol. 29, No. 3, pp. 265-283.

Karpurapu, R. and Bathurst, R. J. (1992). "Numerical Investigation of Controlled Yielding of Soil-Retaining Wall Structures". *Geotextiles and Geomembranes*, Vol. 11, No. 2, pp. 115-131.

Kézdi, A. (1975). "Lateral Earth Pressure". Chapter 5 in *Foundation Engineering Handbook*, H. F. Winterkorn and H.-Y. Fang (ed.), Van Nostrand Reinhold.

Kramer, S. L. (1996). *Geotechnical Earthquake Engineering*. Prentice Hall.

Kulhawy, F. H. and Mayne, P. W. (1990). *Manual on Estimating Soil Properties for Foundation Design*. Final Report - EPRI EL-6800 Project 1493-6, Cornell University, School of Civil Engineering, Geotechnical Engineering Group, Ithaca, NY.

Leshchinsky, D. (2009). "On Global Equilibrium in Design of Geosynthetic Reinforced Walls". *Journal of Geotechnical and Geoenvironmental Engineering*, Vol. 135, No. 3, pp. 309-315

Leshchinsky, D. (2010a). "Geosynthetic Walls and Steep Slopes: Is It Magic?". *Geosynthetics*, Vol. 28, No. 3.

Leshchinsky, D. (2010b). "Is It Magic? Author Offers More Insight". *Geosynthetics*, Vol. 28, No. 5.

Leshchinsky, D., Imamoglu, B., and Meehan, C. L. (2010a). "Exhumed Geogrid-Reinforced Retaining Wall". *Journal of Geotechnical and Geoenvironmental Engineering*, Vol. 136, No. 10, pp. 1311-1323

Leshchinsky, D., Zhu F., and Meehan, C. L. (2010b). "Required Unfactored Strength of Geosynthetic in Reinforced Earth Structures". *Journal of Geotechnical and Geoenvironmental Engineering*, Vol. 136, No. 2, pp. 281-289.

McGown, A., Andrawes, K. Z., and Murray, R. T. (1988). "Controlled Yielding of the Lateral Boundaries of Soil Retaining Structures". *Geosynthetics for Soil Improvement*, American Society of Civil Engineers, pp. 193-210.

McGown, A., Murray, R. T. and Andrawes, K. Z. (1987). *Influence of Wall Yielding on Lateral Stresses in Unreinforced and Reinforced Fills*. Research Report 113, Transport and Road Research Laboratory, Crowthorne, Berkshire, U.K.

Monley, G. J. and Wu, J. T. H. (1993). "Tensile Reinforcement Effects on Bridge-Approach Settlement". *Journal of Geotechnical Engineering*, Vol. 119, No. 4, pp. 749-762.

Monley, G. J. and Wu, J. T. H. (1995). Closure to "Tensile Reinforcement Effects on Bridge-Approach Settlement". *Journal of Geotechnical Engineering*, Vol. 121, No. 1, pp. 96-97.

Murphy, G. (1997). "The Influence of Geofoam Creep on the Performance of a Compressible Inclusion". *Geotextiles and Geomembranes*, Vol. 15, Nos. 1-3, pp. 121-131.

Ostadan, F. (2004). "Seismic Soil Pressure for Building Walls - An Updated Approach". Proceedings of the 11th International Conference on Soil Dynamics and Earthquake Engineering (11th ICSDEE) and the 3rd International Conference on Earthquake Geotechnical Engineering (3rd ICEGE), Berkeley, CA.
www.civil.utah.edu/~bartlett/CVEEN7330/soilpressureonwalls-Ostadan.pdf
Accessed 15 September 2017.

Paik, K. H. and Salgado, R. (2003). "Estimation of Active Earth Pressure Against Rigid Retaining Walls Considering Arching Effects". *Géotechnique*, Vol. 53, No. 7, pp. 643-653; with significant corrections in the attached Errata.
www.civil.utah.edu/~bartlett/CVEEN5305/Handout%2011%20-%20Earth%20pressure%20against%20rigid%20retaining%20walls.pdf
Accessed 5 September 2017.

Partos, A. M. and Kazaniwsky, P. M. (1987). "Geoboard Reduces Lateral Earth Pressures." Proceedings of Geosynthetics '87, Industrial Fabrics Association International, pp. 628-639.

Poulos, H. G. and E. H. Davis (1974). *Elastic Solutions for Soil and Rock Mechanics*. Wiley.

Reeves, J. N. and Filz, G. M. (2000). *Earth Force Reduction by a Synthetic Compressible Inclusion*. Research report issued by Virginia's Center for Innovative Technology, Virginia Tech, Department of Civil Engineering, Blacksburg, VA.

Seed, R. B. and Duncan, J. M. (1983). *Soil-Structure Interaction Effects of Compaction-Induced Stresses and Deflections*. Geotechnical Engineering Research Report UCB/GT/83-06, University of California, Civil Engineering Department, Berkeley, CA.

Sitar, N. and Wagner, N. (2015). "On Seismic Response of Stiff and Flexible Retaining Structures". Proceedings of the 6th International Conference on Earthquake Geotechnical Engineering, Christchurch, New Zealand.
www.ce.berkeley.edu/sites/default/files/assets/users/sitar/Sitar%20and%20Wagner%206ICEGE%20Christchurch%202015.pdf
Accessed 8 September 2017.

Sitar, N., Geraili Mikola, R. G., and Candia, G. (2012). "Seismically Induced Lateral Earth Pressures on Retaining Structures and Basement Walls". Proceedings of GeoCongress 2012, pp. 335-358.
www.researchgate.net/publication/268590571_Seismically_Induced_Lateral_Earth_Pressures_on_Retaining_Structures_and_Basement_Walls
Accessed 18 September 2017.

Spangler, M. G. and Handy, R. L. (1982). *Soil Engineering* (4th edition). Harper & Row.

Stark, T. D., Arellano, D., Horvath, J. S., and Leshchinsky, D. (2004a). *Geofoam Applications in the Design and Construction of Highway Embankments*. National Cooperative Highway Research Project Web Document 65, Transportation Research Board, Washington, DC.

Stark, T. D., Arellano, D., Horvath, J. S., and Leshchinsky, D. (2004b). *Guideline and Recommended Standard for Geofoam Applications in Highway Embankments*. National Cooperative Highway Research Project Report 529, Transportation Research Board, Washington, DC.

Symons, I. F. and Clayton, C. R. I. (1972). "Earth Pressures on Backfilled Retaining Walls". *Ground Engineering*, April, pp. 26-34.

Terzaghi, K. (1943). *Theoretical Soil Mechanics*. Wiley.

Tschebotarioff, G. P. (1973). *Foundations, Retaining and Earth Structures* (2nd edition). McGraw-Hill.

Veletsos, A. S. and Younan, A. H. (1993). *Dynamic Modeling and Response of Soil Wall Systems*. Report BNL-52402, Department of Advanced Technology, Brookhaven National Laboratory, Associated Universities, Inc., Upton, NY.
www.iaea.org/inis/collection/NCLCollectionStore/_Public/25/040/25040746.pdf
Accessed 25 September 2017.

Veletsos, A. S. and Younan, A. H. (1994). "Dynamic Modeling and Response of Soil Wall Systems". *Journal of Geotechnical Engineering*, Vol. 120, No. 12, pp. 2155-2179.

Veletsos, A. S. and Younan, A. H. (1995). "Dynamic Soil Pressures on Vertical Walls". Proceedings of the Third International Conference on Recent Advances in Geotechnical Earthquake Engineering and Soil Dynamics, St. Louis, MO.
scholarsmine.mst.edu/cgi/viewcontent.cgi?article=3283&context=icrageesd
Accessed 25 September 2017.

Virginia Tech (undated). *Instrumented Retaining Wall Facility*. Report issued by the Virginia Polytechnic Institute and State University, Department of Civil Engineering, Geotechnical Engineering program, Blacksburg, VA.

Vokas, C. A. and R. D. Stoll (1987). "Reinforced Elastic Layered Systems". *Transportation Research Record 1153*, Transportation Research Board, Washington, DC, pp. 1-7.

Wagner, N. and Sitar, N. (2016). "Seismic Earth Pressures on Deep Stiff Walls". Proceedings of the Geotechnical and Structural Engineering Congress, American Society of Civil Engineers, pp. 499-508.
www.ce.berkeley.edu/sites/default/files/assets/users/sitar/Wagner%20and%20Sit ar%20-%20ASCE%202016.pdf
Accessed 15 September 2017.

Wang, Y. and Bathurst, R. J. (2008a). "Horizontal Slice Method for Force and Displacement Analysis of EPS Geofoam Seismic Buffers for Rigid Retaining Walls". *China Civil Engineering Journal*, Vol. 41, No. 10, pp. 73-80 [in Chinese with English abstract].

Wang, Y. and Bathurst, R. J. (2008b). "A Numerical Model for EPS Geofoam Seismic Buffers". Proceedings of the 4th Asian Regional Conference on Geosynthetics, Shanghai, China.

Younan, A. H., Veletsos, A. S., and Bandyopadhyay, K. (1997). *Dynamic Response of Flexible Retaining Walls*. Report BNL-52519, Department of Advanced Technology, Brookhaven National Laboratory, Associated Universities, Inc., Upton, NY.
www.osti.gov/scitech/servlets/purl/444031
Accessed 25 September 2017.

Zarnani, S. and Bathurst, R. J. (2005). "Numerical Investigation of Geofoam Seismic Buffers Using FLAC". Proceedings of the North American Geosynthetics Society (NAGS)/GRI19 Conference.

Zarnani, S. and Bathurst, R. J. (2008). "Numerical Modeling of EPS Seismic Buffer Shaking Table Tests". *Geotextiles and Geomembranes*, Vol. 26, No. 5, pp. 371-383.

Zarnani, S. and Bathurst, R. J. (2009a). "Numerical Parametric Study of Expanded Polystyrene (EPS) Geofoam Seismic Buffers". *Canadian Geotechnical Journal*, Vol. 46, pp. 318-338.

Zarnani, S. and Bathurst, R. J. (2009b). "Influence of Constitutive Model on Numerical Simulation of EPS Seismic Buffer Shaking Table Tests". *Geotextiles and Geomembranes*, Vol. 27, pp. 308-312.

Zarnani, S. and Bathurst, R. (2011). "EPS Seismic Buffers for Earthquake Load Attenuation Against Rigid Retaining Walls". Proceedings of Geo-Frontiers 2011, American Society of Civil Engineers, pp. 3166-3176.

Zarnani, S., Bathurst, R. J., and Gaskin, A. (2005). "Experimental Investigation of Geofoam Seismic Buffers Using a Shaking Table". Proceedings of the North American Geosynthetics Society (NAGS)/GRI19 Conference.

This page intentionally left blank.

Lateral Pressure Reduction on Earth-Retaining Structures Using Geofoam
John S. Horvath, Ph.D., P.E., Life Member.ASCE

SUPPLEMENTAL BIBLIOGRAPHY

<u>Note</u>

This section contains a list of reviewed (by the writer), but uncited, publications that are nevertheless deemed relevant to the overall subjects covered in this monograph. The contents of this section are intended to serve as a supplemental bibliography that, when combined with the preceding bibliography of cited references, provide an overall bibliography that is as complete as possible. Towards this end, also included in this section are documents that have not been obtained and reviewed by the writer. However, based on their title and/or author(s) they are believed to be relevant to this monograph and are thus included herein.

Armstrong, R. and Alfaro, M. (2003). "Reduction of Seismic-Induced Pressures on Rigid Retaining Structures Using Compressible Inclusions: A Numerical Study". Proceedings of the 56th Canadian Geotechnical Conference, The Canadian Geotechnical Society, Vol. 2, pp. 500-505.

Athanasopoulos, G. A., Nikolopoulou, C. P., Xenaki, V. C., and Stathopoulou, V. D. (2007). "Reducing the Seismic Earth Pressures on Retaining Walls by EPS Geofoam Buffers - Numerical Parametric Study". Proceedings, Geosynthetics Conference, IFAI.

Bathurst, R. J. and Zarnani, S. (2011). "Recent Research on EPS Geofoam Seismic Buffers". PowerPoint Presentation at EPS 2011 - 4th International Conference on Geofoam Blocks in Construction Applications, Lillestrom, Norway.

Chun, B.-S., Lim, H.-S., Sagong, M., and Kim, K. (2004). "Development of a Hyperbolic Constitutive Model for Expanded Polystyrene (EPS) Geofoam under Triaxial Compression Tests". *Geotextiles and Geomembranes*, Vol. 22, No. 4, pp. 223-237.

Collin, J. G. and Christopher, B. R. (1991). "Finite Element Analysis and Field Instrumentation of Soil/Cement Arch". Proceedings of Geotechnical Engineering Congress 1991, American Society of Civil Engineer, pp. 670-681.

Curtin, W. G., Shaw, G., Parkinson, G. I. and Golding, J. M. (1994). *Structural Foundation Designers' Manual*. Blackwell Scientific Publication.

Dasaka, S. M., Dave, T. N., Gade, V. K., and Chauhan, V. B. (2014). "Seismic Earth Pressure Reduction on Gravity Retaining Walls Using EPS Geofoam". Proceedings of the 8th International Conference on Physical Modeling in Geotechnics, Perth, Australia, pp. 1025-1030.

Gill, S. A. and Bushnell, T. D. (1992). "Reinforced Soil-Cement Embankment". *Stability and Performance of Slopes and Embankments-II*, American Society of Civil Engineers, pp. 1493-1504.

Gnaedinger, J. P. and Gill, S. A. (1991). "Geogrid Reinforced Soil-Cement Arch over Accelerator Ring". Proceedings of Geosynthetics '91, Industrial Fabrics Association International, pp. 917-933.

Hatami, K. and Witthoeft, A. (2008). "A Numerical Study on the Use of Geofoam to Increase the External Stability of Reinforced Soil Walls". *Geosynthetics International*, Vol. 15, No. 6, pp. 452-470.

Hazarika, H. (2000). "Modeling the Application of EPS as Compressible Buffer in Soil-Structure Interaction". Proceedings of the 2nd International Summer Symposium of JSCE, Tokyo, Japan, pp. 261-264.

Hazarika, H. (2001). "Mitigation of Seismic Hazard on Retaining Structures - A Numerical Experiment". Proceedings of the 11th International Offshore and Polar Engineering Conference, International Society of Offshore and Polar Engineers, Cupertino, CA, pp. 459-464.

Hazarika, H. (2006). "Stress-Strain Modeling of EPS Geofoam for Large-Strain Applications". *Geotextiles and Geomembranes*, Vol. 24, No. 2, pp. 79-90.

Hazarika, H., Nakazawa, J., Matsuzawa, H., and Negussey, D. (2001). "On the Seismic Earth Pressure Reduction Against Retaining Structures Using Lightweight Geofoam Fill". Proceedings of the 4th International Conference on Geotechnical Earthquake Engineering and Soil Dynamics, San Diego, CA.

Hazarika, H. and Okuzono, S. (2002). "An Analysis Model for a Hybrid Interactive System Involving Compressible Buffer Material". Proceedings, 12th International Offshore and Polar Engineering Conference (ISOPE-2002), International Society of Offshore and Polar Engineers, Kita Kyushu, Japan, Vol. 2, pp. 622-629.

Hazarika, H. and Okuzono, S. (2004). "Modeling the Behavior of a Hybrid Interactive System Involving Soil, Structure, and EPS Geofoam". *Soils and Foundations*, Vol. 44, No. 5, pp. 149-162.

Hazarika, H., Okuzono, S., and Matsuo, Y. (2003). "Seismic Stability Enhancement of Rigid Non-Yielding Structures". Proceedings of The 13th International Offshore and Polar Engineering Conference, The International Society of Offshore and Polar Engineers, pp. 697-702.

Hazarika, H., Okuzono, S., Matsuo, Y., and Takada, K. (2002). "Evaluation of Lightweight Materials as Geo-Inclusion in Reducing Earth Pressure on Retaining Wall". Proceedings of the 4th International Conference on Ground Improvement Techniques, Kuala Lumpur, Malaysia, Vol. 2, pp. 399-406.

Matsuda, T., Ugai, K., and Gose, S. (1996). "Application of EPS to Backfill of Abutment for Earth Pressure Reduction and Impact Absorption". Proceedings of the International Symposium on EPS Construction Method (EPS Tokyo '96), EPS Construction Method Development Organization, Tokyo, Japan [in Japanese].

345

Matsuda, T., Ugai, K., and Gose, S. (1996). "Application of EPS to Backfill of Abutment for Earth Pressure Reduction and Impact Absorption". Proceedings of the International Symposium on EPS Construction Method (EPS Tokyo '96), EPS Construction Method Development Organization, Tokyo, Japan.

NAVFAC DM-7 (1971). *Design Manual - Soil Mechanics, Foundations, and Earth Structures.* Department of the Navy, Naval Facilities Engineering Command, Washington, DC.

NAVFAC DM-7.2 (1982). *Design Manual 7.2 - Foundations and Earth Structures.* Department of the Navy, Naval Facilities Engineering Command, Alexandria, VA.

Ohbo, N., Horikoshi, K., Ogata, K., Kitamura, Y., Matsumura, K. and Inagaki, M. (2002). "Seismic Earth Pressure on Abutment with Lightweight Geo-Materials for Backfill". Proceedings of the International Workshop on Lightweight Geomaterials, pp. 181-186.

Ostadan, F. (2005). "Seismic Soil Pressure for Building Walls - An Updated Approach". *Journal of Soil Dynamics and Earthquake Engineering*, Vol. 25, pp. 785-793.

Pelekis, P. C., Xenaki, V. C., and Athanasopoulos, G. A. (2000). "Use of EPS Geofoam for Seismic Isolation of Earth-Retaining Structures: Results of a FEM Study". Proceedings of the 2nd European Geosynthetics Conference, Bologna, Italy, pp. 843-846.

Reid, R. A., Soupir, S. P., and Schaefer, V. R. (1998). "Use of Fabric Reinforced Soil Walls for Integral Abutment Bridge End Treatment". Proceedings of the Sixth International Conference on Geosynthetics, Industrial Fabrics Association International, pp. 573-576.

Riad, H. L. and Horvath, J. S. (2004). "Analysis and Design of EPS-Geofoam Embankments for Seismic Loading." Proceedings of Geo-Trans 2004, American Society of Civil Engineers.

Riad, H. L., Ricci, A. L., Osborn, P. W., and Horvath, J. S. (2003). "Expanded Polystyrene (EPS) Geofoam for Road Embankments and Other Lightweight Fills in Urban Environments." Proceedings of Soil and Rock America 2003.

Riad, H. L., Ricci, A. L., Osborn, P. W., Wood, D. C., and Horvath, J. S. (2003). "Innovative Aspects of the Use of Expanded Polystyrene (EPS) on Boston's 'Big Dig'." Preprint paper 03-2823, Transportation Research Board 82nd Annual Meeting, Washington, DC.

Riad, H. L., Ricci, A. L., Osborn, P. W., D'Angelo, D. A., and Horvath, J. S. (2004). "Design of Lightweight Fills for Road Embankments on Boston's Central Artery/Tunnel Project." Proceedings of the Fifth International Conference on Case Histories in Geotechnical Engineering.

Terzaghi, K. (1934). "Large Retaining-Wall Tests". *Engineering News-Record*, Vol. 112, pp. 136-140, 259-262, 316-318, 403-406, 503-508.

Terzaghi, K. (1936). "Stress Distribution in Dry and Saturated Sand Above a Yielding Trap-Door". Proceedings of the 1st International Conference on Soil Mechanics and Foundation Engineering, pp. 35–39.

Trandafir, A. C. and Bartlett, S. F. (2010). "Seismic Performance of Double EPS Geofoam Buffer Systems". Paper 5.10a, Proceedings of the Fifth International Conference on Recent Advances in Geotechnical Earthquake Engineering and Soil Dynamics.

Tsukamoto, Y., Ishihara, K., Nakazawa, H., Kon, H., Matsuo, T., and Hara, K. (2001). "Combined Reinforcement by Means of EPS Blocks and Geogrid for Retaining Wall Structures". *Landmarks in Earth Reinforcement*, pp. 483-487.

Vaziri, H. H. (1997). Closure to "Numerical Study of Parameters Influencing the Response of Flexible Retaining Walls". *Canadian Geotechnical Journal*, Vol. 34, No. 1, p. 167.

Veletsos, A. S. and Younan, A. H. (1992). *Dynamic Soil Pressures on Rigid Vertical Walls*. Report BNL-52357, Brookhaven National Laboratory, Upton, NY.

Veletsos, A. S., and Younan, A. H. (1994). "Dynamic Soil Pressures on Rigid Vertical Walls". *Earthquake Engineering & Structural Dynamics*, Vol. 23, No. 3, pp. 275-301.

Veletsos, A. S., and Younan, A. H. (1994). "Dynamic Soil Pressures on Rigid Cylindrical Vaults". *Earthquake Engineering & Structural Dynamics*, Vol. 23, No. 6, pp. 645-669.

Veletsos, A. S. and Younan, A. H. (1997). "Dynamic Response of Cantilever Retaining Walls". *Journal of Geotechnical Engineering*, Vol. 123, No. 2, pp. 161-172.

Wang, D. and Bathurst, R. J. (2012). "Numerical Analysis of Earthquake Load Mitigation on Rigid Retaining Walls Using EPS Geofoam". *The Open Civil Engineering Journal*, Vol. 6, pp. 21-25.

Wang, Z., Li, Y., and Wang, J. G. (2006). "Numerical Analysis of Attenuation Effect of EPS Geofoam on Stress-Waves in Civil Defense Engineering". *Geotextiles and Geomembranes*, Vol. 24, No. 5, pp. 265-273.

Wood, J. (1973). *Earthquake-Induced Soil Pressures on Structures*. Report EERL 73-05, California Institute of Technology, Pasadena, CA.

Yazdani, M., Azad, A., Farshi, A.-h., and Talatahari, S. (2013). "Extended 'Mononobe-Okabe' Method for Seismic Design of Retaining Walls". *Journal of Applied Mathematics*, Volume 2013.
www.hindawi.com/journals/jam/2013/136132/
Accessed 13 September 2017.

Younan, A. H. and Veletsos, A. S. (2000). "Dynamic Response of Flexible Retaining Walls". *Earthquake Engineering & Structural Dynamics*, Vol. 29, No. 12, pp. 1815-1844.

Zarnani, S. and Bathurst, R. J. (2006). "Application of EPS Geofoam as a Seismic Buffer: Numerical Study Using FLAC". Proceedings of the 59th Canadian Geotechnical Conference, The Canadian Geotechnical Society, Richmond, BC, Canada.

Zarnani, S. and Bathurst, R. J. (2007). "Experimental Investigation of EPS Geofoam Seismic Buffers Using Shaking Table Tests". *Geosynthetics International*, Vol. 14, No. 3, pp. 165-177.

Zarnani, S. and Bathurst, R. J. (2007). "Numerical Parametric Study of EPS Geofoam Seismic Buffers". Proceedings of the 60th Canadian Geotechnical Conference, The Canadian Geotechnical Society, Richmond, BC, Canada.

Zarnani, S. and Bathurst, R. J. (2008). "Influence of Constitutive Model Type on EPS Geofoam Seismic Buffer Simulations Using FLAC". Proceedings of the First Pan American Geosynthetics Conference and Exhibition, Cancun, Mexico, pp. 868-877.

Zou, Y. and Leo, C. H. (1998). "Laboratory Studies on Engineering Properties of Expanded Polystyrene (EPS) Material for Geotechnical Applications". Proceedings of the Second International Conference on Ground Improvement Techniques, Singapore, pp. 581-588.

This page intentionally left blank.

www.ingramcontent.com/pod-product-compliance
Lightning Source LLC
Chambersburg PA
CBHW050238220326
41598CB00047B/7438